MW00989943

A FIRST COURSE IN CONTINUUM MECHANICS

Third Edition

A FIRST COURSE IN CONTINUUM MECHANICS
for Physical and Biological Engineers and Scientists

Y. C. Fung

Professor of Applied Mechanics and Bioengineering
University of California, San Diego

PRENTICE HALL, Englewood Cliffs, New Jersey 07632

Library of Congress Cataloging-in-Publication Data

Fung, Y. C. (Yuan-cheng)
A first course in continuum mechanics : for physical and biological scientists and engineers / Y.C. Fung. — 3rd ed.
p. cm.
Includes bibliographical references and index.
ISBN 0-13-061524-2
1. Continuum mechanics. I. Title.
QA808.F86 1994
531—dc20 92-39910
CIP

Acquisitions editor: Doug Humphrey
Production editor: Bayani Mendoza de Leon
Cover designer: Joe DiDomenico
Copy editor: Brian Baker
Prepress buyer: Linda Behrens
Manufacturing buyer: Dave Dickey
Supplements editor: Alice Dworkin
Editorial assistant: Susan Handy

Printed in the United States of America
20 19 18 17 16 15 14 13 12 11

ISBN 0-13-061524-2

90000

9 780130 615244

ISBN 0-13-061524-2

Prentice-Hall International (UK) Limited, *London*
Prentice-Hall of Australia Pty. Limited, *Sydney*
Prentice-Hall Canada Inc., *Toronto*
Prentice-Hall Hispanoamericana, S.A., *Mexico*
Prentice-Hall of India Private Limited, *New Delhi*
Prentice-Hall of Japan, Inc., *Tokyo*
Editora Prentice-Hall do Brasil, Ltda., *Rio de Janeiro*

Dedicated to

students who would share
my enthusiasm
for the application
of mechanics,

and to

Luna, Conrad, and Brenda.

Contents

Preface to the Third Edition

The objective of this edition is the same as that of the earlier editions: To emphasize the formulation of problems in mechanics, to reduce vague ideas into precise mathematical statements, and to cultivate a habit of questioning, analyzing, designing, and inventing in engineering and science. I have stressed applications even more strongly in this edition. Thus, at the very beginning, continua are defined with regard to real materials. Throughout the book, no discrimination was made against biological materials. Biology is included in science, bioengineering is included in engineering. Mechanics is not limited to physics.

In this book, I often ask the reader to formulate equations, regardless of whether he or she can solve them or not. I have known many students who have worked innumerable exercises without ever formulating a problem of their own. I hope that they will learn things the other way, to generate many problems of their own and then strive to discover the methods of solution and subtleties of the solutions. They are encouraged to observe nature and to think of problems in engineering and science and then to take the first step to write down a possible set of governing equations and boundary conditions. This first step—to derive the basic governing equations—is an objective of this book.

This book is organized as follows. At the outset, the concept of continua is explained. Then follows a thorough treatment of stress and strain. The practical techniques of determining the principal stress and strain, and the concept of compatibility, are given emphasis in two separate chapters. In Chapter 7, idealized specifications of fluids and solids are presented. The important concept of isotropy is described in detail in Chapter 8. The mechanical properties of real fluids and solids are discussed in Chapter 9. In Chapter 10, basic conservation laws of mechanics are given. In Chapters 11 and 12, some features of the flows of ideal and viscous fluids, boundary layer theory, linearized theory of elasticity, theories of bending and torsion, and elastic waves are described. The last Chapter, 13, is new. It deals with long-term changes in material when it is subjected to stress. It deals with active growth and remodeling of materials. It is concerned with the stress-growth law.

Compared with the materials discussed in traditional textbooks of fluid and solid mechanics, some of the materials considered in Chapters 9 and 13 may seem exotic. But they are not. They are certainly not exotic in the sense of being rare. Fluids like blood or paint; solids like heart, lung, muscle, or rubber are certainly common and precious; and it does not hurt to learn something about their constitutive equations.

A large number of problems is dispersed throughout the book. Many are new for this edition. Most are problems for formulation, problems for design, and problems for inventing. A few, especially in Chapters 2 and 10, are exercises that train the reader in applying some of the formulas derived in the text.

If the reader obtains a clear idea about the stress, strain, and constitutive equations from this book, and knows how to use them in formulating scientific and engineering problems, I shall be very happy.

Y. C. Fung
La Jolla, California

Preface to the First Edition

This book is intended for students of science and engineering who are beginning a series of courses in mechanics. At this stage, students normally have had courses in calculus, physics, vector analysis, and elementary differential equations. A course in continuum mechanics then provides a foundation for studies in fluid and solid mechanics, material sciences, and other branches of science and engineering.

It is my opinion that, for a beginner, the approach should be physical rather than mathematical. To engineers and physicists who use continuum mechanics constantly, the primary attraction of the subject lies in its simplicity of conception and concreteness in applications. Therefore, the students should be introduced to the applications as soon as possible.

For the scientist or engineer, the important questions he must find answers to are: How shall I formulate the problem? How shall I state the governing field equations and boundary conditions? How shall I choose alternate hypotheses? What kind of experiments would justify or deny or improve my hypotheses? How exhaustive should the investigation be? Where might errors appear? How much time is required to obtain a reasonable solution? At what cost? These are questions which concern active investigators, and are questions of synthesis, which employ analyses as tools. Complete answers to these questions are beyond the scope of this "first course," but we can make a good beginning. In this book, I often ask the reader to formulate problems, regardless of whether he can solve his equations and understand all the mathematical subtleties. I have known many students who have read many books and worked innumerable exercises without ever formulating a problem of their own. I hope they will learn the other way, to generate many problems of their own and then strive to discover the methods and subtleties of solutions. They should be encouraged to observe nature and to think of problems in engineering and then to take the first step to write down a possible set of governing equations and boundary conditions. This "first step"—to derive the basic governing equations—is the object of this book. Perhaps it is justifiable for a "first course" to be concerned only with this first step. But the preparation required for taking this step is extensive. For such a step to be firm, one would have to understand the basic concepts of mechanics and their mathematical expressions. To be able to use these basic equations with confidence one must know their origins and their derivations. Therefore, the discussions of basic ideas must be thorough. It is for this reason that the first ten chapters of this book are rather comprehensive and detailed.

As for the organization of the book: At the outset, the concept of continua is explained. Then a thorough treatment of the concepts of stress and strain follows. The practical techniques of determining the principal stress and strain, and the concept of compatibility, are given emphasis in two separate chapters. The description of motion is considered. In Chapter 7, an idealized specification of fluids and solids is presented. The important concept of isotropy is described in detail in Chapter 8. Data on the mechanical properties of common fluids and solids appear in Chapter 9. In Chapter 10, a thorough treatment of the basic conservation laws of physics is given. Beginning with Chapter 11, some features of perfect fluids, viscous flow, boundary layer theory, linearized theory of elasticity, theories of bending and torsion, and elastic waves are described briefly. The last two chapters provide a glimpse into the rich fields of fluid and solid mechanics; to treat them comprehensively would require many volumes at a more advanced mathematical level. The introduction given here should prepare the student to enter these fields with greater ease.

If the reader obtains clear ideas about the stress, strain, and constitutive equations from this book, I would consider this introductory text a success. Beyond this, only a sketch of some classical problems is provided. Many discussions are given in the exercises, which should be regarded as an integral part of the text.

I have quoted frequently and borrowed heavily from my previous book, *Foundations of Solid Mechanics*, which can be used for a course following the present one. The material for this "first course" was organized for my class at the University of California, San Diego, where the curriculum offers emphasis on general sciences before specialization. The book should be useful for undergraduates and younger graduate students who have a reasonable background in mathematics and physics.

The writing of this book was a pleasant experience. My wife, Luna, cooperated throughout the task. A mathematician, she gave up her teaching career when I came to La Jolla. Willing to learn some mechanics, she worked through the manuscript very thoroughly. Many passages are clearer because of her declaration that she did not understand. My friend, Chia-Shun Yih, Timoshenko Professor at the University of Michigan, read through the manuscript and gave me many valuable comments. I am also grateful to Drs. Pin Tong of the Massachusetts Institute of Technology and Gilbert Hegemier of the University of California, San Diego, for their comments. Finally, I wish to register my thanks to Nicholas Romanelli of Prentice Hall for editorial assistance, to Mrs. Ling Lin for preparing the index, and to Mrs. Barbara Johnson, whose fast, accurate typing and cheerful good humor made the work a pleasure.

Y. C. Fung
La Jolla, California

A FIRST COURSE
IN CONTINUUM MECHANICS

1 INTRODUCTION

The definition of continua for real-world materials is presented, as are elementary examples through which basic ideas of mechanics evolved.

1.1 THE OBJECTIVE OF THIS COURSE

Our objective is to learn how to formulate problems in mechanics and how to reduce vague questions and ideas into precise mathematical statements, as well as to cultivate a habit of questioning, analyzing, designing, and inventing in engineering and science.

Let us consider a few questions. Suppose an airplane is flying above us. The wings must be under strain in order to support the passengers and freight. How much strain are the wings subjected to? If you were flying a glider, and an anvil cloud appeared, the thermal current would carry the craft higher. Dare you fly into the cloud? Have the wings sufficient strength? Ahead you see the Golden Gate Bridge. Its cables support a tremendous load. How does one design such cables? The cloud contains water and the countryside needs that water. If the cloud were seeded, would that produce rain? And would the rain fall where needed? Would the amount of rainfall be adequate and not produce a flood? In the distance there is a nuclear reactor power station. How is the heat transported in the reactor? What kind of thermal stresses are there in the reactor? How does one assess the safety of the power station against earthquakes? What happens to the earth in an earthquake? Thinking about the globe, you may wonder how the continents float, move, or tear apart. And how about ourselves: How do we breathe? What changes take place in our lungs if we do a yoga exercise and stand on our heads?

Interestingly, all these questions are concerned with force, motion, flow, deformation, energy, properties of matter, external interaction between bodies, or internal interaction between one part of a body and another part; and changes in matter, temporarily or permanently, reversibly or irreversibly. These changes, together with the axioms of continuum mechanics, can be reduced to certain differential equations and boundary conditions. By solving such equations, we obtain precise quantitative information. In this book, we deal with the fundamental principles that underlie these differential equations and boundary conditions. Although it would be nice to solve these equations once they are formulated, we shall not become involved in discussing their solutions in detail. Our objective is formulation: the formal reduction of general ideas to a mathematical form.

1.2 APPLICATIONS TO SCIENCE AND TECHNOLOGY

The mathematical approach taken in this book will be aimed at serving science and technology. I want the applications to be apparent to the student; hence, the examples and the problems to be solved are often stated in terms of scientific research or engineering design. A person's frame of mind with regard to designing and inventing things, devices, methods, theories, and experiments can be strengthened by constant practice—by forming a habit.

1.3 WHAT IS MECHANICS?

Mechanics is the study of the motion (or equilibrium) of matter and the forces that cause such motion (or equilibrium). Mechanics is based on the concepts of time, space, force, energy, and matter. A knowledge of mechanics is needed for the study of all branches of physics, chemistry, biology, and engineering.

1.4 A PROTOTYPE OF A CONTINUUM: THE CLASSICAL DEFINITION

The classical concept of a continuum is derived from mathematics. We say that the real number system is a *continuum*. Between any two distinct real numbers there is another distinct real number, and therefore, there are infinitely many real numbers between any two distinct real numbers. Intuitively, we feel that *time* can be represented by a real number system t and that a three-dimensional *space* can be represented by three real number systems x, y, z. Thus, we identify *time* and *space* together as a four-dimensional continuum.

Extending the concept of a continuum to *matter*, we speak of a continuous distribution of matter in space. This may be best illustrated by considering the concept of *density*. Let the amount of matter be measured by its *mass*, and let us assume that a certain matter permeates a certain space $\mathcal{V}_0$, as in Fig. 1.1. Let us consider a point P in $\mathcal{V}_0$ and a sequence of subspaces $\mathcal{V}_1, \mathcal{V}_2, \ldots,$ converging on P:

$$\mathcal{V}_n \subset \mathcal{V}_{n-1}, \qquad P \in \mathcal{V}_n, \qquad (n = 1, 2, \ldots). \tag{1.4–1}$$

Let the volume of $\mathcal{V}_n$ be V_n and the mass of the matter contained in $\mathcal{V}_n$ be M_n. We form the ratio M_n/V_n. Then if the limit of M_n/V_n exists as $n \to \infty$ and $V_n \to 0$, the

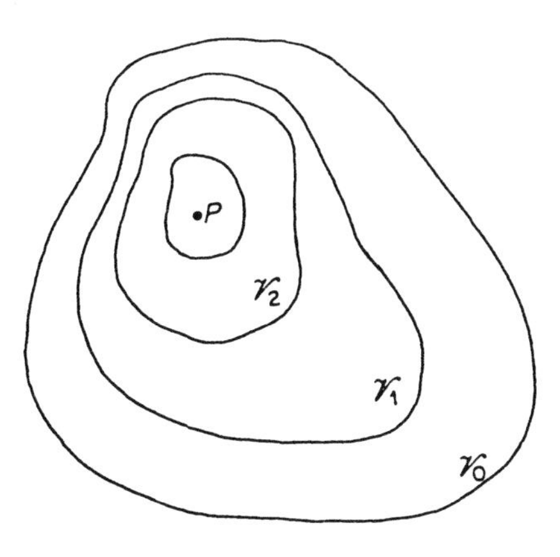

Figure 1.1 A sequence of spatial domains converging on P.

limiting value is defined as the *density of the mass distribution at the point P* and is denoted by $\rho(P)$:

$$\rho(P) = \lim_{\substack{n\to\infty \\ V_n\to 0}} \frac{M_n}{V_n}. \tag{1.4–2}$$

If the density is well defined everywhere in $\mathcal{V}_0$, the mass is said to be *continuously distributed.*

A similar consideration can be used to define the density of momentum, the density of energy, and so on. *A material continuum is a material for which the densities of mass, momentum, and energy exist in the mathematical sense. The mechanics of such a material continuum is continuum mechanics.*

This is the usual definition of a material continuum. However, if we adhere rigorously to it, it will be of no use to science and technology, because the set of real-world systems satisfying such a definition is empty. For example, no gas would satisfy Eq. (1.4–2) when V_n becomes smaller than the mean free path. And no fluid would satisfy the equation when V_n becomes atomic sized. No polycrystalline metal or fiber composite structure, no ceramic, and no polymer plastic can meet this requirement; no living organism, no tissue of any animal, no single cell, and no cell aggregate can either.

1.5 OUR DEFINITION OF A CONTINUUM

We shall define a material as a continuum in a way similar to the classical approach presented in the preceding section, except that the size of V_n will be bounded below and the material particles will not be required to have a one-to-one isomorphism with the real number system. The material particles may be discrete and have voids between them. For the concept of the density of the material, consider a point P in a space $\mathcal{V}_0$. Consider also a sequence of subspaces $\mathcal{V}_1, \mathcal{V}_2, \ldots, \mathcal{V}_n$ in $\mathcal{V}_0$, with volumes $V_1, V_2, \ldots, V_n$ respectively, each enclosing the next one and all enclosing

P. As $n \to \infty$, *the limit of* V_n *tends to a finite positive number* ω. Let the mass of the material enclosed in $\mathcal{V}_n$ be M_n. The sequence of the ratios M_n/V_n is said to have a *limit* ρ *with an acceptable variability* ϵ if

$$\left|\rho - \frac{M_n}{V_n}\right| < \epsilon$$

as $n \to \infty$. The quantity ρ is then said to be the *density of the material at* P *with an acceptable variability* ϵ *in a defining limit volume* ω.

We define the momentum of the material particles per unit volume and the energy per unit volume similarly, each associated with an acceptable variability and a defining volume. Later (see Sec. 1.6), we shall deal with the force acting on a surface of a material body, and it would be necessary to consider whether a limit of force per unit area exists at any point on the surface with an acceptable variability in a defining limit area. If it does exist, then the limit is called the *traction* or *stress*, and the collective entity of tractions in every orientation of the surface is called the *stress tensor*. Further, in Chap. 5 we shall consider the change of spacing between particles and define the *strain tensor*. The existence of strain components will be associated with an acceptable variability and a defining limit length.

If, with a clear understanding of acceptable variabilities and defining limit lengths, areas, and volumes, the density, momentum, energy, stress, and strain can be defined at every point in the space $\mathcal{V}_0$, and if they are all continuous functions of spatial coordinates in $\mathcal{V}_0$, then we say that the material in $\mathcal{V}_0$ is a *continuum*.

If a material is a continuum in the classical sense, then it is also a continuum in our sense. For a classical continuum the acceptable variability and the defining limit length, area, and volume are zero.

In other books on continuum mechanics, the authors say or imply that to decide whether continuum mechanics is applicable to science and technology is a matter for the experimenters in each discipline to decide. I say instead that every experimenter knows that the classical theory does not apply; hence, it is the responsibility of the theorist to refine the theory to fit the real world. Our approach does fit many fields of science and technology; the need to specify acceptable variabilities and defining dimensions is the price we pay.

1.6 THE CONCEPT OF STRESS IN OUR DEFINITION OF A CONTINUUM

Consider a material B occupying a spatial region V (Fig. 1.2). Imagine a closed surface S within B, and consider the interaction between the material outside S and that in the interior. Let ΔS be a small surface element on S. Let us draw, from a point on ΔS, a unit vector $\boldsymbol{\nu}$ normal to ΔS, with its direction pointing outward from the interior of S. Then we can distinguish the two sides of ΔS according to the direction of $\boldsymbol{\nu}$. Let the side to which this normal vector points be called the positive side. Consider the part of material lying on the positive side. This part exerts a force $\Delta\mathbf{F}$ on the other part, which is situated on the negative side of the

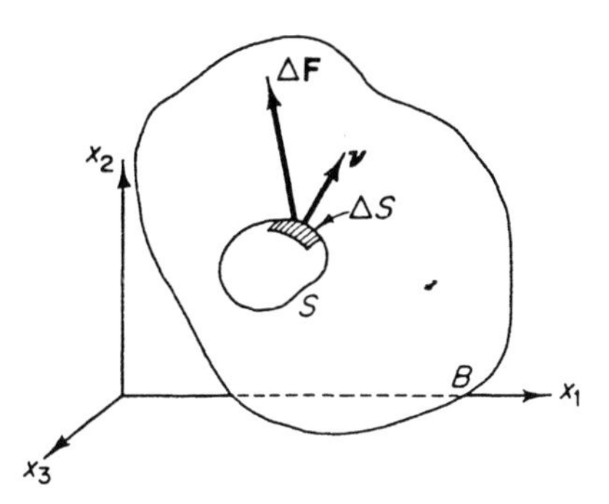

Figure 1.2 Stress principle.

normal. The force $\Delta\mathbf{F}$ depends on the location and size of the area and the orientation of the normal. We introduce the assumption that *as ΔS tends to a small but bounded size α, the ratio $\Delta\mathbf{F}/\Delta S$ tends to a definite limit $d\mathbf{F}/dS$ with an acceptable variability ϵ, and the moment of the force acting on the surface ΔS about any point within the area vanishes in the limit of small but bounded area α with an acceptable variability*. The limiting vector will be written as

$$\overset{\nu}{\mathbf{T}} = \frac{d\mathbf{F}}{dS},$$

where the superscript $\boldsymbol{\nu}$ is introduced to denote the direction of the normal $\boldsymbol{\nu}$ of the surface ΔS. The limiting vector $\overset{\nu}{\mathbf{T}}$ is called the *traction*, or the *stress vector*, and represents the force per unit area acting on the surface.

The assertion that there is defined upon any imagined closed surface S in the interior of a continuum a stress vector field whose action on the material occupying the space interior to S is equipollent to the action of the exterior material upon it is the *stress principle of Euler and Cauchy*. We accept this principle as an axiom; however, it is no more than a basic simplification. For example, there is no *a priori* justification why the interaction of the material on the two sides of the surface element ΔS must be momentless. Indeed, some people who do not like the restrictive idea that the moment of the forces acting on the surface ΔS about any point within the area vanishes in the limit have proposed a generalization of the stress principle of Euler and Cauchy, namely, that across any small surface element in a material, the action of the exterior material upon the interior is equipollent to *a force and a couple*. The resulting theory requires the concept of *couple stress* and is much more complex than the conventional theory. So far, no real application has been found for the couple-stress theory; hence, we shall not discuss it further in this book.

1.7 ABSTRACT COPY OF A REAL CONTINUUM

Once it is decided that a material body can be regarded as a continuum, one can make an abstract copy of the real material according to the classical definition. The abstract copy is isomorphic with the real number system: It is an idealization of the real material. The rules of idealization are as follows: The mass density of the idealized system is the same as the real one in the range of its definition. When a set of forces is applied to both the real material and the abstract copy, the stress and strain of the two systems are the same, except that the calculus of the idealized system can be carried out rigorously, whereas that of the real material would have a limitation on the lower bound of sizes and a statistical variability that must be evaluated. The constitutive equation of the real material is used to describe the mechanical properties of the idealized copy. The real system satisfies the equations of motion (or equilibrium), continuity, and balance of energy of the idealized system with errors whose bounds can be calculated. Often, a library full of results concerning the abstract copy exists and can be borrowed. The known, acceptable variability and defining dimensions will allow us to evaluate the difference between the real material and the abstraction and let us know something about the real system.

The constitutive equation of an abstract copy of a real material in a certain range of sizes may differ from the constitutive equation of another copy of the same material in a different range of sizes. If the constitutive equations of the real material were the same for consecutive ranges of sizes, then the abstract copy has one constitutive equation in the total range. If the constitutive equations of the real material are different in different ranges of sizes, then the abstract copies in successive ranges of sizes have different constitutive equations. This is indeed a useful feature of our system, because it allows us to recognize different structures of an object at different dimensions of observation, to ask different questions at different levels of sizes, and to gain a better understanding of the whole.

The hierarchy of the constitutive equations at different dimensions of observation of a material is related to the similarity or dissimilarity of the structure of the material at different dimensions. The structure of a material may be *fractal*, i.e., self-similar in successive ranges of sizes; or it may not be fractal. For example, the geometric pattern of the airways of the lung, from the largest bronchi to the smallest bronchioles, is fractal in structure, so the bronchi in this range of sizes can be expected to obey the same constitutive equation. The structural pattern of the alveolar ducts, from the respiratory bronchiole to the alveolar sacs, is another fractal; hence, a different constitutive equation is expected to hold in this range. The pulmonary alveoli—the alveolar walls—are not fractal at all; neither are the collagen and elastin fibers in the alveolar walls. Hence, their mechanical properties need entirely different descriptions.

1.8 WHAT IS CONTINUUM MECHANICS ABOUT?

We shall call a continuum enclosed in a closed surface a *body*. The surface may be real, like the skin of a man or the shell of an airplane. But it may also be imaginary, visualized to enclose a bit of space.

Real-world material objects are subjected to forces acting on their bodies (such as gravitational and electromagnetic forces) and forces acting on their surfaces (such as atmospheric pressure, wind and rain, burdens to be carried, and loads to be transmitted to a remote place). If the body is a continuum in the sense described in the preceding section, then we would want to know how the material in the body reacts to the external forces. The determination of the internal condition of a body in response to external forces is what continuum mechanics is about.

1.9 AXIOMS OF CONTINUUM MECHANICS

The axioms of physics are taken as the axioms of continuum mechanics. In particular, we use Newton's laws of motion and the first and second laws of thermodynamics in this book.

There are three additional axioms of continuum mechanics. First, *a material continuum remains a continuum under the action of forces*. Hence, two particles that are neighbors at one time remain neighbors at all times. We do allow bodies to be breakable (i.e., they can be fractured); but the surfaces of fracture must be identified as newly created external surfaces. In living bodies, we allow new growth (e.g., cellular or extracellular mass increased, new cells migrated into an area of the body, or proliferated from existing cells by division) and resorption (cellular or extracellular mass reduced, cells migrated out of an area of the body, or cells died and subsequently washed away by blood, or absorbed in tissue). Each newly added or resorbed cell creates a new surface in the body.

The second axiom of continuum mechanics is that stress, as described in Sec. 1.6, and strain, as described in Chap. 5, can be defined everywhere in the body. The third axiom of continuum mechanics is that the stress at a point is related to the strain and the rate of change of strain with respect to time at the same point. This axiom is a great simplifying assumption. It asserts that the stress at any point in the body depends only on the deformation in the immediate neighborhood of that point. This stress-strain relationship may be influenced by other parameters, such as temperature, electric charges, nerve impulses, muscle contraction, ion transport, etc., but these influences can be studied separately.

1.10 A BIOLOGICAL EXAMPLE OF A HIERARCHY OF CONTINUA DEPENDING ON THE SIZE OF THE OBJECT INVOLVED IN A SCIENTIFIC INQUIRY

We are familiar with telescopic views of the sky and microscopic views of cells, tissues, metals, and ceramics. As scales of observation change, the object appears different. An example is the human lung. Figure 1.3 shows that the lung may be considered to be composed of three trees: an airway tree, an arterial tree, and a venous tree. The airway tree is for ventilation. The trachea is divided into bronchi, which enter the lung and subdivide repeatedly (in humans, 23 times, statistically speaking) into smaller and smaller branches and, finally, into the smallest units, called pulmonary alveoli. Figure 1.4 shows a photograph of human alveoli, as seen

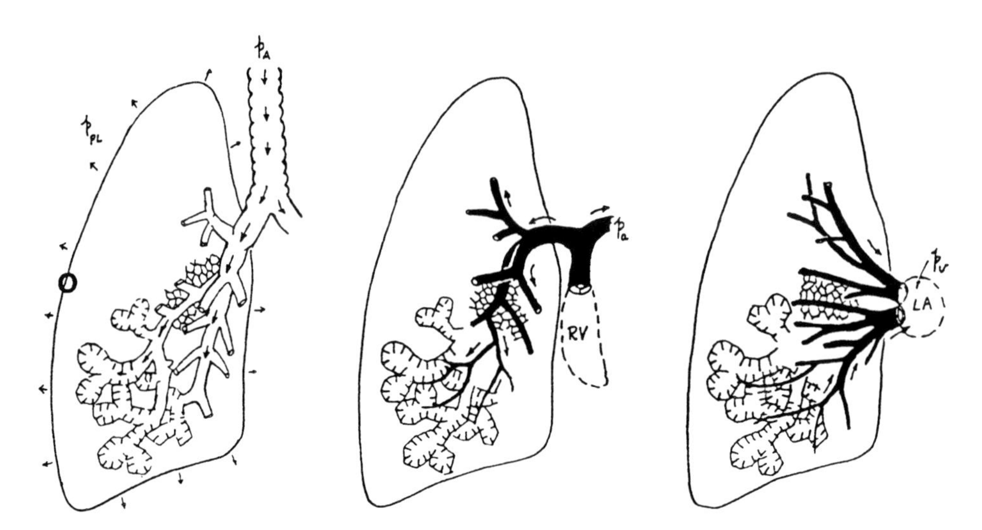

Figure 1.3 A conceptual sketch showing the lung as composed of three trees: the airway tree (trachea–bronchi–bronchioles–alveolar ducts–alveoli), shown on the left; the pulmonary arterial tree (arteries–arterioles–capillaries), shown at the center; and the pulmonary venous tree (venules–veins–left atrium), shown on the right. Total height on the order of 40 cm.

in a microscope. The photo covers a small area within the little circle on the left border of the leftmost drawing in Fig. 1.3. The walls of the alveoli are capillary blood vessels. Every wall of each alveolus is exposed to gas on both sides. The pulmonary artery also bifurcates again and again until it becomes capillary blood vessels that lie in the alveolar wall. The venous tree begins with the capillaries in the alveolar wall. The veins converge repeatedly until they become pulmonary veins, which enter the left atrium of the heart. The main function of the lung takes place in the alveoli. The venous blood takes up oxygen from the gas in the alveoli and releases carbon dioxide to the alveolar gas. The gas exchange takes place across the alveolar wall. Figure 1.5 shows a micrograph of the alveolar walls enclosed in a small circle on the left border of Fig. 1.4. Figure 1.6 shows the collagen fibers in the alveolar walls. The collagen fibers were stained with silver and appear as black bundles. The collagen fibers are formed by fibrils. Fibrils are formed by collagen molecules. One could go on to smaller hierarchies consisting of molecules, atoms, nuclei, and quarks.

Depending on what property of the lung we wish to investigate, we may consider the lung as a continuum at successive hierarchies. For example, if one is interested in comparing the difference in strain in the upper part of the human lung from that in the lower part, then the individual alveoli can be considered infinitesimal and one can speak of deformation averaged over volumes that are large compared with the volume of a single alveolus, but small compared with the whole lung. Such an approximation would be appropriate in studying the interaction

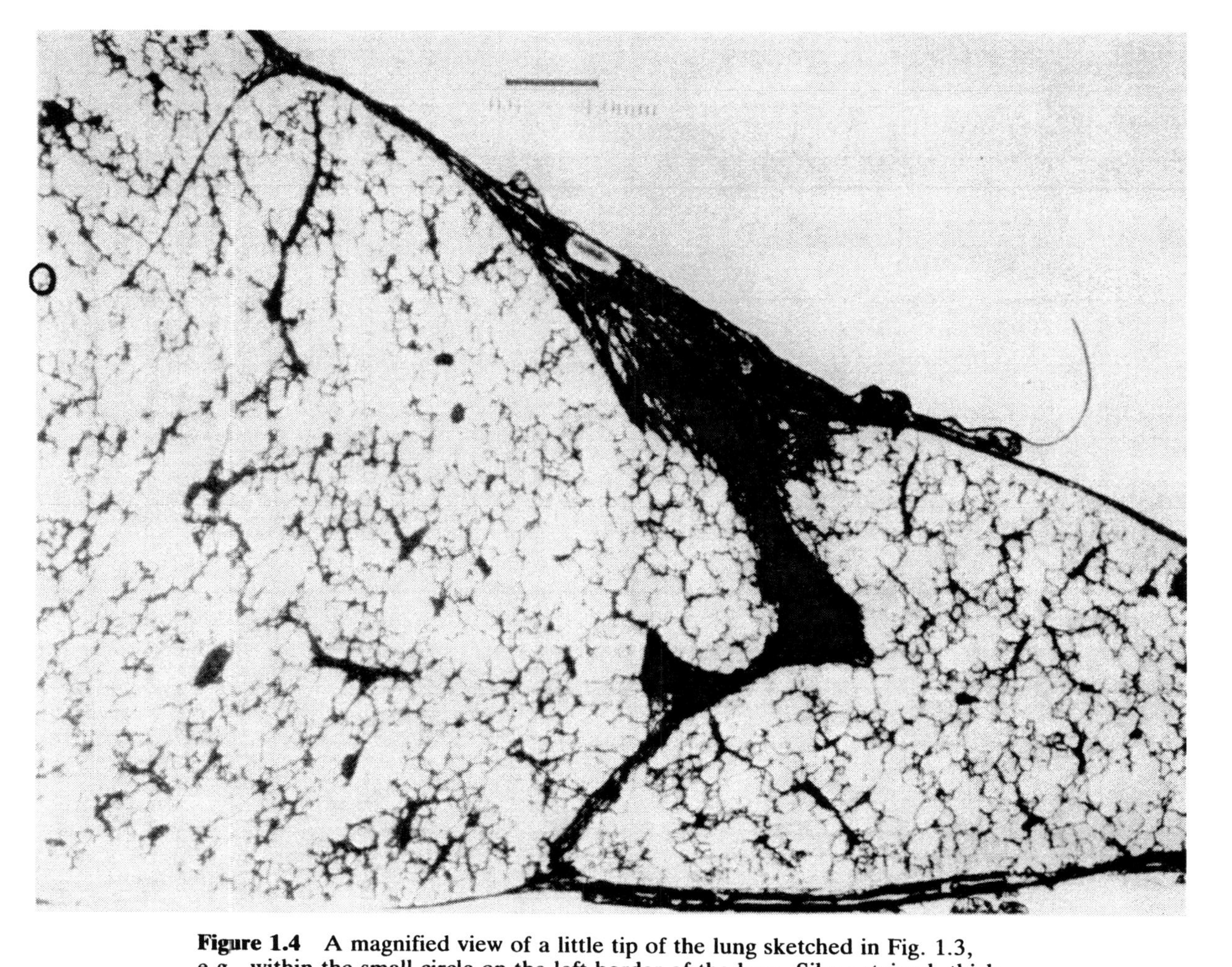

Figure 1.4 A magnified view of a little tip of the lung sketched in Fig. 1.3, e.g., within the small circle on the left border of the lung. Silver-stained, thick (150 μm) section of lung tissue. *Black lines*, collagen fibers. *Gray regions*, alveolar walls. From Matsuda, M., Fung, Y.C., and Sobin, S.S., "Collagen and Elastin Fibers in Human Pulmonary Alveolar Mouths and Ducts," *J. Appl. Physiology* 63(3): 1185–1194, 1987. Reproduced by permission.

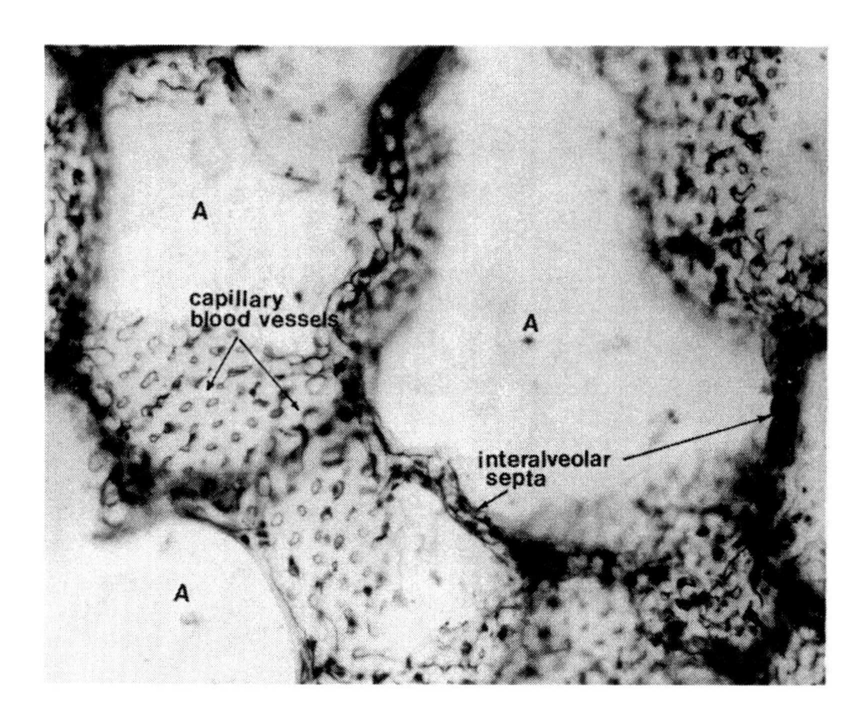

Figure 1.5 A magnified view of the alveolar walls enclosed in a small circle on the left border of Fig. 1.4, showing the capillary blood vessels in the walls (also called the interalveolar septa). A = alveolar gas space.

between the lung and the chest wall, the distribution of pleural pressure, or the distribution of ventilation in the whole lung (because ventilation is related to alveolar size, which is proportional to the strain). For these problems, a hierarchy with a minimum dimension on the order of 1 cm will suffice. On the other hand, if one is interested in the blood flow in the large pulmonary artery, then the blood can be regarded as a homogeneous fluid and the blood vessel wall can be regarded as a continuum, with a minimum dimension for the definition of stress and strain on the order of 10 μm. If one is interested instead in the stress in a single alveolar wall (with a plane area on the order of $100 \times 100\ \mu m^2$ and a thickness of about 10 μm), then even the individual collagen and elastin fibers in the wall cannot be ignored, and the wall must be considered a composite structure made of several different materials and constructed in a special way. What kinds of averages are useful depend on what the purpose of one's investigation is. Engineers, biologists, and physicists are concerned about these questions. We bend the classical continuum mechanics in this direction to make it useful for dealing with practical problems.

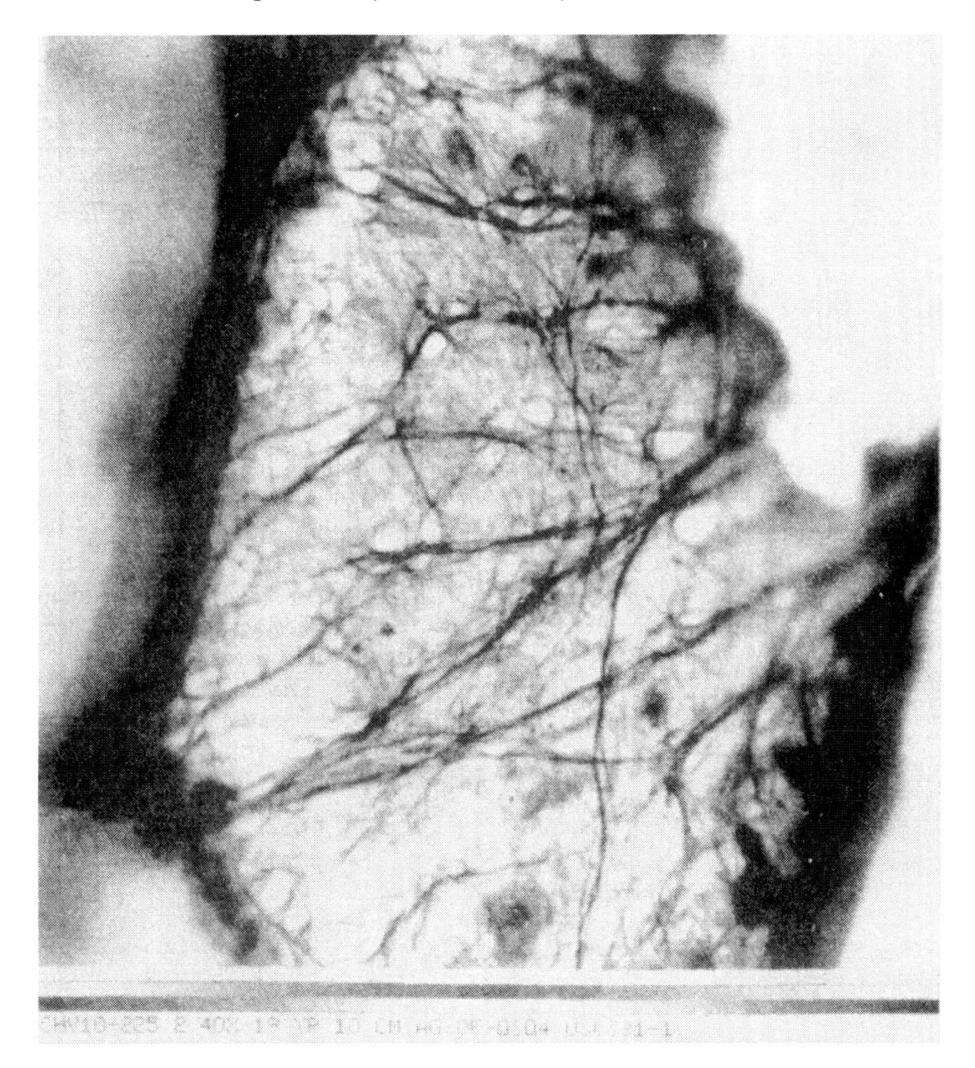

Figure 1.6 Collagen fibers in a pulmonary alveolar wall of a human lung inflated to a transpulmonary pressure (alveolar gas pressure minus pleural pressure) of 10 cm H_2O. Scale is marked on border; 800 pixels are equal to 200 μm in tissue. OsO_4 fixed. Silver stained. Black lines are collagen fibers. Larger ones are about 1 μm wide. From Sobin, S.S., Fung, Y.C., and Tremer, H.M., "Collagen and Elastin Fibers in Human Pulmonary Alveolar Walls," *J. Appl. Physiology* 64(4): 1659–1675, 1988. Reproduced by permission.

1.11 ELEMENTARY TOPICS THROUGH WHICH BASIC IDEAS EVOLVED

As an introduction to the rest of the book, let us consider some elementary topics that are simple and useful and that have been fundamental in the history of mechanics. These include Newton's laws of motion, the equations of equilibrium, the use of free-body diagrams, the analyses of a truss, a beam, a block, a plate, and a shell, and the classical beam theory. If you are familiar with these topics, you can go over them quickly. If some points are new to you, I can assure you that learning them would not be a waste of your time.

Newton's Laws of Motion

Newton's laws are stated with respect to material particles in a three-dimensional space that obeys Euclidean geometry. A material particle has a unique, positive measure, the *mass* of the particle. The location of the particle can be described with respect to a rectangular Cartesian frame of reference. It is assumed that an *inertial frame of reference* exists, with respect to which the Newtonian equations of motion are valid. It can be shown that any frame of reference moving with a uniform velocity with respect to an inertial frame is again inertial. Consider a particle of mass m. Let the position, velocity, and acceleration of this particle be denoted by the vectors $\mathbf{x}$, $\mathbf{v}$, and $\mathbf{a}$, respectively, all defined in an inertial frame of reference. By definition,

$$\mathbf{v} = \frac{d\mathbf{x}}{dt}, \qquad \mathbf{a} = \frac{d\mathbf{v}}{dt}. \tag{1.11–1}$$

Let $\mathbf{F}$ be the total force acting on the particle. If $\mathbf{F} = 0$, then *Newton's first law* states that

$$\mathbf{v} = \text{constant}. \tag{1.11–2}$$

If $\mathbf{F} \neq 0$, then *Newton's second law* states that

$$\frac{d}{dt} m\mathbf{v} = \mathbf{F}, \quad \text{or} \quad \mathbf{F} = m\mathbf{a}. \tag{1.11–3}$$

When Eq. (1.11–3) is written as

$$\mathbf{F} + (-m\mathbf{a}) = 0, \tag{1.11–4}$$

it appears as an equation of equilibrium of two forces. The term $-m\mathbf{a}$ is called the *inertial force*. Equation (1.11–4) states that the sum of the external force acting on a particle and the inertial force vanishes; i.e., the inertial force balances the external force. The Newtonian equation of motion stated in this way is called *D'Alembert's principle*.

Now, consider a system of particles that interact with each other. Every particle is influenced by all the other particles in the system. Let an index I denote the Ith particle. Let $\mathbf{F}_{IJ}$ denote the force of interaction exerted by particle number J on particle number I and $\mathbf{F}_{JI}$, that of particle I on particle J. Then *Newton's third law* states that

$$\mathbf{F}_{IJ} = -\mathbf{F}_{JI} \quad \text{or} \quad \mathbf{F}_{IJ} + \mathbf{F}_{JI} = 0. \tag{1.11–5}$$

If $I = J$, then we set $\mathbf{F}_{II} = 0$, (I not summed) in agreement with Eq. (1.11–5). Let K be the total number of particles in the system. The force $\mathbf{F}_I$ that acts on the Ith particle consists of an external force $\mathbf{F}_I^{(e)}$, such as gravity, and an internal force that is the resultant of mutual interaction between particles. Thus,

$$\mathbf{F}_I = \mathbf{F}_I^{(e)} + \sum_{J=1}^{K} \mathbf{F}_{IJ}. \tag{1.11–6}$$

The equation of motion of the Ith particle is, therefore,

$$\frac{d}{dt} m_I \mathbf{v}_I = \mathbf{F}_I^{(e)} + \sum_{J=1}^{K} \mathbf{F}_{IJ}, \qquad (I = 1, 2, \ldots, K). \tag{1.11–7}$$

Each particle is described by such an equation. The totality of K equations describes the motion of the system.

To make further progress, we must specify how the forces of interaction $\mathbf{F}_{IJ}$ can be computed. Such a specification is a statement of the material property of the system of particles and is referred to as a *constitutive equation of the material system*.

Equilibrium

A special motion is *equilibrium*, i.e., one in which there is no acceleration for any particles of the system.

At equilibrium, Eq. (1.11–7) becomes

$$\mathbf{F}_I^{(e)} + \sum_{J=1}^{K} \mathbf{F}_{IJ} = 0, \qquad (I = 1, 2, \ldots, K). \tag{1.11–8}$$

Summing over I from 1 to K, we obtain

$$\sum_{I=1}^{K} \mathbf{F}_I^{(e)} + \sum_{I=1}^{K} \sum_{J=1}^{K} \mathbf{F}_{IJ} = 0. \tag{1.11–9}$$

In the last sum, whenever $\mathbf{F}_{IJ}$ appears, $\mathbf{F}_{JI}$ appears also; they add up to zero, according to Eq. (1.11–5). Therefore, Eq. (1.11–9) reduces to

$$\sum_{I=1}^{K} \mathbf{F}_I^{(e)} = 0. \tag{1.11–10}$$

That is, *for a body in equilibrium, the summation of all external forces acting on the body is zero*.

Next, let us consider the tendency of a body to rotate. If a body is pivoted at a point O and is acted on by a force $\mathbf{F}_I$, then the *moment* of the force about O that tends to cause the body to rotate about O is given by the vector product $\mathbf{r}_I \times \mathbf{F}_I$, where $\mathbf{r}_I$ is a radius vector from O to any point on the line of action of the force $\mathbf{F}_I$. Forming a vector product of $\mathbf{r}_I$ with Eq. (1.11–8), i.e., with every term of that equation, setting $I = 1, 2, \ldots K$, adding the results together, i.e., summing over I from 1 to K, and using Eq. (1.11–5) to simplify the grand total, we obtain

$$\sum_{j=1}^{k} \mathbf{r}_I \times \mathbf{F}_I^{(e)} = O. \qquad (1.11\text{–}11)$$

The choice of the point O is arbitrary. Hence, we obtain the second condition of equilibrium of the body: *The summation of the moments of all the external forces acting on the body about any point is zero.*

Use of Free-Body Diagram in the Analysis of Problems

The word *body* or the phrase *a system of particles* used in the previous section can be interpreted in the most general way. If a machine is in equilibrium, every part of it is in equilibrium. By a proper selection of the parts to be examined, a variety of information can be obtained. This method is like a surgeon's exploration of a diseased organ by biopsy. With imagined sections, we cut free certain parts of the body and examine their conditions of equilibrium. A diagram of the part with all the external forces acting on it clearly indicated is called a *free-body diagram*. The method we use is therefore called the *free-body method*.

Example 1. Analysis of a Truss

Trusses are frame structures commonly seen in bridges, buildings, lifts in construction sites, TV towers, radio astronomical antennas, etc. Figure 1.7(a) shows a typical truss of a small railway bridge. It is made of steel members *ab*, *bc*, *ac*, . . . , bolted together. The joints at *a*, *b*, *c*, . . . may be considered pin joints, meaning that the members are joined together with pins and are free to rotate relative to each other. The whole truss is "simply supported" at the ends *a* and *l*, which anchor the truss but impose no moment on the truss. The support at *l* rests on a roller so that the horizontal reaction from the foundation is eliminated.

Trusses are made with slender members. The weights of these members are small compared with the load carried by the truss. Hence, as a first approximation, we may ignore the weights of the members.

Since each member is pin jointed and is considered weightless, the condition of equilibrium of the member requires that the pair of forces coming from the joints must be equal and opposite. Hence, each member can transmit forces only along its axis.

Let the truss be loaded with a weight W at the center (point g). We would like to know the load acting in various members of the truss.

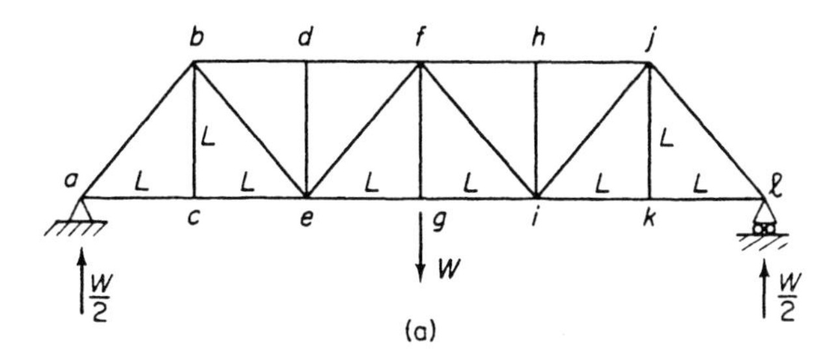

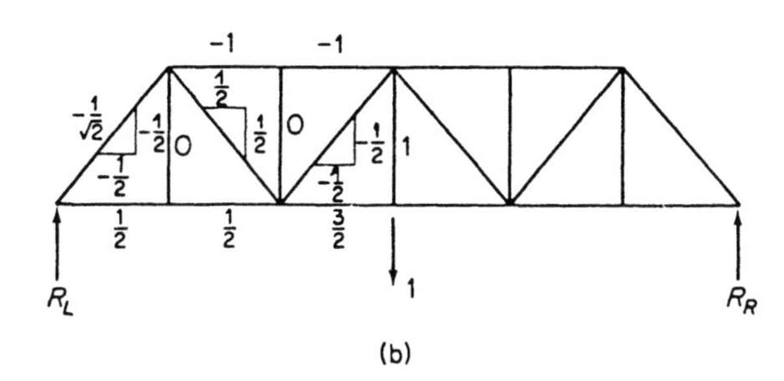

Figure 1.7 A simply-supported, pin-jointed truss loaded by a force W. (a) Nomenclature of the joints, $(a, b, \ldots)$, and the lengths of the members, (L). (b) Forces in members when $W = 1$. + for tension, − for compression. The triangles on members ab, be, ef show the horizontal and vertical components and the resultant of the load in these members.

Let us first compute the reactions at the two supports. Consider the whole truss as a free body. It is subjected to three external forces: W, R_L, and R_R [Fig. 1.7(b)]. The conditions of equilibrium are:

(1) Summation of vertical forces is zero:

$$W - R_L - R_R = 0.$$

(2) Summation of moments about the point a is zero:

$$W \cdot 3L - R_R \cdot 6L = 0.$$

The solutions are $R_R = R_L = W/2$.

Next, we wish to know the tension in the members ab and ac. For this purpose, we cut through ab and ac with an imaginary plane and consider the portion $ab'c'$ as a free body [see Fig. 1.7(c)]. At the cut and exposed end b', the tension $\mathbf{F}_{ab}$ acts in the member ab. The tension $\mathbf{F}_{ac}$ acts at the cut c' of ac. At the support, a force $W/2$ acts (the reaction R_L just computed). Now, summing all the forces in the vertical direction, and letting F denote the magnitude of $\mathbf{F}$, we obtain

$$\frac{W}{2} + F_{ab} \sin \theta = 0.$$

Since $\theta = 45°$, $\sin \theta = \sqrt{2}/2$, and we obtain $F_{ab} = -W/\sqrt{2}$.

Summing all the forces in the horizontal direction yields

$$F_{ab} \cos \theta + F_{ac} = 0.$$

Hence, for $\theta = 45°$ and $F_{ab} = -W/\sqrt{2}$, we obtain $F_{ac} = W/2$.

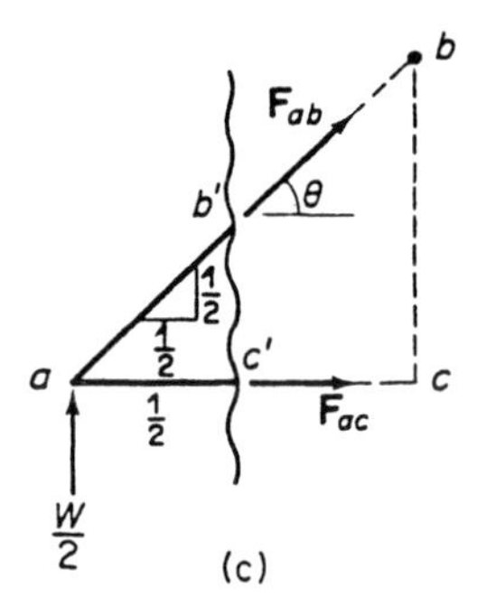

Figure 1.7 (Cont.) (c) Free-body diagram of the members ab' and ac'.

Next, we compute the tensions in the members df, ef, and eg. We pass a cut through these members and consider the left portion of the truss as a free body [see Fig. 1.7(d)]. For convenience, we resolve the tension in the member ef into two components: the horizontal H_{ef} and the vertical V_{ef}. All the external forces acting on this free body are shown in Fig. 1.7(d). The equilibrium conditions are:

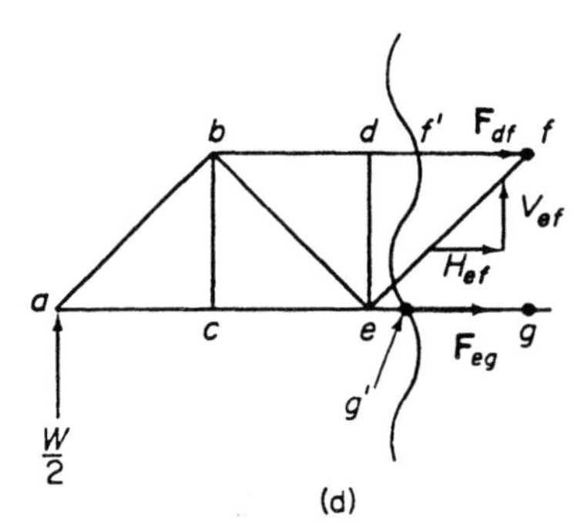

Figure 1.7 (Cont.) (d) Free-body diagram of a part of the truss to the left of the cutting surface passing through f' and g'.

(1) Summation of all horizontal forces vanish:

$$F_{df} + H_{ef} + F_{eg} = 0.$$

(2) Summation of all vertical forces vanish:

$$\frac{W}{2} + V_{ef} = 0.$$

(3) Sum of moments about the point e vanish:

$$\frac{W}{2}\cdot 2L + F_{df}\cdot L + 0\cdot F_{ef} + 0\cdot F_{eg} = 0.$$

(4) Sum of moments about the point f vanish:

$$\frac{W}{2}\cdot 3L - F_{eg}\cdot L = 0.$$

Hence, from condition 3, we obtain $F_{df} = -W$; from condition 4, we obtain

$F_{eg} = 3W/2$; from condition 2, we obtain $V_{ef} = -W/2$; and, finally, from condition 1, we obtain $H_{ef} = -W/2$. A similar calculation can be done for other members of the truss.

The results can be presented as in Fig. 1.7(b). Since the load in every member is proportional to W, we may express the load in each member in units of W and set W equal to 1. For the truss design, it is important to know whether a member is subjected to tension or compression. (A rod pushed at both ends is said to subject to *compression*; a rod pulled at both ends is said to be subject to *tension*.) The design of a steel member in tension is different from that in compression. A member in tension may fail by plastic yielding; a member in compression may fail by elastic buckling. Whereas the signs of F_{ac}, F_{ab}, V_{ef}, H_{ef}, etc., in the preceding equations depend on the directions of the vectors we draw on the free-body diagram (which is done arbitrarily), the tension-compression character of the stress in each member is fixed by the load W. We present the final result in Fig. 1.7(b), with the *convention* that *if a member is in tension, we give the load a positive sign; if the member is in compression, we give the load a negative sign*. Thus, in Fig. 1.7(b), we see that the member *ab* is in compression; the members *be*, *ac*, and *eg* are in tension; and the member *ef* is in compression.

Example 2. A Simply Supported Beam

A beam is a solid member that resists lateral load by bending. Figure 1.8(a) illustrates a simply supported beam. Its function is similar to the truss discussed in Example 1. However, whereas the truss resists the load by tension or compression in the members, the beam resists it by continuously distributed tensile and compressive stresses.

The ends of the beam shown in Fig. 1.8(a) are supported on pins that do not resist moment. The reactions at the supports are obviously $W/2$.

Let us ask how the beam resists the external load. For this purpose, let us make a cut with an imaginary plane perpendicular to the beam at a distance x from the left end [Fig. 1.8(b)]. Consider the free-body diagram of the left portion of the beam, as shown in Fig. 1.8(b). At the cut surface, there acts a "shear force" S tangential to the cut, an "axial force" H perpendicular to the cut, and a couple M, called the *bending moment* in the beam. The conditions of equilibrium are

(1) Sum of all forces in the horizontal direction vanish:

$$H = 0.$$

(2) Sum of all forces in the vertical direction vanish:

$$S = \frac{W}{2}.$$

(3) Sum of the moments of all forces about the left end support vanish:

$$M = Sx = \left(\frac{W}{2}\right)x.$$

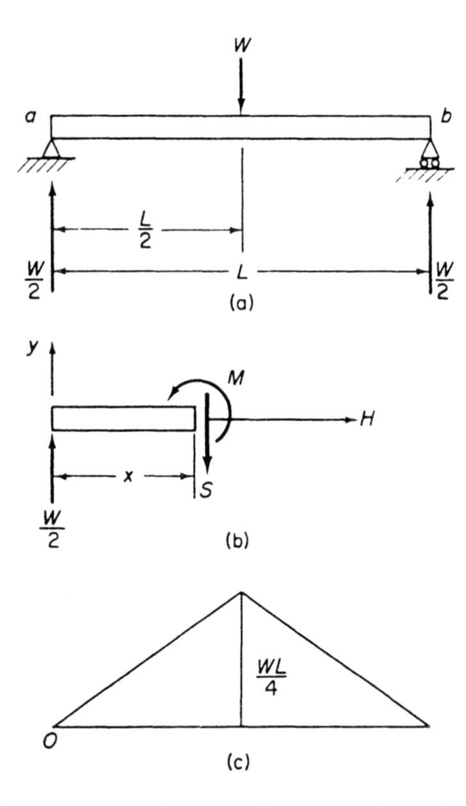

Figure 1.8 A simply supported beam. (a) A sketch of the beam of length L, loaded by a force W. (b) Free-body diagram of a part of the beam to the left of a plane perpendicular to the beam axis at a distance x from the left end. The shear, S, tension, H, and bending moment, M, act in the cross section at x. (c) The bending moment diagram showing the bending moment, M, in the ordinate and the position of the cross section, x, in the abscissa. The length of the base is L. The maximum bending moment is $WL/4$ and acts in the cross section located at $x = L/2$.

Thus, in a cross section at a distance x from the left end, the stresses in the beam are equipollent to a shear force $S = W/2$ and a moment $M = Wx/2$.

As x varies, the moment varies, as shown in Fig. 1.8(c). Such a figure is called a *bending moment diagram* of a beam subjected to a specific loading. Knowing the bending moment, we can compute the stresses acting in a beam. [See Eq. (1.11–31) infra.] Beams are generally designed on the basis of the maximum bending moments they have to resist.

Example 3. Stresses in a Block

Consider a block of solid material compressed by a load W acting on its end, as shown in Fig. 1.9(a). The block is a rectangular parallelepiped. We wish to know the stress in the block.

Let us assume that at a distance sufficiently far away from the ends, the stresses are uniform in the block, i.e., everywhere the same. Let us erect a set of rectangular Cartesian coordinates x, y, z, as shown in the figure, with the z-axis parallel to the axis of the block. Let us pass an imaginary plane $z = 0$ through the block and consider the free-body diagram of the upper part of the block, Fig. 1.9(b). The stresses acting on the surface $z = 0$ must have a resultant force and a resultant moment. Applying the conditions of equilibrium as before, we find at once that the horizontal component of the resultant force vanishes, that the vertical

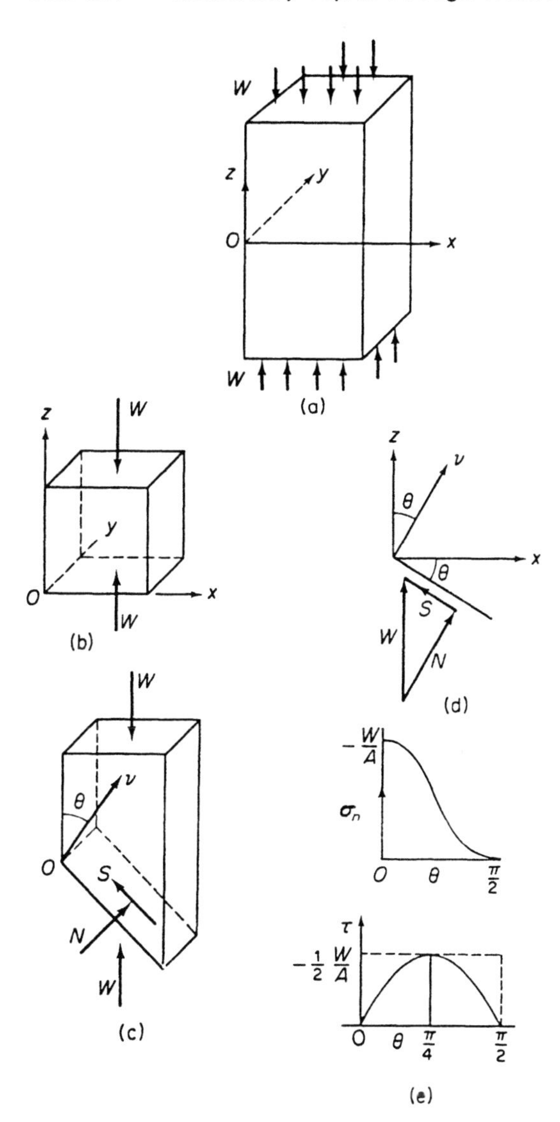

Figure 1.9 Stresses in a block. (a) The frame of reference x, y, z, and the loading, W, acting on the block. (b) Free-body diagram of the part of the block above the plane $z = 0$. (c) Free-body diagram of the upper part of the block cut by a plane whose normal vector $\boldsymbol{\nu}$ makes an angle θ with the z-axis. (d) The force acting on the inclined plane with normal vector $\boldsymbol{\nu}$ is resolved into a normal force, N, and a shear force, S. (e) The variations of the normal and shear stresses acting on the inclined plane, σ_n and τ, respectively, are plotted against the angle of inclination, θ.

component of the resultant force is W, and that the resultant moment is zero. In this case, we say that the *stress* acting on a plane $z = 0$ is a *compressive, normal* stress with a magnitude

$$\sigma = -\frac{W}{A}, \tag{1.11–12}$$

where A is the normal cross-sectional area of the block (cut by the plane $z = 0$ normal to the axis of the block). The stress is *compressive* because the material is compressed in that direction. It is *normal* because σ is a force (per unit area) perpendicular (normal) to the surface $z = 0$. We indicate the compressive nature of the stress by giving it a negative value.

Next, let us make a cut with a plane that is inclined at an angle θ to the xy-plane. The simplest way to express the orientation of a plane is to specify the normal vector of the plane. Let $\boldsymbol{\nu}$ be the *unit* normal vector (of unit length) of the inclined plane, and $\mathbf{z}$ be a unit vector in the direction of the z-axis; then $\boldsymbol{\nu}\cdot\mathbf{z} = \cos(\boldsymbol{\nu}, \mathbf{z}) = \cos\theta$. Consider the upper half of the block as a free body, as shown in Fig. 1.9(c). The balance of forces requires that the resultant force acting on the plane $\boldsymbol{\nu}$ (a plane whose unit normal vector is $\boldsymbol{\nu}$) is exactly equal to $-W$. This resultant can be resolved into two components, one normal and one tangential to the surface, as shown in Fig. 1.9(c). Let these be N and S, respectively; then [see Fig. 1.9(d)]

$$N = -W\cos\theta, \qquad S = -W\sin\theta. \tag{1.11–13}$$

The cross-sectional area of the block cut by the plane $\boldsymbol{\nu}$ is $A/\cos\theta$, where A is the normal cross-sectional area. Dividing N and S by the area of the surface on which they act and denoting the results by σ_n and τ, we obtain

$$\sigma_n = -\cos^2\theta W/A, \qquad \tau = -\sin\theta\cos\theta W/A. \tag{1.11–14}$$

These are the *normal stress* and *shear stress*, respectively, acting on the inclined surface $\boldsymbol{\nu}$. We give the normal stress σ_n a negative value to indicate that it is a *compressive* stress. If the load W is reversed so that the block is *pulled*, then the material on the two sides of the plane $\boldsymbol{\nu}$ tends to be pulled apart. We say in that case that the stress is in *tension* and indicate that fact by assigning σ_n a positive numerical value.

The sign convention of the shear stress will be discussed in Sec. 3.1.

The normal and shear stresses σ_n and τ vary with the angle θ. If we plot them as a function of θ, we obtain the curves shown in Fig. 1.9(e). We see that σ_n is a maximum when $\theta = 0$, whereas the shear τ reaches a maximum when $\theta = 45°$, and that the maximum shear is $\tau_{\max} = \frac{1}{2}W/A$.

The principal lesson that we learn from this example is that there are two components of stress, normal and shear, whose values at any given point in a body depend on the direction of the surface on which the stress acts. Thus, stress is a vector (σ_n, τ) associated with another vector $(\boldsymbol{\nu})$. To specify a stress, we have to specify two vectors. To specify fully the *state of stress at a given point in a continuum*, we must know the stresses acting on all possible planes $\boldsymbol{\nu}$ (i.e., sections oriented in all possible directions). A quantity such as the state of stress is called a *tensor*. Thus, this example tells us that *stress is a tensorial quantity*.

In the International System of Units (SI Units), the basic unit of force is the *newton* (n) and that of length is the *meter* (m). Thus, the basic unit of stress is *newton per square meter* (n/m^2), or *pascal* (Pa, in honor of Pascal). We also have 1 MPa = 1 n/mm^2. A force of 1 n can accelerate a body of mass 1 kg to 1 m/sec^2. A force of 1 dyne can accelerate a body of mass 1 gram to 1 cm/sec^2. Hence, 1 dyne = 10^{-5} newton. Following are some conversion factors:

1 kilogram force (kgf) = 9.806 65 newton
1 pound force ≐ 4.448 221 newtons
1 pound mass avoirdupois ≐ 0.453 592 kg
1 pound per square inch (psi) ≐ 6.894 757 kPa
1 dyne/cm^2 = 0.100 n/m^2
1 atmosphere ≐ 1.013 25 × 10^5 n/m^2 = 1.013 25 bar
1 mm Hg at 0°C ≐ 133.322 n/m^2 = 1 torr ~ $\frac{1}{7.5}$ kPa
1 cm H_2O at 4°C ≐ 98.063 8 n/m^2
1 poise (viscosity) = 0.1 newton sec/m^2 = 0.1 Pa·sec
1 cp = 0.001 Pa·sec

The notion of stress has practical value. If you have large blocks and small blocks of the same material, obviously the large ones can take larger loads, and the small ones can take smaller loads; but both will break at the same critical state of stress. Hence, engineers look at stresses.

Example 4. Stresses in a Plate

Consider a thin rectangular plate of uniform thickness and homogeneous material. As shown in Figs. 1.10(a) and (b), the plate is subjected to a uniformly distributed load acting on the surfaces $x = \pm a$ and $y = \pm b$ and no load on the surface $z = \pm h/2$. In Fig. 1.10(b), it is shown that the stress acting on the edge $x = a$ is of a magnitude σ_{xx} per unit area. (σ_{xx} is equal to the total load on the edge $x = a$ divided by the cross-sectional area of the plate cut by the plane $x = a$.) The stress acting on the edge $y = b$ is of magnitude σ_{yy} per unit area. In Fig. 1.10(c), it is shown that the plate is subjected to a shear stress τ_{xy} on the edge $x = a$ (τ_{xy} is equal to the total shear load acting on the edge $x = a$ in the direction of the y-axis, divided by the cross-sectional area of the section $x = a$) and a shear stress τ_{yx} on the edge $y = b$. σ_{xx}, σ_{yy}, τ_{xy}, and τ_{yx} are called stresses because they are all in units of force per unit area.

Applying the equations of equilibrium to the plate shown in Fig. 1.10(b), we see that the stress σ_{xx} acting on the edge $x = -a$ is equal to that acting on the edge $x = a$. Applying the equations of equilibrium to the plate shown in Fig. 1.10(c), we see that τ_{xy} on $x = -a$ is equal to τ_{xy} on $x = a$, that τ_{yx} on $y = b$ and $y = -b$ are also equal, and further, that by taking the moment of all forces (stresses × cross-sectional area) about the origin 0, we obtain

$$2a \cdot \tau_{xy} \cdot 2bh - 2b \cdot \tau_{yx} \cdot 2ah = 0, \quad \text{or} \quad \tau_{xy} = \tau_{yx}.$$

The state of stress in the plate shown in Fig. 1.10(b) is specified by σ_{xx} and σ_{yy}. The state of stress in the plate shown in Fig. 1.10(c) is specified by $\tau_{xy} = \tau_{yx}$. If a plate is subjected to both the *normal stresses* σ_{xx}, σ_{yy} and the *shear stresses* τ_{xy}, τ_{yx} [a superposition of the condition shown in Fig. 1.10(b) and (c)], then the state of stress is specified by the four numbers σ_{xx}, σ_{yy}, τ_{xy}, τ_{yx} ($\tau_{yx} = \tau_{xy}$). To clarify this

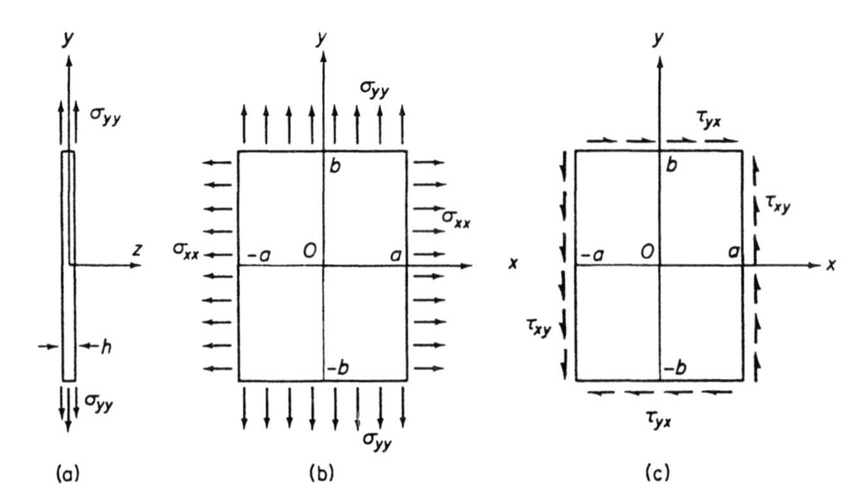

Figure 1.10 Stresses in a plate. (a) Side view of a plate in the yz-plane, showing the stress σ_{yy} acting in cross sections perpendicular to the y-axis. (b) Plan view of a plate in the xy-plane, showing the normal stress σ_{xx} acting on a cross section normal to the x-axis and in the direction of the x-axis; and the normal stress σ_{yy} acting on cross-sectional plane $\perp$ to y in the direction of y. (c) The shear stress τ_{xy} is a force per unit area acting in a plane cross section $\perp$ to the x-axis, and in the direction of y-axis. If the outer normal of the cross section points in the positive direction of the x-axis, then a positive τ_{xy} is a stress pointing in the positive direction of the y-axis. If the outer normal points to $-x$ direction then a positive τ_{xy} is a stress vector pointing in $-y$ direction. Similarly, τ_{yx} is a stress vector acting in cross section $\perp$ to y and in the direction of x.

double subscript notation, we specify the *rule* that *the first index of the stress denotes the plane on which the stress acts, whereas the second index denotes the direction in which the force acts*. Thus, the tensorial character of stress mentioned at the end of Example 3 becomes even clearer in the present example.

Example 5. A Pressurized Spherical Shell

The wall of an inflated balloon shown in Fig. 1.11(a) is in tension. We would like to know the tensile stress in the wall. For this purpose, it is simplest to cut the sphere with a diametrical plane and consider the hemisphere as a free body, as shown in Fig. 1.11(b). Let the inner radius of the shell be r_i, the outer radius be r_o, and the thickness of the wall be $h = r_o - r_i$. The internal pressure p_i acts on the inner wall. The resultant pressure force acting on a hemisphere is $\pi r_i^2 p_i$. The normal stress in the wall of the shell is not uniform; the calculation of this must await a general formulation (Chap. 10, et seq.), but it is easy to calculate the average tensile stress in the wall. Let $\langle\sigma\rangle$ be the average normal stress acting on a surface normal to the wall (i.e., passing through the center of the sphere). The area of the wall on the diametrical plane is $\pi r_o^2 - \pi r_i^2$. The resultant tensile force

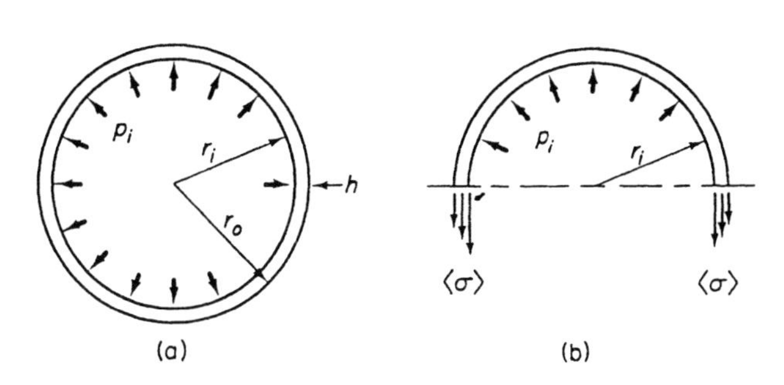

Figure 1.11 A pressurized spherical shell. (a) A diametrical cross section of the shell in a plane passing through the center of the shell, showing the inner radius, r_i, the outer radius, r_o, the wall thickness, h, and the internal pressure, p_i. (b) A free-body diagram of a thin slice of the shell cut by two parallel planes at a small distance apart, one on each side of the center of the sphere, and a third plane normal to the first two and passing through the center of the sphere. The circumferential stress σ is a stress acting on, and normal to the last mentioned cross-sectional plane. σ is not uniform across the wall of the shell. The average value of σ is $\langle\sigma\rangle$. The value of $\langle\sigma\rangle$ is computed in the text.

due to wall stress is $\pi(r_o^2 - r_i^2)\langle\sigma\rangle$. The balance of the forces in equilibrium requires, therefore, that

$$\pi(r_o^2 - r_i^2)\langle\sigma\rangle = \pi r_i^2 p_i \tag{1.11-15}$$

or

$$\langle\sigma\rangle = p_i \frac{r_i^2}{r_o^2 - r_i^2} = \frac{r_i^2 p_i}{h(r_o + r_i)}. \tag{1.11-16}$$

This is a useful formula that is valid for thick-walled, as well as thin-walled, spherical shells.

If a pressure p_o acts on the outside of the shell, as in Fig. 1.12, the resulting normal stress in the wall will be

$$\langle\sigma\rangle = -\frac{r_o^2 p_o}{h(r_o + r_i)}. \tag{1.11-17}$$

If the shell is subjected to both an internal and an external pressure, and the wall

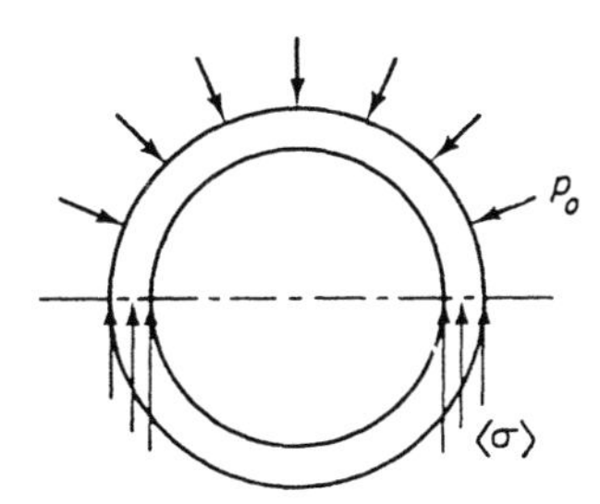

Figure 1.12 A spherical shell subjected to external pressure.

of the sphere is very thin, then $r_o - r_i = h$; $r_o \cong r_i = r$, and the foregoing equations reduce to

$$\langle \sigma \rangle = \frac{rp_i}{2h} - \frac{rp_o}{2h}. \tag{1.11–18}$$

Example 6. Pressurized Circular Cylindrical Tanks

Consider a cylindrical shell subjected to an internal pressure p_i, as shown in Fig. 1.13(a). Let us pass two planes perpendicular to the axis of the cylinder to cut the shell into a ring, pass another plane through the axis of the cylinder to cut the ring into two halves, and isolate the semicircular ring as a free body, as shown in Fig. 1.13(b). The stress acting on the radial cut CD is normal to the surface and is directed in the direction of increasing polar angle θ in polar coordinates; hence, it will be denoted by σ_θ. As in Example 5, we do not know the exact distribution of σ_θ in the cross section, but if $\langle \sigma_\theta \rangle$ denotes the average value of σ_θ over the cross section, then $\langle \sigma_\theta \rangle$ multiplied by the area $(r_o - r_i)L$ is the resultant force acting in the cross section CD. Similarly, the tensile force in the section EF [Fig. 1.13(b)] is also $\langle \sigma_\theta \rangle (r_o - r_i)L$. The resultant of pressure acting on the inside is $2r_i L p_i$. The

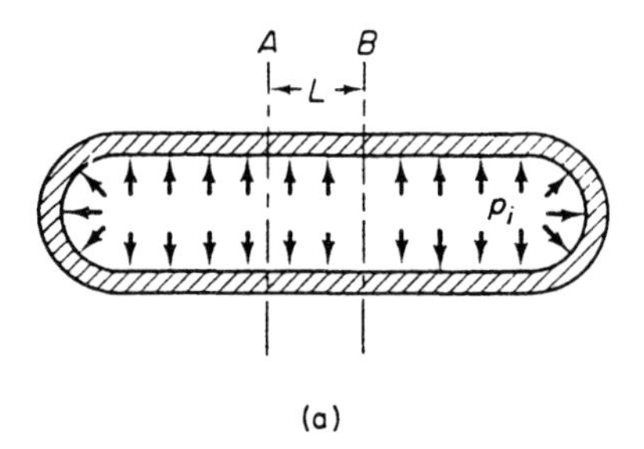

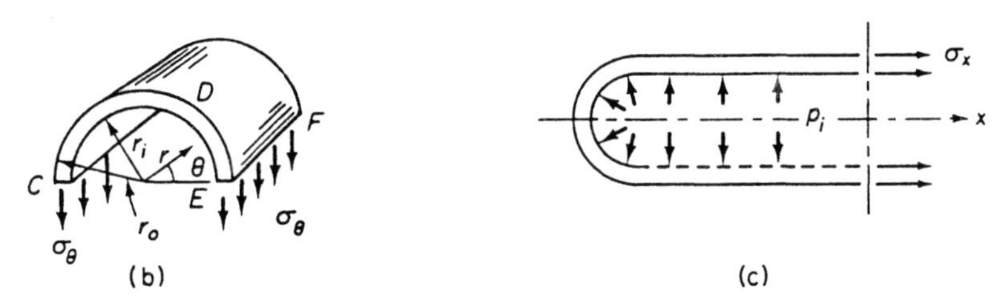

Figure 1.13 A pressurized cylindrical tank. (a) The shell seen in a cross section passing through the central axis of the shell. (b) A free-body diagram of a part of the shell cut by two planes perpendicular to the central axis and a third plane containing the central axis. (c) A free-body diagram of a part of the shell to the left of a plane perpendicular to the central axis.

balance of forces acting on the semicircular ring in the vertical direction requires that

$$2\langle\sigma_\theta\rangle(r_o - r_i)L = 2r_i L p_i. \tag{1.11–19}$$

Hence,

$$\langle\sigma_\theta\rangle = \frac{r_i p_i}{r_o - r_i}, \tag{1.11–20}$$

which is another very useful exact formula.

If we cut the cylinder by a plane perpendicular to its axis and consider the left half of the tank as a free body, as shown in Fig. 1.13(c), we can examine the average value of the axial stress σ_x that acts in the axial direction x on a cross section perpendicular to x. We note that the area on which σ_x acts is $\pi(r_o^2 - r_i^2)$. On the other hand, the surface on which the internal pressure p_i acts has a projected area in the axial direction equal to πr_i^2. Hence, the balance of forces in the axial direction yields

$$\pi r_i^2 p_i = \langle\sigma_x\rangle\pi(r_o^2 - r_i^2) \tag{1.11–21}$$

or

$$\langle\sigma_x\rangle = \frac{r_i^2 p_i}{r_o^2 - r_i^2}. \tag{1.11–22}$$

If the shell wall is very thin, so that $r_o - r_i = h$ and thus, $r_o \cong r_i = r$, then these equations are simplified to

$$\langle\sigma_\theta\rangle = \frac{r p_i}{h}, \qquad \langle\sigma_x\rangle = \frac{r p_i}{2h}. \tag{1.11–23}$$

Simple Beam Theory

Consider a prismatic beam of a uniform isotropic Hookean material with a rectangular cross section subjected to a pair of bending moments of magnitude M acting on its ends, as shown in Fig. 1.14(a). If the cross section of the beam is symmetric with respect to the plane containing the end moments, the beam will deflect into a circular arc in the same plane, as shown in Fig. 1.14(b). The deflection curve must be a circular arc because of symmetry, since every cross section is subjected to the same stress and strain. Let us assume that the deflection is small (compared with the length of the beam). We choose a rectangular frame of reference $x\ y\ z$, with the x-axis pointing in the direction of the longitudinal axis of the beam, y perpendicular to x but in the plane of bending, and z normal to x and y. [See Fig. 1.14(a).] The origin of the coordinates will be chosen at the centroid of one cross section, for reasons that will become clear shortly.

The deflection of the beam can be described by the deflection of the centroidal surface (the plane $y = 0$ when the beam is in the undeflected configuration) and

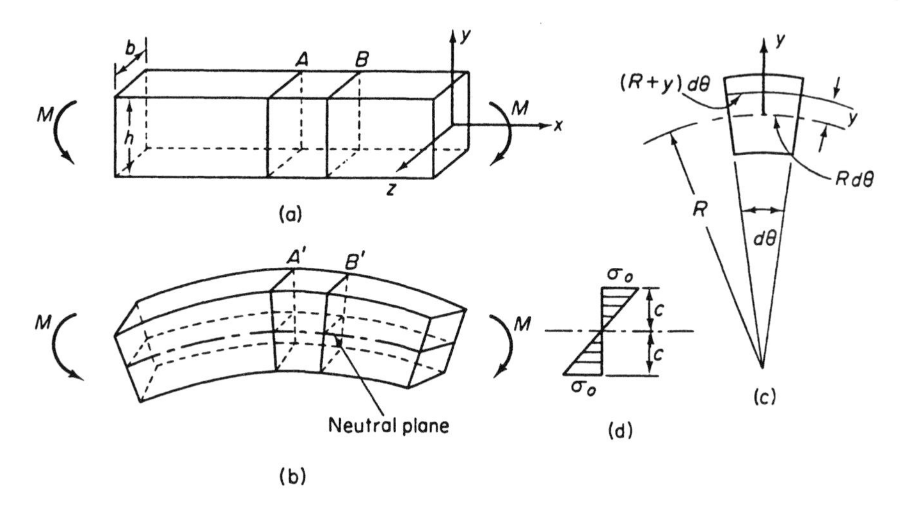

Figure 1.14 Bending of a prismatic beam. (a) The configuration of the beam at zero-stress state. (b) The beam bent by a uniform bending moment M. (c) Under the hypotheses of the classical beam theory, the deformed form of a small segment of the beam bounded by two cross-sectional planes A and B in (a) and A', B' in (b) is shown. R is the radius of curvature of the neutral surface on which the origin of the coordinates system x, y, z is located. (d) The distribution of the bending stress in a cross section of the beam. The bending stress is zero on the neutral surface, and is a linear function of y under the assumption that plane cross sections perpendicular to the neutral axis remain plane in bending deformation.

any displacements relative to this surface. Consider two neighboring cross sections A and B that are perpendicular to the plane $y = 0$ when the beam is unloaded. When the beam is bent into a circular arc, the two planes A and B are deformed into planes A' and B' that remain normal to the arc. [See Fig. 1.14(c).] That A' and B' are planes is because of symmetry. That they are perpendicular to the centroidal arc is also because of symmetry. Let the radius of curvature of the centroidal arc be R. When the cross sections A' and B' are bent to a relative angle of $d\theta$, the centroidal arc length is $Rd\theta$; whereas a line at a distance y above the centroidal line will have a length $(R + y)d\theta$. The change in length is $y\, d\theta$. A division by its original length $R\, d\theta$ yields the strain

$$e_{xx} = \frac{y}{R}. \tag{1.11–24}$$

In response to the strain e_{xx}, there will be a stress σ_{xx}. We now make the assumption that σ_{xx} is the only nonvanishing component of stress, whereas $\sigma_{yy} = \sigma_{zz} = \tau_{xy} = \tau_{yz} = \tau_{zx} = 0$. Then according to Hooke's law, we have

$$\sigma_{xx} = Ee_{xx} = E\frac{y}{R}. \tag{1.11–25}$$

Since only pure bending moments act on the beam, the resultant axial force must vanish. That is,

$$\int_A \sigma_{xx} dA = 0, \tag{1.11–26}$$

where A is the cross section, dA is an element of area in the cross section, and the integration extends over the entire cross section. Substituting Eq. (1.11–25) into Eq. (1.11–26) yields

$$\int_A y\, dA = 0, \tag{1.11–27}$$

which says that the origin ($y = 0$) must be the *centroid* of the cross section. This explains our original choice of the centroid as origin. The centroidal plane $y = 0$ is unstressed during bending [according to Eq. (1.11–25)]. Material particles on it are not strained in the axial direction. This plane is therefore called the *neutral surface* of the beam.

The resultant moment of the bending stress σ_{xx} about the z-axis must be equal to the external moment M. A force $\sigma_{xx}\, dA$ acting on an element of area dA in a cross section has a moment arm y; hence, the bending moment is

$$M = \int_A y\sigma_{xx}\, dA. \tag{1.11–28}$$

Substituting Eq. (1.11–25) into this equation yields

$$M = \frac{E}{R}\int_A y^2\, dA. \tag{1.11–29}$$

We now define the last integral as the *area moment of inertia of the cross section and denote it by I*:

$$I = \int_A y^2\, dA. \tag{1.11–30}$$

Then the foregoing equations may be written as*

$$\frac{M}{EI} = \frac{1}{R}, \qquad \sigma_{xx} = \sigma_0 \frac{y}{c}, \qquad \sigma_0 = \frac{Mc}{I}, \qquad \blacktriangle \tag{1.11–31}$$

where c is the largest distance from the neutral surface to the edge of the cross section. [See Fig. 1.14(d).] The stress σ_0 is the largest bending stress in the beam. It is called the *outer fiber stress* because it is associated with the outer edge of the beam cross section. I is a property of the cross-sectional geometry. For a rectangular cross section with depth h and width b, as shown in Fig. 1.14(a), we have $c = h/2$ and $I = \frac{1}{12}bh^3$.

*Thirty important formulas in this book are marked by black triangles on the right. These are thirty lines worthy of memorizing.

These formulas give us the stress and strain in a prismatic beam when it is subjected to pure bending. Can we use them for a prismatic beam subjected to a general loading, such as the one shown in Fig. 1.15? Or to a beam with variable cross sections? The answer is that although the solution is then no longer exactly correct, it is found empirically to be surprisingly good. The basic reason is that the shear stresses which must exist in the general case cause a deflection which usually is negligible compared with that due to the bending moment. Therefore, in general, the hypothesis that plane sections remain plane is very good, and Eqs. (1.11–24) through (1.11–31) can be considered *locally* true along the beam.

Figure 1.15 A beam subjected to a distributed loading.

Deflection of Beams

Based on such an empirical observation, we can analyze the deflection of a beam under a lateral load. For example, consider the beams illustrated in Fig. 1.16. Let the beam deflection curve (deflection of the neutral surface) be $y(x)$. When $y(x)$ is small (much less than the length of the beam), its curvature can be approximated by d^2y/dx^2, and the use of Eq. (1.11–31) leads to the basic equation

$$\frac{d^2y}{dx^2} = \frac{1}{EI} M(x). \qquad \blacktriangle \quad (1.11\text{–}32)$$

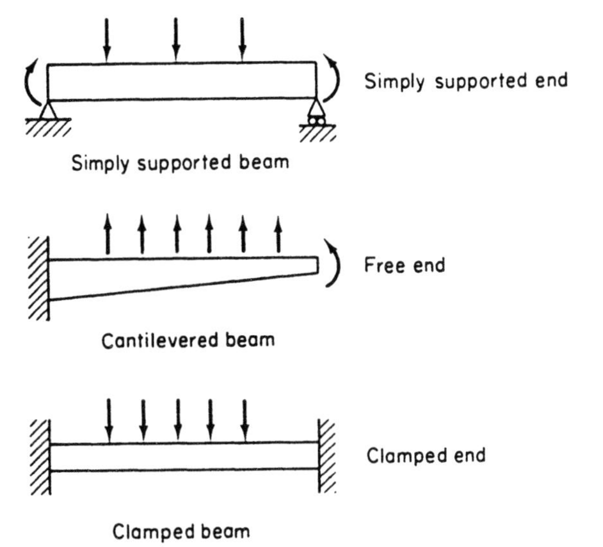

Figure 1.16 End conditions of beams.

The beam deflection $y(x)$ can be obtained by solving this equation with appropriate boundary conditions, which are:

Simply supported end (deflection and moment vanish):

$$y = 0, \quad \frac{d^2y}{dx^2} = 0. \tag{1.11–33}$$

Clamped end (deflection and slope vanish):

$$y = 0, \quad \frac{dy}{dx} = 0. \tag{1.11–34}$$

Free end (moment and shear specified):

$$EI\frac{d^2y}{dx^2} = M, \qquad EI\frac{d^3y}{dx^3} = S. \tag{1.11–35}$$

(See Fig. 1.16.) All these are pretty evident except the last one, for whose explanation we should examine Prob. 1.14 infra, which shows that the bending moment M, the transverse shear S, and the lateral load per unit length w are related by the equations

$$\frac{dM}{dx} = S, \qquad \frac{dS}{dx} = w. \qquad \blacktriangle \tag{1.11–36}$$

But since $M = EI\, d^2y/dx^2$, we must have $S = EI\, d^3y/dx^3$ as in Eq. (1.11–35).

If the curvature is small (so that the preceding analysis can be valid), but the slope is finite, then we should use the exact expression for $1/R$, which leads to the following equation in place of Eq. (1.11–32):

$$\frac{d^2y}{dx^2}\left[1 + \left(\frac{dy}{dx}\right)^2\right]^{-3/2} = \frac{M(x)}{EI}. \tag{1.11–37}$$

As an example, consider the small deflection of a cantilever beam clamped at the left end, as shown in Fig. 1.17, and subjected to a constant bending moment. The right-hand side of Eq. (1.11–32) is constant in this case, so that equation can be integrated to obtain

$$y(x) = \frac{M}{EI}\frac{x^2}{2} + Ax + B, \tag{1.11–38}$$

Figure 1.17 Bending of a cantilever beam.

where A and B are arbitrary constants. The boundary conditions $y = dy/dx = 0$ at $x = 0$ then yield $A = B = 0$, so that the solution is

$$y = \frac{M}{EI}\frac{x^2}{2}. \tag{1.11–39}$$

In this special case, the boundary conditions on the free end are also satisfied because $M = \text{const}$.

Can we, however, satisfy all the boundary conditions in general? After all, our beam has two ends with two conditions each, so that we have four boundary conditions, whereas our differential equation (1.11–32) is only of the second order. Are we going to have a sufficient number of arbitrary constants to satisfy all boundary conditions? The answer, as it stands, is no. A further reflection tells us, however, that for a general loading, the differential equation must be obtained by combining Eq. (1.11–36) with Eq. (1.11–32). Thus, the general equation must be

$$w = \frac{dS}{dx} = \frac{d^2M}{dx^2} = \frac{d^2}{dx^2}\left(EI\frac{d^2y}{dx^2}\right), \qquad \blacktriangle \quad (1.11–40)$$

which is a fourth-order differential equation, able to handle four boundary conditions. In the case of a uniform beam, we have

$$EI\frac{d^4y}{dx^4} = w(x). \qquad \blacktriangle \quad (1.11–41)$$

Equation (1.11–40) is an approximate equation, exact only in the pure bending of a prismatic beam, but it is used often to describe beam deflection in the general case, even for beams of variable cross section. In general, for a slender beam, it yields close approximations. Significant deviation occurs only when the beam is not slender or for sandwich constructions with very soft core material in which shear deflection becomes significant.

PROBLEMS

1.1 Ice melts into water with a slight reduction in volume. The molecules of water in ice rearrange themselves to achieve this feature. Construct some macroscopic examples that can do the same, i.e., change a solid structure into one that can be deformed easily.

1.2 When a truck of premixed concrete pours the mixture into a mold at a construction site, the mixture can be treated as a fluid continuum. Similarly, rice flowing in a grain chute of a silo can be considered as a fluid. Solar flares, sunspots, and the lava flow after a volcanic explosion are other examples. Name 10 more examples in which an aggregate of solid bodies flows like a fluid and to which the continuum concept can be applied in some sense.

1.3 Consider a spacecraft that you would like to bring back to the earth. You are faced with the problem of frictional heating upon reentry into the earth's atmosphere. You know that for a gas, the length of the mean free path between collisions is a measure of the average distance between molecules. For air at 1 km above the ground, the mean free path is 8×10^{-6} cm; at 100 km, it is 9.5 cm; and at 200 km, it is 3×10^4 cm. To analyze the flow of air around your spacecraft as it reenters the atmosphere, would it be permissible to use the method of continuum mechanics? At what level and for what purpose can you consider air a continuum? What kind of problems do you have to solve to bring the spacecraft back to earth safely?

1.4 Suppose that you are a surgeon, and you have a patient who has a dime-shaped and dime-sized piece of skin you would like to remove because it is cancerous. After cutting away the diseased tissue, you would like to sew up the healthy skin to cover up the hole. Here is a chance to do some engineering planning. Invent a way to do the job so as to obtain the best results. First define what you mean by "best." How would the healthy skin survive the surgery? What kind of healing process would you expect? What final result do you want? Can you treat the skin as a continuum? In which way would you use the continuum concept to deal with this problem? Does the location of the cancerous lesion (e.g., on the face, the hand, the back, or the abdomen) make a difference to your method?

1.5 A 100-story building is to be built in town, and you are challenged to design a lift to elevate heavy material in the construction process. Make several alternative designs, and then make a choice from among them. Explain why your choice is a good one.

1.6 An engineer looked at the simple truss shown in Fig. P1.6(a) and felt that he could improve the safety of the truss by adding another member AB, as shown in Fig. P1.6(b). Then, if one of the members AB, CD, AC, BD, and AD were broken by some accident, the truss would not immediately collapse.

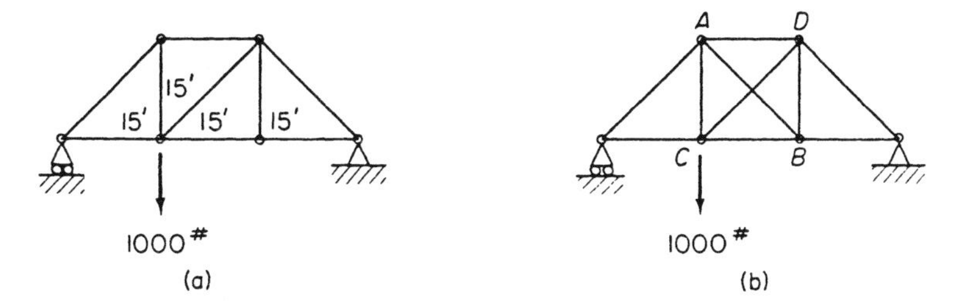

Figure P1.6 (a) A statically determinate truss. (b) A statically indeterminate truss.

Now, fail-safe construction is a great idea, especially for critical public structures such as airplanes, bridges, and ships. But introducing another member AB changes the character of the truss. To understand this, find first the load in every member of the truss in Fig. P1.6(a). Then, determine the load in every member of the truss of Fig. P1.6(b) also. You'll find that you cannot do it. Why? What additional information is needed? With the additional information, how do you determine the loads in all the members? What cost do you pay for the additional safety?

Note: A truss like that shown in Fig. 1.6(b) is called a *statically indeterminate* structure because the tension in the bars cannot be determined by statics alone.

1.7 Galileo, in his *Two New Sciences*, posed the following problem: A column of marble rested on two supports, in the manner of a simply supported beam. The citizens of

Rome were worried about the safety of the column and sought to increase the support. They inserted a third support in the middle of the span, as in Fig. P1.7. The column broke. Why?

Figure P1.7 Galileo's problem of a fallen column.

1.8 It is pretty easy to demonstrate that the tension in an Achilles tendon is considerable when we stand on tiptoe or when we poise for a jump. A tension gauge can be built using the same principle we would have used to measure the tension in the string of a bow or in a rubber slingshot. Design such a gauge.

Hint. If we pull on a bow as shown in Fig. P1.11, and if a lateral force F induces a deflection angle θ, show that $T = F/(2 \sin \theta)$.

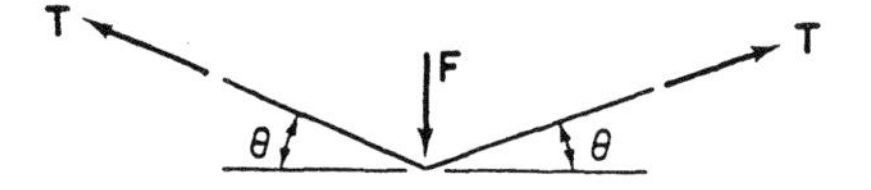

Figure P1.8 Measurement of tension in a string.

1.9 Compare the bending moment acting on the spinal column at the level of a lumbar vertebra for the following cases:

(a) A secretary bends down to pick up a book on the floor (i) with knees straight and (ii) with knees bent.

(b) A water skier skiis (i) with arms straight and (ii) with elbows hugging the waist.

Discuss these cases quantitatively with proper free-body diagrams.

1.10 Your doctor always trys to "feel your pulse" by putting his fingers on your radial artery. To understand what he can find by doing this, let us consider a simpler case. A small balloon is inflated by air at a pressure of p pascals. I press my finger on the balloon. (See Fig. P1.10.) I assume that the bending resistance of the wall is negligible. How far down should I press so that the pressure acting on my finger is exactly p?

Hint. Consider the free-body diagram of a small piece of the balloon under the finger. Consider the condition when the spot under the finger becomes flattened into a plane surface.

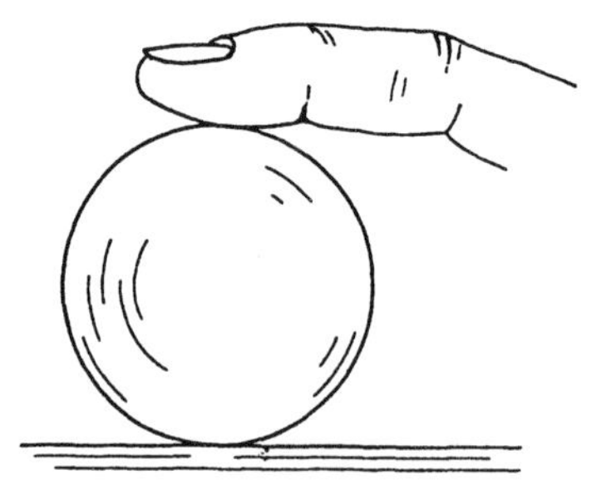

Figure P1.10 Feeling the pressure in a balloon.

1.11 One man is twice as tall as another man. Assume that they are completely similar and doing exactly the same stationary gymnastic maneuver. Are they subjected to the same stresses in their bones and muscles?

Answer. The ratio of the linear dimensions is 2. The ratio of the mass of the corresponding organs is 8. The ratio of the corresponding areas is 4. The ratio of the stresses is 2.

1.12 The gas pressure P in a soap bubble is related to the surface tension σ and the radius R by the equation $P = 4\sigma/R$. Derive this equation, which is known as Laplace's law.

Take a pippette, put a valve in the middle, close it, and blow two bubbles, one at each end. (See Fig. P1.12.) One bubble is large and one is small. Now open the middle valve so that the gas in the two bubbles can move. In which way will the bubble diameters change? Explain in detail.

Answer. The small bubble will disappear.

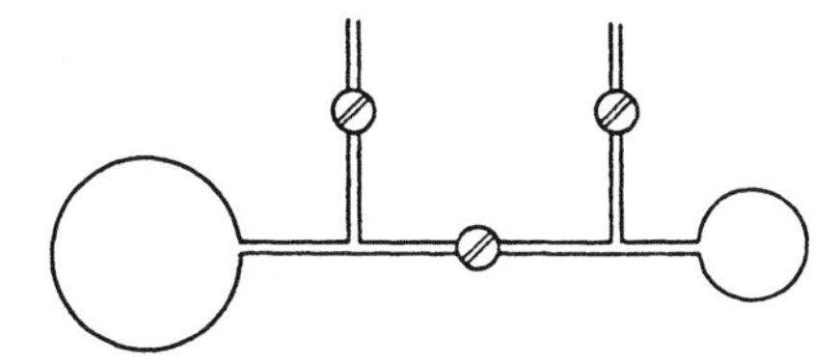

Figure P1.12 Collapse of a small soap bubble into a large one.

1.13 When a person breathes, air enters the mouth, nose, trachea, bronchi, bronchioles, and alveolar ducts and ends in the alveoli, which are the final units of respiration. Most textbooks of physiology liken each alveolus to the gas bubble considered in Prob. 1.12 and state that a human lung consists of 300 million bubbles ventilated in parallel to the atmosphere. Now, apply the results of Prob. 1.12 to this statement. One cannot help reaching the conclusion that all the alveoli will collapse except for the largest one. So the lung would consist of one open alveolus. This is obviously absurd. What went wrong? What is the correct answer?

Hint. Mammalian lungs are so well packed that each wall of an alveolus serves as a wall to two neighboring alveoli. Hence, the walls are more accurately called interalveolar septa, and the gas bubble analogy is incorrect for the pulmonary alveoli.

1.14 Let M denote bending moment in a beam, S the shear, and w the load. Show that, according to the free-body diagram in Fig. P1.21,

$$S = \frac{dM}{dx}, \qquad w = \frac{dS}{dx};$$

hence,

$$\frac{d^2M}{dx^2} = w.$$

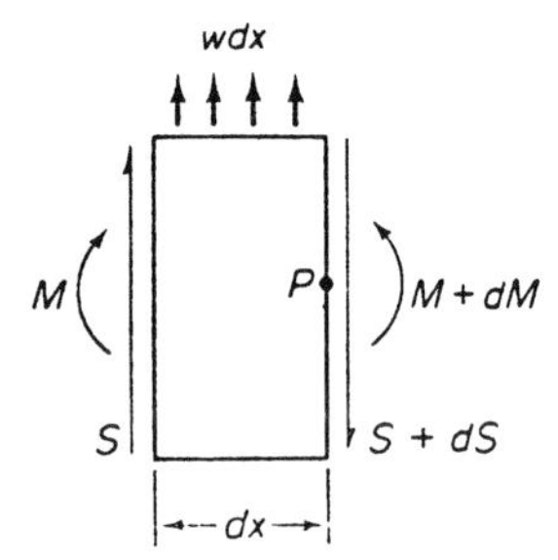

Figure P1.14 Equilibrium of a beam segment.

1.15 Using the differential equation derived in Prob. 1.14, find the bending moment distribution in the beams shown in Figs. P1.15(a) and (b) under a loading per unit length of

$$w = a \sin \frac{\pi x}{L}.$$

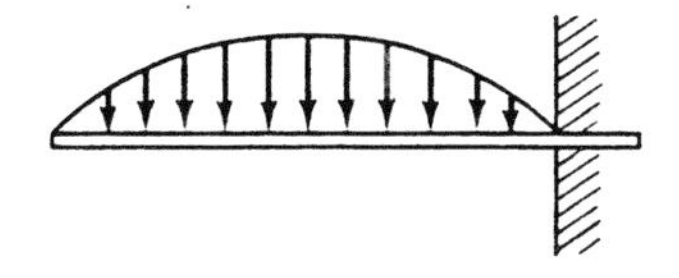

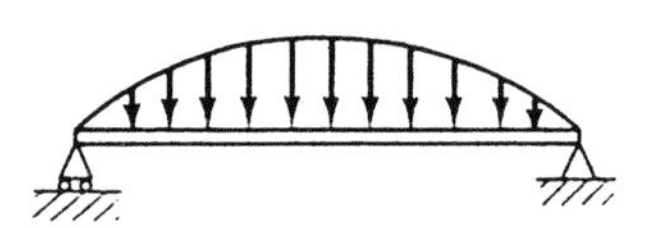

Figure P1.15 Bending of beams under a sinusoidally distributed load. (a) Beam is cantilevered, i.e., free at one end and clamped at the other. (b) Beam is simply supported.

1.16 A person weighing W pounds tries to walk over a plank that is simply supported at two ends across a river. (See Fig. P1.16.) The plank will break whenever the bending moment exceeds M_{cr}. At what place (x) will the plank break and the person fall into the river?

Answer. $x = \frac{1}{2}[L \pm (L^2 - 4K)^{1/2}]$, where $K = LM_{cr}/W$.

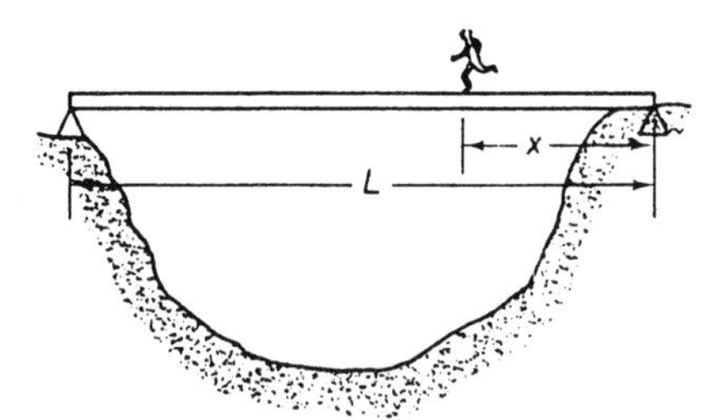

Figure P1.16 A person walking over a plank.

1.17 A hinge is added to the left end of a cantilever beam that is loaded by a constant force P, as shown in Fig. P1.17. How would you determine the bending moment in the beam?

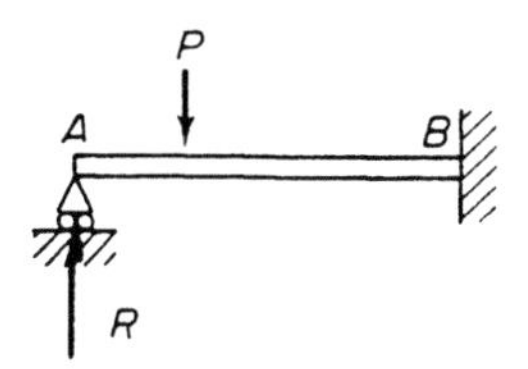

Figure P1.17 Redundant supports.

Solution. A beam clamped at B and simply supported at A is *statically indeterminate* because the reaction at A cannot be computed by statics alone. To solve the problem, we must consider the elasticity of the beam.

One method is as follows. Take the support of A away. Then the beam becomes a cantilever beam. We can find the deflection at A due to the load P. Let this be $\delta_A^{(P)}$, which is proportional to P.

Next, consider the same cantilever beam loaded by a force R at the tip. This produces a deflection $\delta_A^{(R)}$ at the end A. In reality, the end A does not move. Hence, $\delta_A^{(R)} + \delta_A^{(P)} = 0$. From this equation, we can compute R. With R known, we can then complete the moment diagram.

1.18 A beam (Fig. P1.18) rests on three hinges which, unlike Galileo's rocks discussed in Prob. 1.7, are so rigidly attached to the foundation that both push and pull can be sustained. Sketch a method with which the bending moment distribution in the beam can be calculated.

Solution. First, withdraw one of the supports, so that the problem becomes statically determinate. Compute the deflection at the location of the withdrawn support due to the load P.

Next, apply a force R at the location of the withdrawn support, and compute the displacement at this point.

The condition that the net displacement at all the supports must vanish provides an equation to compute the reaction R. Then all the forces are known, and the moment diagram can be completed.

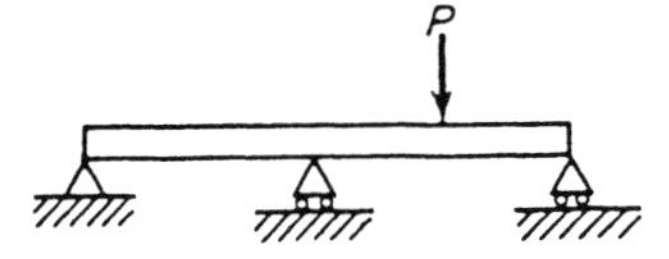

Figure P1.18 Redundant supports.

1.19 A strong wind blows on a palm tree. (See Fig. P1.19.) The wind load on the trunk is $w = kD$ per unit length of the trunk, where D is the local diameter of the trunk and k is a constant. How should the diameter vary with the height so that the tree is uniformly strong from top to bottom with respect to bending in wind? Note that the area moment of inertia of the cross section of the tree trunk is proportional to D^4, and the outer fiber stress due to bending is proportional to $Mc/I \sim MD^{-3}$, where M is the bending moment. Ignore the bending moment contributed by the leaves.

Hint. Let x and ξ be measured downward from the treetop. The bending moment at x is

$$M(x) = \int_0^x (x - \xi)kD(\xi)\,d\xi.$$

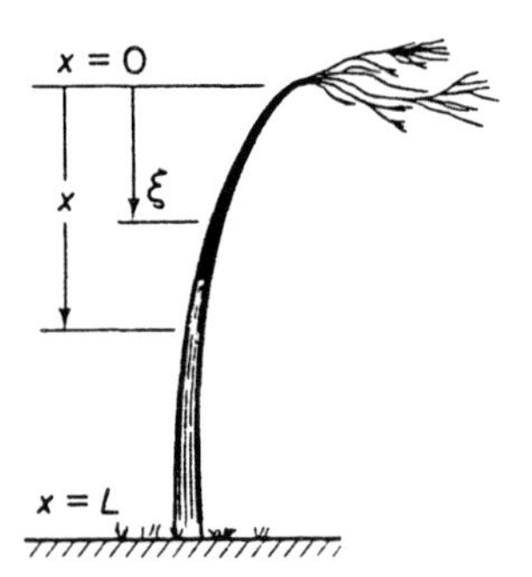

Figure P1.19 A strong wind blowing on a palm tree.

The maximum bending stress at x is proportional to $M(x)/D^3(x)$. The problem is to determine $D(x)$ so that $M(x)/D^3(x)$ is constant. Try a power law, such as $D(x) = \text{const.} \cdot x^m$, and show that $m = 1$. The tree trunk should look like a slender cone.

1.20 One of the most beautiful results in aerodynamic theory says that the best design for the minimum induced drag of an airplane (the air resistance to forward motion due to lifting the weight of the airplane) is one with elliptic loading. By loading is meant the aerodynamic lift force per unit span. By elliptic loading is meant that the lift distribution from wing tip to wing tip is an ellipse. Let x be the distance along the wing span measured from the centerline of the airplane. Let b be the semispan of the wing (the distance from the centerline to the wing tip). Then the theorem says that the minimum induced drag is achieved if the lift is distributed according to the formula

$$\ell(x) = k(1 - x^2/b^2)^{1/2}$$

where k is a constant. This lift distribution, shown in Fig. P1.20, yields the best fuel economy.

Assume that the airplane has an elliptic loading. Consider the wing as a cantilever beam. Compute the bending moment $M(x)$ at x in the wing. Plot the bending moment diagram to show the moment at every station in the wing due to aerodynamic load.

If the lift distribution is approximated by $\ell(x) = A \cos(\pi x/2L)$, and the wing's bending rigidity is $EI(x)$, find the wing tip deflection relative to the wing root.

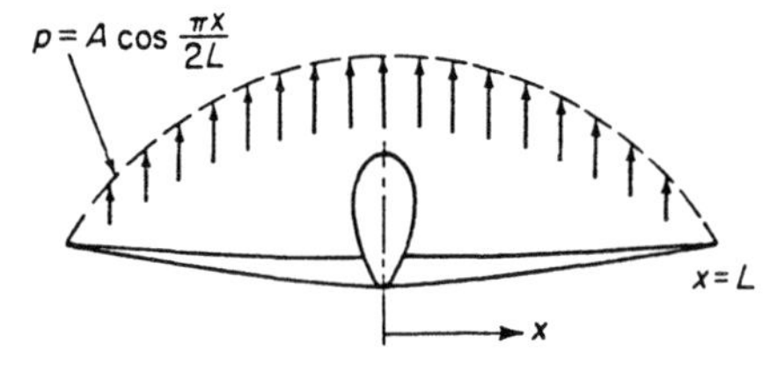

Figure P1.20 The optimal elliptic aerodynamic loading acting on an airplane wing.

1.21 A red blood cell is an axisymmetric shell with a very thin wall in the shape of a doughnut without a hole. (See Fig. P1.21.) People describe it as a biconcave disk. It is filled with a Newtonian fluid, and it floats in another Newtonian fluid. By considering suitable free-body diagrams of the red blood cell, we can compute the difference between the internal pressure p_i and external pressure p_e under the assumption that the bending rigidity of the cell membrane can be neglected. What conclusion do you reach? What is the physiological significance of this conclusion?

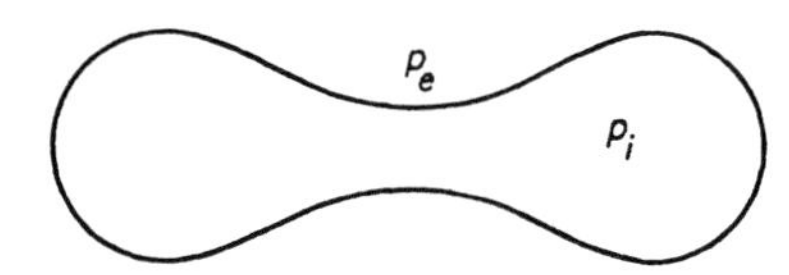

Figure P1.21 A red blood cell suspended in a buffered isotonic plasma is an axisymmetric body of revolution with a central cross section as shown.

1.22 Figure P1.22(a) shows a person working. Figure P1.22(b) shows a free-body diagram of the upper body, with a cross section passing through an intervertebral disc in the lumbar region. The structure of the lumbar spine is sketched in Fig. P1.22(c). The discs serve as pivots of rotation: They cannot resist bending and torsion moments. In resisting the external load, the vertebra, the discs, and the muscles are stressed. The major muscles behind the centroidal line of the discs are the erector spinae, whose centroid is located about 22% of the depth of the trunk behind the center of the disc. For a load W acting with a moment arm of L, what is the bending moment of the external load about the disc? How large is the tension in the erector spinae muscle for a person of your size?

Low back pain is so common that lots of attention has been given to this problem. The loads acting on the discs have been measured with strain gages in some cases. It was found that no agreement with prediction can be obtained if we do not take into account the fact that when one lifts a heavy weight, one tenses up the abdominal muscles so that the pressure in the abdomen is increased. Show that it helps for a lifter to have a large abdomen and strong abdominal muscles. *Ref.* Schultz, A.B., and Ashton-Miller, J.A.: "Biomechanics of human spine." In *Basic Orthopaedic Biomechanics*, ed. by V.C. Mow and W.C. Hayes, Raven Press, New York, 1991, pp. 337–364.

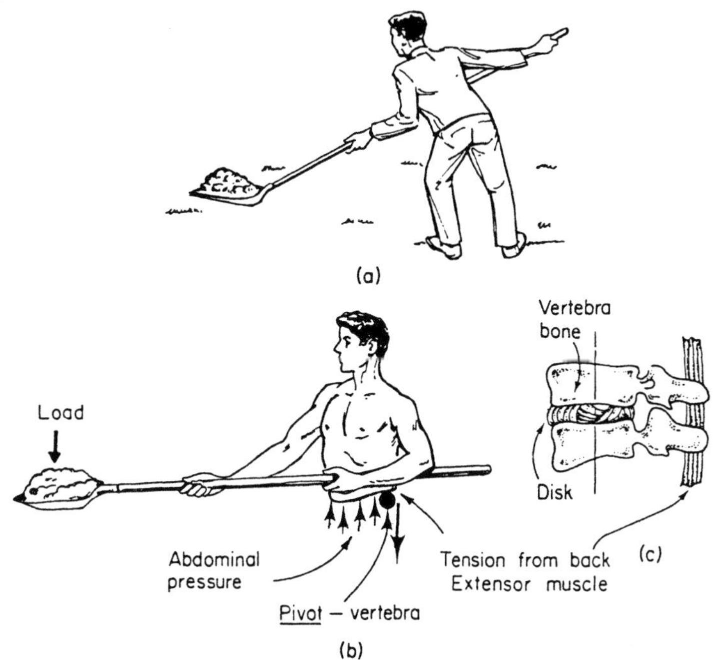

Figure P1.22 Loads in the spine when a man shoves a weight.

1.23 Figure P1.23 is a classic from a book by Giovanni Alphonso Borelli (1608–1679) entitled "De Motu Animalium" (On the Movement of Animals), published in 1680 (first part) and 1681 (second part), recently translated by P. Maquet, Springer-Verlag, New York, 1989. The figure shows a person carrying a heavy load. Several parts are cut open to show how bones and muscles work in this effort. Further clarification can be obtained, of course, by use of more detailed free-body diagrams. Use them to estimate how large is the load acting on the hip joint when a 70 kg person walks carrying a 30 kg globe on the shoulder.

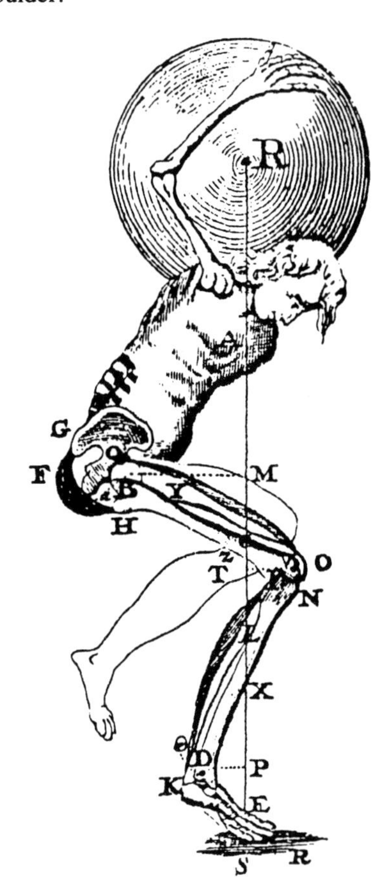

Figure P1.23 A figure from Table VI, Fig. 1 of Borelli's book.

2 VECTORS AND TENSORS

A beautiful story needs a beautiful language to tell. Tensor is the language of mechanics.

2.1 VECTORS

A vector in a three-dimensional Euclidean space is defined as a directed line segment with a given magnitude and a given direction. We shall denote vectors by $\overrightarrow{AB}$, $\overrightarrow{PQ}$, . . . , or by boldface letters, **u, v, F, T, ...**

Two vectors are *equal* if they have the same direction and same magnitude. A *unit vector* is a vector of magnitude 1. The *zero vector*, denoted by **0**, is a vector of zero magnitude. We use the symbols $|\overrightarrow{AB}|$, $|\mathbf{u}|$, and v to represent the magnitudes of $\overrightarrow{AB}$, **u**, and **v**, respectively.

The sum of two vectors is another vector obtained by the "parallelogram law," and we write, for example, $\overrightarrow{AB} + \overrightarrow{BC} = \overrightarrow{AC}$. Vector addition is commutative and associative.

A vector multiplied by a number yields another vector. If k is a positive real number, $k\mathbf{a}$ represents a vector having the same direction as **a** and a magnitude k times as large. If k is negative, $k\mathbf{a}$ is a vector whose magnitude is $|k|$ times as large and whose direction is opposite to **a**. If $k = 0$, we have $0 \cdot \mathbf{a} = \mathbf{0}$.

The subtraction of vectors can be defined by

$$\mathbf{a} - \mathbf{b} = \mathbf{a} + (-\mathbf{b}).$$

If we let $\mathbf{e}_1$, $\mathbf{e}_2$, $\mathbf{e}_3$ be the unit vectors in the directions of the positive x_1, x_2, x_3 axes, respectively, we can show that every vector in a three-dimensional Euclidean space with coordinate axes x_1, x_2, x_3 may be represented by a linear combination of $\mathbf{e}_1$, $\mathbf{e}_2$, and $\mathbf{e}_3$. Furthermore, if the vector **u** is represented by the linear combination

$$\mathbf{u} = u_1\mathbf{e}_1 + u_2\mathbf{e}_2 + u_3\mathbf{e}_3, \tag{2.1–1}$$

then u_1, u_2, u_3 are the components of **u**, and **u** can be represented by a matrix (u_1, u_2, u_3).

The magnitude $|\mathbf{u}|$ is then given by

$$|\mathbf{u}| = \sqrt{u_1^2 + u_2^2 + u_3^2}, \tag{2.1–2}$$

and therefore $\mathbf{u} = \mathbf{0}$ if and only if $u_1 = u_2 = u_3 = 0$.

The *scalar* (or *dot*) *product* of $\mathbf{u}$ and $\mathbf{v}$, denoted by $\mathbf{u}\cdot\mathbf{v}$, is defined by the formula

$$\mathbf{u}\cdot\mathbf{v} = |\mathbf{u}||\mathbf{v}| \cos\theta \qquad (0 \leq \theta \leq \pi), \tag{2.1–3}$$

where θ is the angle between the given vectors. This represents the product of the magnitude of one vector and the component of the second vector in the direction of the first; that is,

$$\mathbf{u}\cdot\mathbf{v} = (\text{magnitude of } \mathbf{u})(\text{component of } \mathbf{v} \text{ along } \mathbf{u}). \tag{2.1–4}$$

If

$$\mathbf{u} = u_1\mathbf{e}_1 + u_2\mathbf{e}_2 + u_3\mathbf{e}_3 \qquad \mathbf{v} = v_1\mathbf{e}_1 + v_2\mathbf{e}_2 + v_3\mathbf{e}_3$$

the scalar product of these two vectors can also be expressed in terms of the components:

$$\mathbf{u}\cdot\mathbf{v} = u_1v_1 + u_2v_2 + u_3v_3. \tag{2.1–5}$$

Whereas the scalar product of two vectors is a scalar quantity, the *vector* (or *cross*) *product* of two vectors $\mathbf{u}$ and $\mathbf{v}$ produces another vector $\mathbf{w}$; and we write $\mathbf{w} = \mathbf{u} \times \mathbf{v}$. The magnitude of $\mathbf{w}$ is defined as

$$|\mathbf{w}| = |\mathbf{u}||\mathbf{v}| \sin\theta \qquad (0 \leq \theta \leq \pi), \tag{2.1–6}$$

where θ is the angle between $\mathbf{u}$ and $\mathbf{v}$, and the direction of $\mathbf{w}$ is defined as perpendicular to the plane determined by $\mathbf{u}$ and $\mathbf{v}$, in such a way that $\mathbf{u}$, $\mathbf{v}$, $\mathbf{w}$ form a right-handed system. Vector products satisfy the following relations:

$$\begin{aligned}
&\mathbf{u} \times \mathbf{v} = -(\mathbf{v} \times \mathbf{u}) \\
&\mathbf{u} \times (\mathbf{v} + \mathbf{w}) = \mathbf{u} \times \mathbf{v} + \mathbf{u} \times \mathbf{w} \\
&\mathbf{u} \times \mathbf{u} = 0 \\
&\mathbf{e}_1 \times \mathbf{e}_1 = \mathbf{e}_2 \times \mathbf{e}_2 = \mathbf{e}_3 \times \mathbf{e}_3 = \mathbf{0} \\
&\mathbf{e}_1 \times \mathbf{e}_2 = \mathbf{e}_3 \qquad \mathbf{e}_2 \times \mathbf{e}_3 = \mathbf{e}_1 \qquad \mathbf{e}_3 \times \mathbf{e}_1 = \mathbf{e}_2 \\
&k\mathbf{u} \times \mathbf{v} = \mathbf{u} \times k\mathbf{v} = k(\mathbf{u} \times \mathbf{v}).
\end{aligned} \tag{2.1–7}$$

Using these relations, the vector product can be expressed in terms of the components as follows:

$$\mathbf{u} \times \mathbf{v}\ (u_2v_3 - u_3v_2)\mathbf{e}_1 + (u_3v_1 - u_1v_3)\mathbf{e}_2 + (u_1v_2 - u_2v_1)\mathbf{e}_3. \tag{2.1–8}$$

PROBLEMS

2.1 Given vector $\mathbf{u} = -3\mathbf{e}_1 + 4\mathbf{e}_2 + 5\mathbf{e}_3$, find a unit vector in the direction of $\mathbf{u}$.

Answer: $(\sqrt{2}/10)\mathbf{u}$.

2.2 If $\overrightarrow{AB} = -2\mathbf{e}_1 + 3\mathbf{e}_2$, and the midpoint of the segment $\overrightarrow{AB}$ has coordinates $(-4, 2)$, find the coordinates of A and B.

Answer: $(-3, \frac{1}{2}), (-5, \frac{7}{2})$.

2.3 Prove that, for any two vectors $\mathbf{u}, \mathbf{v}$, $|\mathbf{u} - \mathbf{v}|^2 + |\mathbf{u} + \mathbf{v}|^2 = 2(|\mathbf{u}|^2 + |\mathbf{v}|^2)$.

2.4 Find the magnitude and direction of the resultant force of three coplanar forces of 10 lb each acting outward on a body at the origin and making angles of 60°, 120°, and 270°, respectively, with the x-axis.

Answer: $10(\sqrt{3} - 1)$, $\perp x$.

2.5 Find the angles between $\mathbf{u} = 6\mathbf{e}_1 + 2\mathbf{e}_2 - 3\mathbf{e}_3$ and $\mathbf{v} = -\mathbf{e}_1 + 8\mathbf{e}_2 + 4\mathbf{e}_3$.

Answer: $\cos^{-1}(-\frac{2}{63})$.

2.6 Given $\mathbf{u} = 3\mathbf{e}_1 + 4\mathbf{e}_2 - \mathbf{e}_3$, $\mathbf{v} = 2\mathbf{e}_1 + 5\mathbf{e}_3$, find the value of α so that $\mathbf{u} + \alpha\mathbf{v}$ is orthogonal to $\mathbf{v}$.

Answer: $-\frac{1}{29}$.

2.7 Given $\mathbf{u} = 2\mathbf{e}_1 + 3\mathbf{e}_2$, $\mathbf{v} = \mathbf{e}_1 - \mathbf{e}_2 + 2\mathbf{e}_3$, $\mathbf{w} = \mathbf{e}_1 - 2\mathbf{e}_3$, evaluate $\mathbf{u}\cdot(\mathbf{v} \times \mathbf{w})$ and $(\mathbf{u} \times \mathbf{v})\cdot\mathbf{w}$.

Answer: 16.

2.8 $(\mathbf{u} \times \mathbf{v})\cdot\mathbf{w}$ is called the *scalar triple product* of $\mathbf{u}, \mathbf{v}, \mathbf{w}$. Show that $(\mathbf{u} \times \mathbf{v})\cdot\mathbf{w} = \mathbf{u}\cdot(\mathbf{v} \times \mathbf{w})$.

2.9 Find the equation of the plane through $A(1, 0, 2)$, $B(0, 1, -1)$, and $C(2, 2, 3)$.

Answer: $7x - 2y - 3z - 1 = 0$.

2.10 Find the area of $\triangle ABC$ in Prob. 2.9.

Answer: $\sqrt{62}/2$.

2.11 Find a vector perpendicular to both $\mathbf{u} = 2\mathbf{e}_1 + 3\mathbf{e}_2 - \mathbf{e}_3$ and $\mathbf{v} = \mathbf{e}_1 - 2\mathbf{e}_2 + 3\mathbf{k}$.

Answer: $7\mathbf{e}_1 - 7\mathbf{e}_2 - 7\mathbf{e}_3$.

2.2 VECTOR EQUATIONS

The spirit of vector analysis is to use symbols to represent physical or geometric quantities and to express a physical relationship or a geometric fact by an equation. For example, if we have a particle on which the forces $\mathbf{F}^{(1)}, \mathbf{F}^{(2)}, \ldots, \mathbf{F}^{(n)}$ act, then we say that the condition of equilibrium for this particle is

$$\mathbf{F}^{(1)} + \mathbf{F}^{(2)} + \cdots + \mathbf{F}^{(n)} = \mathbf{0}. \tag{2.2–1}$$

As another example, we say that the following equation for the vector **r** represents a plane if **n** is a unit vector and p is a constant:

$$\mathbf{r}\cdot\mathbf{n} = p. \tag{2.2–2}$$

By this statement, we mean that the locus of the end point of a radius vector **r** satisfying the preceding equation is a plane. The geometric meaning is again clear. The vector **n**, called the *unit normal vector* of the plane, is specified. The scalar product $\mathbf{r}\cdot\mathbf{n}$ represents the scalar projection of **r** on **n**. Equation (2.2–2) then states that if we consider all radius vectors **r** whose component on **n** is a constant p, we shall obtain a plane. (See Fig. 2.1.)

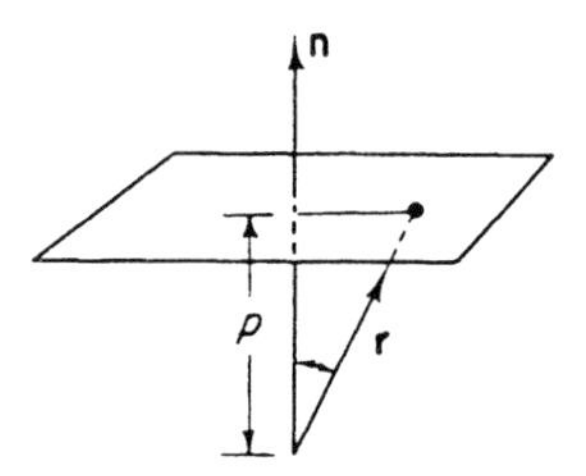

Figure 2.1 Equation of a plane, $\mathbf{r}\cdot\mathbf{n} = p$.

On the other hand, elegant as they are, vector equations are not always convenient. Indeed, when Descartes introduced analytic geometry in which vectors are expressed by their components with respect to a fixed frame of reference, it was a great contribution. Thus, with reference to a set of rectangular Cartesian coordinate axes O-xyz, Eqs. (2.2–1) and (2.2–2) may be written, respectively, as

$$\sum_{i=1}^{n} F_x^{(i)} = 0, \qquad \sum_{i=1}^{n} F_y^{(i)} = 0, \qquad \sum_{i=1}^{n} F_z^{(i)} = 0, \tag{2.2–3}$$

$$ax + by + cz = p, \tag{2.2–4}$$

where $F_x^{(i)}$, $F_y^{(i)}$, $F_z^{(i)}$ represent the components of the vector $\mathbf{F}^{(i)}$ with respect to the frame of reference O-xyz; x, y, z represent the components of **r**; and a, b, c represent those of the unit normal vector **n**.

Why is the analytic form preferred? Why are we willing to sacrifice the elegance of the vector notation? The answer is compelling: We like to express physical quantities in numbers. To specify a radius vector, it *is* convenient to specify a triple of numbers (x, y, z). To specify a force **F**, it *is* convenient to define the three components F_x, F_y, F_z. In fact, in practical calculations, we use Eqs. (2.2–3) and (2.2–4) much more frequently than Eqs. (2.2–1) and (2.2–2).

PROBLEMS

2.12 Express the basic laws of elementary physics—e.g., Newton's law of motion, Coulomb's law for the attraction or repulsion between electric charges, and Maxwell's equation for the electromagnetic field—in the form of vector equations.

For example, to express Newton's law of gravitation in vector form, let m_1 and m_2 be the masses of two particles. Let the position vector from particle 1 to particle 2 be $\mathbf{r}_{12}$. Then the force produced on particle 1 due to the gravitational attraction between 1 and 2 is

$$\mathbf{F}_{12} = G\frac{m_1 m_2}{|\mathbf{r}_{12}|^2}\frac{\mathbf{r}_{12}}{|\mathbf{r}_{12}|}$$

where G is the gravitational constant.

2.13 Consider a particle constrained to move in a circular orbit at a constant speed. Let $\mathbf{v}$ be the velocity at any instant. What is the acceleration of the particle; i.e., what is the vector $d\mathbf{v}/dt$?

Answer. The velocity vector $\mathbf{v}$ may be represented in polar coordinates as follows. Let $\hat{\mathbf{r}}$, $\hat{\boldsymbol{\theta}}$, $\hat{\mathbf{z}}$, be, respectively, the unit vectors with origin at P in the directions of the radius, the tangent, and the polar axis perpendicular to the plane of the orbit. (See Fig. P2.13.) Then $\mathbf{v} = v\hat{\boldsymbol{\theta}}$, where v is the absolute value of $\mathbf{v}$. Hence, by differentiation,

$$\frac{d\mathbf{v}}{dt} = v\frac{d\hat{\boldsymbol{\theta}}}{dt} + \frac{dv}{dt}\hat{\boldsymbol{\theta}}.$$

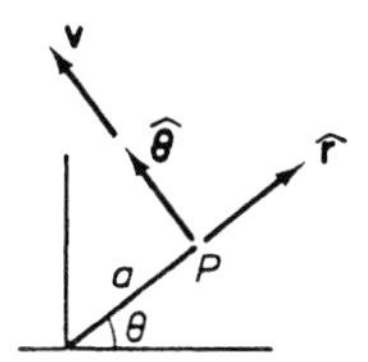

Figure P2.13 Velocity vector of a particle moving in a circular orbit.

The last term vanishes because v is a constant. To evaluate $d\hat{\boldsymbol{\theta}}/dt$, we note that $\hat{\boldsymbol{\theta}}$ is a unit vector; hence, it can only change direction. $d\hat{\boldsymbol{\theta}}/dt$ is, therefore, perpendicular to the vector $\hat{\boldsymbol{\theta}}$, i.e., parallel to $\hat{\mathbf{r}}$. Let ω be the angular velocity of the particle about the center of the orbit. Obviously, $\hat{\boldsymbol{\theta}}$ is turning at a rate of $\omega = v/a$. Hence, $d\hat{\boldsymbol{\theta}}/dt = -(v/a)\hat{\mathbf{r}}$, and $d\mathbf{v}/dt = -(v^2/a)\hat{\mathbf{r}}$.

2.14 A particle is constrained to move along a circular helix of radius a and pitch h at a constant speed v. What is the acceleration of the particle? If the particle is located at P, as shown in Fig. P2.14, express the velocity and acceleration vectors in terms of

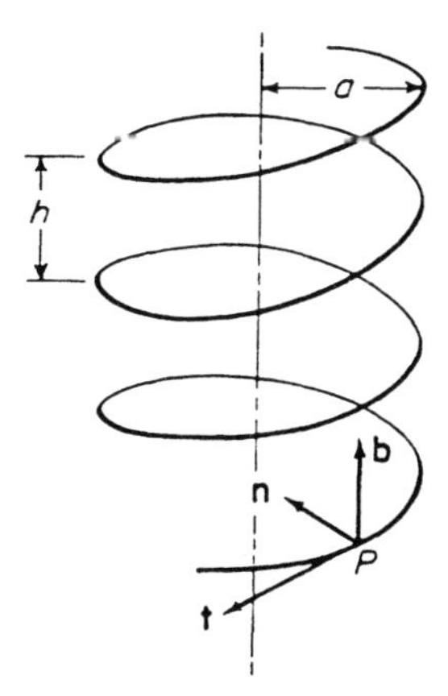

Figure P2.14 A helical orbit.

unit vectors $\mathbf{t}$, $\mathbf{n}$, and $\mathbf{b}$ that are, respectively, tangent, normal, and binormal to the helix at P.

Answer. The velocity vector is parallel to $\mathbf{t}$ and has a magnitude v. Hence, $\mathbf{v} = v\mathbf{t}$. By differentiation, and noting that v is a constant, we have $d\mathbf{v}/dt = v\, d\mathbf{t}/dt$. But since $\mathbf{t}$ has a constant length of unity, $d\mathbf{t}/dt$ must be perpendicular to $\mathbf{t}$ and, hence, must be a combination of $\mathbf{n}$ and $\mathbf{b}$. That is,

$$\frac{d\mathbf{t}}{dt} = \kappa\mathbf{n} + \tau\mathbf{b}$$

where κ and τ are constants. If the particle moves with unit velocity, the constants κ and τ are called the *curvature* and the *torsion* of the space curve, respectively.

It is convenient to use polar coordinates for this problem. Let the unit vectors in the direction of the radial, circumferential, and axial directions be $\hat{\mathbf{r}}$, $\hat{\boldsymbol{\theta}}$, and $\hat{\mathbf{z}}$, respectively. Then

$$\mathbf{v} = u\hat{\boldsymbol{\theta}} + w\hat{\mathbf{z}}$$

where u and w are the circumferential and axial velocities, respectively. Hence, $d\mathbf{v}/dt = (du/dt)\hat{\boldsymbol{\theta}} + u\, d\hat{\boldsymbol{\theta}}/dt + (dw/dt)\hat{\mathbf{z}} + w(d\hat{\mathbf{z}}/dt) = u\, d\hat{\boldsymbol{\theta}}/dt = -(u^2/a)\hat{\mathbf{r}}$. The velocities u and w are related to v as follows: In the time interval $\Delta t = 2\pi a/u$, the axial position z is changed by h. Hence, $w = h/\Delta t = hu/2\pi a$, and $v = u[1 + h^2/(4\pi^2 a^2)]^{1/2}$.

2.3 THE SUMMATION CONVENTION

For further development, an important matter of notation must be mastered.

A set of n variables $x_1, x_2, \ldots, x_n$ is usually denoted as x_i, $i = 1, \ldots, n$. When written singly, the symbol x_i stands for *any one* of the variables $x_1, x_2, \ldots, x_n$. The *range* of i must be indicated in every case; the simplest way is to write, as illustrated here, $i = 1, 2, \ldots, n$. The symbol i is an *index*. An index may be either a subscript or a superscript. A system of notations using indices is said to be an *indicial notation*.

Consider an equation describing a plane in a three-dimensional space referred to a rectangular Cartesian frame of reference with axes x_1, x_2, x_3, i.e.,

$$a_1x_1 + a_2x_2 + a_3x_3 = p, \tag{2.3–1}$$

where a_i and p are constants. This equation can be written as

$$\sum_{i=1}^{3} a_ix_i = p. \tag{2.3–2}$$

However, we shall introduce the *summation convention* and write the preceding equation in the simple form

$$a_ix_i = p. \tag{2.3–3}$$

The convention is as follows: *The repetition of an index in a term will denote a summation with respect to that index over its range.* The *range* of an index i is the set of n integers 1 to n. An index that is summed over is called a *dummy index*; one that is not summed is called a *free index*.

Since a dummy index indicates summation, it is immaterial which symbol is used. Thus, $a_i x_i$ is the same as $a_j x_j$, etc. This is analogous to the dummy variable in an integral, e.g.,

$$\int_a^b f(x)\,dx = \int_a^b f(y)\,dy.$$

Examples

The use of the index and summation convention may be illustrated by other examples. Consider a unit vector $\mathbf{v}$ in a three-dimensional Euclidean space with rectangular Cartesian coordinates x, y, and z. Let the direction cosines α_i be defined as

$$\alpha_1 = \cos(\boldsymbol{\nu}, x), \qquad \alpha_2 = \cos(\boldsymbol{\nu}, y), \qquad \alpha_3 = \cos(\boldsymbol{\nu}, z),$$

where $(\boldsymbol{\nu}, x)$ denotes the angle between $\boldsymbol{\nu}$ and the x-axis, and so forth. The set of numbers $\alpha_i (i = 1, 2, 3)$ represents the components of the unit vector on the coordinate axes. The fact that the length of the vector is unity is expressed by the equation

$$(\alpha_1)^2 + (\alpha_2)^2 + (\alpha_3)^2 = 1,$$

or, simply,

$$\alpha_i \alpha_i = 1. \tag{2.3–4}$$

As another illustration, consider a line element with components dx, dy, dz in a three-dimensional Euclidean space with rectangular Cartesian coordinates x, y, and z. The square of the length of the line element is

$$ds^2 = dx^2 + dy^2 + dz^2. \tag{2.3–5}$$

If we define

$$dx_1 = dx, \qquad dx_2 = dy, \qquad dx_3 = dz, \tag{2.3–6}$$

and

$$\begin{aligned} &\delta_{11} = \delta_{22} = \delta_{33} = 1, \\ &\delta_{12} = \delta_{21} = \delta_{13} = \delta_{31} = \delta_{23} = \delta_{32} = 0, \end{aligned} \tag{2.3–7}$$

then Eq. (2.3–5) may be written as

$$ds^2 = \delta_{ij}\,dx_i\,dx_j, \qquad \blacktriangle \tag{2.3–8}$$

with the understanding that the range of the indices i and j is 1 to 3. Note that there are two summations in this expression, one over i and one over j. The symbol δ_{ij}, as defined in Eq. (2.3–7), is called the *Kronecker delta*.

Matrices and Determinants

The rules of matrix algebra and the evaluation of determinants can be expressed more simply with the summation convention. An $m \times n$ *matrix* $\mathbf{A}$ is an ordered rectangular array of mn elements. We denote

$$\mathbf{A} = (a_{ij}) = \begin{pmatrix} a_{11} & a_{12} & \cdots & a_{1n} \\ a_{21} & a_{22} & \cdots & a_{2n} \\ a_{m1} & a_{m2} & \cdots & a_{mn} \end{pmatrix} \tag{2.3–9}$$

so that a_{ij} is the element in the ith row and jth column of the matrix $\mathbf{A}$. The index i takes the values 1, 2, . . . , m, and the index j takes the values 1, 2, . . . , n. A transpose of $\mathbf{A}$ is another matrix, denoted by $\mathbf{A}^T$, whose elements are the same as those of $\mathbf{A}$, except that the row numbers and column numbers are interchanged. Thus,

$$\mathbf{A}^T = (a_{ij})^T = \begin{pmatrix} a_{11} & a_{21} & \cdots & a_{m1} \\ a_{12} & a_{22} & \cdots & a_{m2} \\ a_{1n} & a_{2n} & \cdots & a_{mn} \end{pmatrix} \tag{2.3–10}$$

The product of two 3 × 3 matrices $\mathbf{A} = (a_{ij})$, $\mathbf{B} = (b_{ij})$ is a 3 × 3 square matrix defined as

$$\begin{aligned} \mathbf{A}\cdot\mathbf{B} &= \begin{pmatrix} a_{11} & a_{12} & a_{13} \\ a_{21} & a_{22} & a_{23} \\ a_{31} & a_{32} & a_{33} \end{pmatrix} \begin{pmatrix} b_{11} & b_{12} & b_{13} \\ b_{21} & b_{22} & b_{23} \\ b_{31} & b_{32} & b_{33} \end{pmatrix} \\ &= \begin{pmatrix} a_{11}b_{11} + a_{12}b_{21} + a_{13}b_{31} & \cdots \\ a_{21}b_{11} + a_{22}b_{21} + a_{23}b_{31} & \cdots \\ a_{31}b_{11} + a_{32}b_{21} + a_{33}b_{31} & \cdots \end{pmatrix} \end{aligned} \tag{2.3–11}$$

whose element in the ith row and jth column can be written, with the summation convention, as

$$(\mathbf{A}\cdot\mathbf{B})_{ij} = (a_{ik}b_{kj}) \tag{2.3–12}$$

A vector $\mathbf{u}$ may be represented by a row matrix (u_i), and Eq. (2.1–2) can be written

$$|\mathbf{u}|^2 = (u_i)\cdot(u_j)^T = u_1^2 + u_2^2 + u_3^2 = u_i u_i. \tag{2.3–13}$$

By this rule, the scalar product of two vectors $\mathbf{u}\cdot\mathbf{v}$, Eq. (2.1–3), can be written as

$$\mathbf{u}\cdot\mathbf{v} = (u_i)(v_i)^T = u_1v_1 + u_2v_2 + u_3v_3 = u_i v_i. \tag{2.3–14}$$

The *determinant* of a square matrix is a number that is the sum of all the products of the elements of the matrix, taken one from each row and one from each column, and no two or more from any row or column, and with sign specified by a rule given shortly. For example, the determinant of a 3×3 matrix $\mathbf{A}$ is written as det $\mathbf{A}$ and is defined as

$$\det \mathbf{A} = \det (a_{ij}) = \begin{vmatrix} a_{11} & a_{12} & a_{13} \\ a_{21} & a_{22} & a_{23} \\ a_{31} & a_{32} & a_{33} \end{vmatrix} \tag{2.3–15}$$

$$= a_{11}a_{22}a_{33} + a_{12}a_{23}a_{31} + a_{13}a_{21}a_{32} - a_{11}a_{23}a_{32} - a_{12}a_{21}a_{33} - a_{13}a_{22}a_{31}$$

The special rule of signs is as follows: Arrange the first index in the order 1, 2, 3. Then check the order of the second index. If they permute as 1, 2, 3, 1, 2, 3, . . . , then the sign is positive; otherwise the sign is negative.

Let us introduce a special symbol, ϵ_{rst}, called the *permulation symbol* and defined by the equations

$$\begin{aligned} &\epsilon_{111} = \epsilon_{222} = \epsilon_{333} = \epsilon_{112} = \epsilon_{121} = \epsilon_{211} = \epsilon_{221} = \epsilon_{331} = \cdots = 0, \\ &\epsilon_{123} = \epsilon_{231} = \epsilon_{312} = 1, \\ &\epsilon_{213} = \epsilon_{321} = \epsilon_{132} = -1. \end{aligned} \tag{2.3–16}$$

In other words, ϵ_{ijk} vanishes whenever the values of any two indices coincide; $\epsilon_{ijk} = 1$ when the subscripts permute as 1, 2, 3; and $\epsilon_{ijk} = -1$ otherwise. Then the determinant of the matrix (a_{ij}) can be written as

$$\det (a_{ij}) = \epsilon_{rst}a_{r1}a_{s2}a_{t3} \tag{2.3–17}$$

Using the symbol ϵ_{rst}, we can write Eq. (2.1–8) defining the vector product $\mathbf{u} \times \mathbf{v}$ as

$$\mathbf{u} \times \mathbf{v} = \epsilon_{rst}u_s v_t \mathbf{e}_r \tag{2.3–18}$$

The ϵ-δ Identity

The Kronecker delta and the permutation symbol are very important quantities that will appear again and again in this book. They are connected by the identity

$$\epsilon_{ijk}\epsilon_{ist} = \delta_{js}\delta_{kt} - \delta_{jt}\delta_{ks}. \qquad \blacktriangle \tag{2.3–19}$$

This ϵ-δ identity is used frequently enough to warrant special attention here. It can be verified by actual trial.

Differentiation

Finally, we shall extend the summation convention to differentiation formulas. Let $f(x_1, x_2, \ldots, x_n)$ be a function of n variables $x_1, x_2, \ldots, x_n$. Then its differential shall be written as

$$df = \frac{\partial f}{\partial x_1} dx_1 + \frac{\partial f}{\partial x_2} dx_2 + \cdots + \frac{\partial f}{\partial x_n} dx_n = \frac{\partial f}{\partial x_i} dx_i. \tag{2.3–20}$$

PROBLEMS

2.15 Write Eq. (2.2–1) or (2.2–3) in the index form. Let the components of $\mathbf{F}^{(i)}$ be written as $F_k^{(i)}$, $k = 1, 2, 3$; i.e., $F_x = F_1$, etc.

Answer. $\sum_{i=1}^{n} F_k^{(i)} = 0.$

2.16 Show that
(a) $\delta_{ii} = 3$
(b) $\delta_{ij}\delta_{ij} = 3$
(c) $\epsilon_{ijk}\epsilon_{jki} = 6$
(d) $\epsilon_{ijk}A_j A_k = 0$
(e) $\delta_{ij}\delta_{jk} = \delta_{ik}$
(f) $\delta_{ij}\epsilon_{ijk} = 0$

2.17 Write Eqs. (2.1–1) and (2.1–5) in the index form, e.g., $\mathbf{u}\cdot\mathbf{v} = u_i v_i$.

Note. For Eq. (2.1–1), we may do the following: Define three unit vectors $\boldsymbol{\nu}^{(1)} = \mathbf{e}_1$, $\boldsymbol{\nu}^{(2)} = \mathbf{e}_2$, $\boldsymbol{\nu}^{(3)} = \mathbf{e}_3$; then $\mathbf{u} = u_i \cdot \boldsymbol{\nu}^{(i)}$.

2.18 Use the index form of vector equations to solve Probs. 2.5 through 2.9.

2.19 The vector product of two vectors $\mathbf{u} = (u_1, u_2, u_3)$ and $\mathbf{v} = (v_1, v_2, v_3)$ is the vector $\mathbf{w} = \mathbf{u} \times \mathbf{v}$ whose components are

$$w_1 = u_2 v_3 - u_3 v_2, \qquad w_2 = u_3 v_1 - u_1 v_3, \qquad w_3 = u_1 v_2 - u_2 v_1.$$

Show that this can be shortened by writing

$$w_i = \epsilon_{ijk} u_j v_k.$$

2.20 Express Eqs. (2.1–7) in the index form.

2.21 Derive the vector identity connecting three arbitrary vectors **A**, **B**, **C** by the method of vector analysis:

$$\mathbf{A} \times (\mathbf{B} \times \mathbf{C}) = (\mathbf{A}\cdot\mathbf{C})\mathbf{B} - (\mathbf{A}\cdot\mathbf{B})\mathbf{C}.$$

Solution. Since $\mathbf{A} \times (\mathbf{B} \times \mathbf{C})$ is perpendicular to $\mathbf{B} \times \mathbf{C}$, it must lie in the plane of **B** and **C**. Hence, we may write $\mathbf{A} \times (\mathbf{B} \times \mathbf{C}) = a\mathbf{B} + b\mathbf{C}$, where a, b are scalar quantities. But $\mathbf{A} \times (\mathbf{B} \times \mathbf{C})$ is a linear function of **A**, **B**, and **C**; hence, a must be a linear scalar combination of **A** and **C**, and b must be a linear scalar combination of **A** and **B**. Accordingly, a, b are proportional to $\mathbf{A}\cdot\mathbf{C}$ and $\mathbf{A}\cdot\mathbf{B}$, respectively, and we may write

$$\mathbf{A} \times (\mathbf{B} \times \mathbf{C}) = \lambda(\mathbf{A}\cdot\mathbf{C})\mathbf{B} + \mu(\mathbf{A}\cdot\mathbf{B})\mathbf{C}$$

where λ, μ are pure numbers, independent of **A**, **B**, and **C**. We can, therefore, evaluate λ, μ by special cases, e.g., if **i**, **j**, **k** are the unit vectors in the directions of the x-, y-, and z-axes (a right-handed rectangular Cartesian coordinate system), respectively, we may put $\mathbf{B} = \mathbf{i}$, $\mathbf{C} = \mathbf{j}$, $\mathbf{A} = \mathbf{i}$ to show that $\mu = -1$; and $\mathbf{B} = \mathbf{i}$, $\mathbf{C} = \mathbf{j}$, $\mathbf{A} = \mathbf{j}$ to show that $\lambda = 1$.

2.22 Write the equation in Prob. 2.21 in the index form, and prove its validity by means of the ϵ-δ identity (2.3–19).

Note. Since the equation in Prob. 2.21 is valid for arbitrary vectors **A**, **B**, **C**, this proof may be regarded as a proof of the ϵ-δ identity.

Solution. $[\mathbf{A} \times (\mathbf{B} \times \mathbf{C})]_l = \epsilon_{lmn}a_m(\mathbf{B} \times \mathbf{C})_n = \epsilon_{lmn}a_m\epsilon_{njk}b_jc_k = \epsilon_{nlm}\epsilon_{njk}a_mb_jc_k$. By the ϵ-δ identity, Eq. (2.3–19), this becomes $(\delta_{lj}\delta_{mk} - \delta_{lk}\delta_{mj})a_mb_jc_k$. Hence, it is $\delta_{lj}a_mc_mb_j - \delta_{lk}a_mb_mc_k = a_mc_mb_l - a_mb_mc_l = (\mathbf{A}\cdot\mathbf{C})(\mathbf{B})_l - (\mathbf{A}\cdot\mathbf{B})(\mathbf{C})_l$.

2.4 TRANSLATION AND ROTATION OF COORDINATES

Two-Dimensional Space

Consider two sets of rectangular Cartesian frames of reference O-xy and O'-$x'y'$ on a plane. If the frame of reference O'-$x'y'$ is obtained from O-xy by a shift of origin without a change in orientation, then the transformation is a *translation*. If a point P has coordinates (x, y) and (x', y') with respect to the old and new frames of reference, respectively, and if the coordinates of the new origin O' are (h, k) relative to O-xy, then

$$\begin{cases} x = x' + h \\ y = y' + k \end{cases} \quad \text{or} \quad \begin{cases} x' = x - h \\ y' = y - k. \end{cases} \tag{2.4–1}$$

If the origin remains fixed, and the new axes are obtained by rotating Ox and Oy through an angle θ in the counterclockwise direction, then the transformation of axes is a *rotation*. Let P have coordinates (x, y), (x', y') relative to the old and new frames of reference, respectively. Then (see Fig. 2.2),

$$\begin{aligned} x &= x' \cos\theta - y' \sin\theta \\ y &= x' \sin\theta + y' \cos\theta. \end{aligned} \tag{2.4–2}$$

$$\begin{aligned} x' &= x \cos\theta + y \sin\theta \\ y' &= -x \sin\theta + y \cos\theta. \end{aligned} \tag{2.4–3}$$

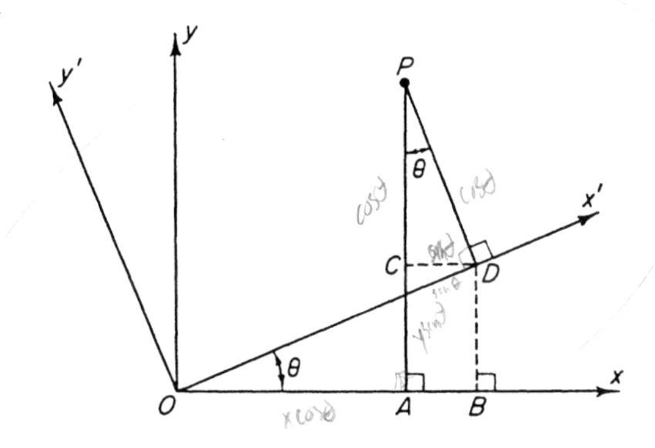

Figure 2.2 Rotation of coordinates.

Using the index notion, we let x_1, x_2 replace x, y and x'_1, x'_2 replace x', y'. Then obviously, a rotation specified by Eq. (2.4–3) can be represented by the equation

$$x'_i = \beta_{ij}x_j, \qquad (i = 1, 2) \tag{2.4–4}$$

where β_{ij} are elements of the square matrix

$$(\beta_{ij}) = \begin{pmatrix} \beta_{11} & \beta_{12} \\ \beta_{21} & \beta_{22} \end{pmatrix} = \begin{pmatrix} \cos\theta & \sin\theta \\ -\sin\theta & \cos\theta \end{pmatrix}. \tag{2.4–5}$$

The inverse transform of Eq. (2.4–4) is

$$x_i = \beta_{ji}x'_j, \qquad (i = 1, 2) \tag{2.4–6}$$

where, according to Eq. (2.4–2), β_{ji} is the element in the jth row and ith column of the matrix (β_{ij}). It is clear that the matrix (β_{ji}) is the *transpose* of the matrix (β_{ij}), i.e.,

$$(\beta_{ji}) = (\beta_{ij})^T. \tag{2.4–7}$$

On the other hand, from the point of view of the solution of the set of simultaneous linear equations (2.4–4), the matrix (β_{ji}) in Eq. (2.4–6) must be identified as the *inverse* of the matrix (β_{ij}), i.e.,

$$(\beta_{ji}) = (\beta_{ij})^{-1}. \tag{2.4–8}$$

Thus, we obtain a fundamental property of the transformation matrix (β_{ij}) that defines a rotation of rectangular Cartesian coordinates:

$$(\beta_{ij})^T = (\beta_{ij})^{-1}. \tag{2.4–9}$$

A matrix (β_{ij}), $i, j = 1, 2, \ldots, n$, that satisfies Eq. (2.4–9) is called an *orthogonal* matrix. A transformation is said to be orthogonal if the associated matrix is orthogonal. The matrix of Eq. (2.4–5) defining a rotation of coordinates is orthogonal.

For an orthogonal matrix, we have

$$(\beta_{ij})(\beta_{ij})^T = (\beta_{ij})(\beta_{ij})^{-1} = (\delta_{ij}),$$

where δ_{ij} is the Kronecker delta. Hence,

$$\beta_{ik}\beta_{jk} = \delta_{ij}. \tag{2.4–10}$$

To clarify the geometric meaning of this important equation, we rederive it directly for the rotation transformation as follows. A unit vector issued from the origin along the x_i'-axis has direction cosines β_{i1}, β_{i2} with respect to the x_1-, x_2-axes, respectively. The fact that its length is unity is expressed by the equation

$$(\beta_{i1})^2 + (\beta_{i2})^2 = 1, \qquad (i = 1, 2). \tag{2.4–11}$$

The fact that a unit vector along the x_i'-axis is perpendicular to a unit vector along the x_j'-axis if $j \neq i$ is expressed by the equation

$$\beta_{i1}\beta_{j1} + \beta_{i2}\beta_{j2} = 0, \qquad (i \neq j). \tag{2.4–12}$$

Combining Eqs. (2.4–11) and (2.4–12), we obtain Eq. (2.4–10).

Note: Alternatively, since we know what the β_{ij}'s are from Eq. (2.4–5), we can verify Eq. (2.4–10) by direct computation.

Three-Dimensional Space

Obviously, the preceding discussion can be extended to three dimensions without much ado. The range of indices i, j can be extended to 1, 2, 3. Thus, consider two right-handed rectangular Cartesian coordinate systems x_1, x_2, x_3 and x_1', x_2', x_3', with the same origin O. Let $\mathbf{x}$ denote the position vector of a point P with components x_1, x_2, x_3 or x_1', x_2', x_3'. Let $\mathbf{e}_1, \mathbf{e}_2, \mathbf{e}_3$ be unit vectors in the directions of the positive x_1, x_2, x_3-axes. They are called *base vectors* of the x_1, x_2, x_3 coordinate system. Let $\mathbf{e}_1', \mathbf{e}_2', \mathbf{e}_3'$ be the base vectors of the x_1', x_2', x_3' coordinate system. Note that since the coordinates are orthogonal, we have

$$\mathbf{e}_i \cdot \mathbf{e}_j = \delta_{ij}, \qquad \mathbf{e}_i' \cdot \mathbf{e}_j' = \delta_{ij}. \tag{2.4–13}$$

In terms of the base vectors, the vector $\mathbf{x}$ may be expressed as follows:

$$\mathbf{x} = x_j \mathbf{e}_j = x_j' \mathbf{e}_j'. \tag{2.4–14}$$

A scalar product of both sides of Eq. (2.4–14) with $\mathbf{e}_i$ gives

$$x_j(\mathbf{e}_j \cdot \mathbf{e}_i) = x_j'(\mathbf{e}_j' \cdot \mathbf{e}_i). \tag{2.4–15}$$

But

$$x_j(\mathbf{e}_j \cdot \mathbf{e}_i) = x_j \delta_{ij} = x_i;$$

therefore,

$$x_i = (\mathbf{e}_j' \cdot \mathbf{e}_i) x_j'. \tag{2.4–16}$$

Now, define

$$(\mathbf{e}_j' \cdot \mathbf{e}_i) \equiv \beta_{ji}; \tag{2.4–17}$$

then,

$$x_i = \beta_{ji} x'_j, \qquad (j = 1, 2, 3). \tag{2.4–18}$$

Next, dot both sides of Eq. (2.4–14) with $\mathbf{e}'_i$. This gives

$$x_j(\mathbf{e}_j \cdot \mathbf{e}'_i) = x'_j(\mathbf{e}'_j \cdot \mathbf{e}'_i).$$

But $(\mathbf{e}'_j \cdot \mathbf{e}'_i) = \delta_{ij}$ and $(\mathbf{e}_j \cdot \mathbf{e}'_i) = \beta_{ij}$; therefore, we obtain

$$x'_i = \beta_{ij} x_j, \qquad (i = 1, 2, 3). \tag{2.4–19}$$

Equations (2.4–18) and (2.4–19) are generalizations of Eqs. (2.4–4) and (2.4–6) to the three-dimensional case.

Equation (2.4–17) shows the geometric meaning of the coefficient β_{ij}. That Eqs. (2.4–7) and (2.4–8) hold for $i, j = 1, 2, 3$ is clear because Eqs. (2.4–18) and (2.4–19) are inverse transformations of each other. Then, Eqs. (2.4–9) and (2.4–10) follow.

Now, the numbers x_1, x_2, x_3 that represent the coordinates of the point P in Fig. 2.3 are also the components of the radius vector $\mathbf{A}$. A recognition of this fact

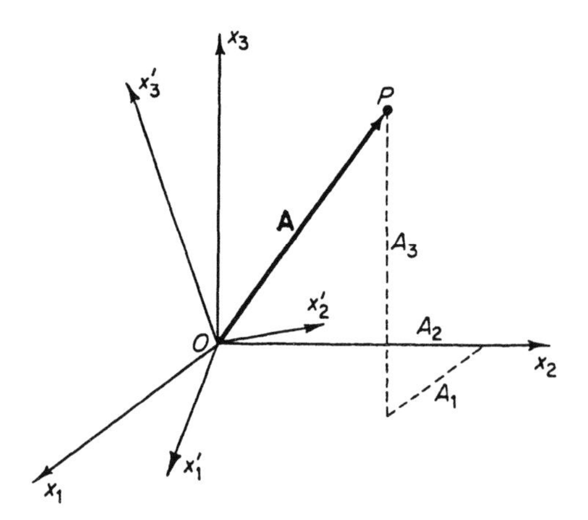

Figure 2.3 Radius vector and coordinates.

gives us immediately the law of transformation of the components of a vector in rectangular Cartesian coordinates:

$$A'_i = \beta_{ij} A_j, \qquad A_i = \beta_{ji} A'_j, \tag{2.4–20}$$

in which β_{ij} represents the cosine of the angle between the axes Ox'_i and Ox_j.

Finally, let us point out that the three unit vectors along x'_1, x'_2, x'_3 form the edges of a cube with volume 1. The volume of a parallelepiped having any three vectors $\mathbf{u}$, $\mathbf{v}$, $\mathbf{w}$ as edges is given either by the triple product $\mathbf{u} \cdot (\mathbf{v} \times \mathbf{w})$ or by its negative; the sign is determined by whether the three vectors $\mathbf{u}$, $\mathbf{v}$, $\mathbf{w}$, in this order, form a right-handed screw system or not. If they are right handed, then the volume is equal to the determinant of their components:

$$\text{Volume} = (\mathbf{u} \times \mathbf{v}) \cdot \mathbf{w} = \begin{vmatrix} u_1 & u_2 & u_3 \\ v_1 & v_2 & v_3 \\ w_1 & w_2 & w_3 \end{vmatrix}. \tag{2.4-21}$$

Let us assume that $\mathbf{x}_1, \mathbf{x}_2, \mathbf{x}_3$ and $\mathbf{x}'_1, \mathbf{x}'_2, \mathbf{x}'_3$ are right handed. Then it is clear that the determinant of β_{ij} represents the volume of a unit cube and hence has the value 1:

$$|\beta_{ij}| \equiv \begin{vmatrix} \beta_{11} & \beta_{12} & \beta_{13} \\ \beta_{21} & \beta_{22} & \beta_{23} \\ \beta_{31} & \beta_{32} & \beta_{33} \end{vmatrix} = 1. \tag{2.4-22}$$

PROBLEMS

2.23 Write out Eq. (2.4–10) in extenso, and interpret the geometric meaning of the six resulting equations; $i = 1, 2, 3$.

Solution. Let the index i stand for 1, 2, 3.

$$\text{If } i = 1, j = 1\text{: then } \beta_{11}\beta_{11} + \beta_{12}\beta_{12} + \beta_{13}\beta_{13} = 1. \tag{1}$$

$$\text{If } i = 1, j = 2\text{: then } \beta_{11}\beta_{21} + \beta_{12}\beta_{22} + \beta_{13}\beta_{23} = 0. \tag{2}$$

Equation (1) means that the length of the vector $(\beta_{11}, \beta_{12}, \beta_{13})$ is 1. Equation (2) means that the vectors $(\beta_{11}, \beta_{21}, \beta_{13})$, $(\beta_{21}, \beta_{22}, \beta_{23})$ are orthogonal to each other.

Other combinations of i, j are similar.

2.24 Derive Eq. (2.4–10) by the following alternative procedure. Differentiate both sides of Eq. (2.4–4) with respect to x'_j. Then use Eq. (2.4–6) and the fact that $\partial x_i/\partial x_j = \delta_{ij}$ to simplify the results.

Solution. Differentiating Eq. (2.4–4) with respect to x'_j, we obtain $\delta_{ij} = \beta_{ik}\partial x_k/\partial x'_j$. But $x_i = \beta_{ji}x'_j$. On changing the index i to k and differentiating, we have $\partial x_k/\partial x'_j = \beta_{jk}$. Combining these results yields $\delta_{ij} = \beta_{ik}\beta_{jk}$.

2.5 COORDINATE TRANSFORMATION IN GENERAL

A set of independent variables x_1, x_2, x_3 specifies the coordinates of a point in a frame of reference. A set of equations

$$\bar{x}_1 = f_i(x_1, x_2, x_3), \qquad (i = 1, 2, 3) \tag{2.5-1}$$

describes a transformation from x_1, x_2, x_3 to a set of new variables $\bar{x}_1, \bar{x}_2, \bar{x}_3$. The inverse transformation

$$x_i = g_i(\bar{x}_1, \bar{x}_2, \bar{x}_3), \qquad (i = 1, 2, 3) \tag{2.5-2}$$

proceeds in the reverse direction. In order to ensure that such a reversible transformation exists and is in one-to-one correspondence in a certain region R of the variables (x_1, x_2, x_3)—i.e., in order that each set of numbers $(\bar{x}_1, \bar{x}_2, \bar{x}_3)$ defines a unique set of numbers (x_1, x_2, x_3), for (x_1, x_2, x_3) in the region R, and vice versa—it is sufficient that

(1) The functions f_i are single valued, are continuous, and possess continuous first partial derivatives in the region R.

(2) The *Jacobian determinant* $J = \det(\partial\bar{x}_i/\partial x_j)$ does not vanish at any point of the region R. That is,

$$J = \det\left(\frac{\partial\bar{x}_i}{\partial x_j}\right) \equiv \begin{vmatrix} \dfrac{\partial\bar{x}_1}{\partial x_1} & \dfrac{\partial\bar{x}_1}{\partial x_2} & \dfrac{\partial\bar{x}_1}{\partial x_3} \\ \dfrac{\partial\bar{x}_2}{\partial x_1} & \dfrac{\partial\bar{x}_2}{\partial x_2} & \dfrac{\partial\bar{x}_2}{\partial x_3} \\ \dfrac{\partial\bar{x}_3}{\partial x_1} & \dfrac{\partial\bar{x}_3}{\partial x_2} & \dfrac{\partial\bar{x}_3}{\partial x_3} \end{vmatrix} \neq 0. \tag{2.5–3}$$

Coordinate transformations with the properties 1 and 2 are called *admissible transformations*. If the Jacobian is positive everywhere, then a right-hand set of coordinates is transformed into another right-hand set, and the transformation is said to be *proper*. If the Jacobian is negative everywhere, a right-hand set of coordinates is transformed into a left-hand one, and the transformation is said to be *improper*. *In this book, we shall* tacitly assume *that our transformations are admissible and proper*.

Significance of the Jacobian Determinant

To appreciate the significance of the *Jacobian determinant*, let us assume that we have found that (x_1^0, x_2^0, x_3^0) corresponds to $(\bar{x}_1^0, \bar{x}_2^0, \bar{x}_3^0)$, i.e., they satisfy Eq. (2.5–1), and ask whether we can find an inverse transformation in a small neighborhood of this point. We differentiate Eq. (2.5–1) to obtain

$$d\bar{x}_i = \frac{\partial f_i}{\partial x_j}\, dx_j \qquad (i = 1, 2, 3) \tag{2.5–4}$$

and evaluate the partial derivatives $\partial f_i/\partial x_j$ at the point (x_1^0, x_2^0, x_3^0). The Eq. (2.5–4) defines a linear transformation of the vector dx_j to a vector $d\bar{x}_i$. If we solve the set of linear equations (2.5–4) for dx_j, we know that the solution exists only if the determinant of the coefficients does not vanish:

$$\det\left(\frac{\partial f_i}{\partial x_j}\right) \neq 0. \tag{2.5–5}$$

Thus, an inverse exists in the neighborhood of (x_1^0, x_2^0, x_3^0) only if Eq. (2.5–3) is valid. Further, when $J \neq 0$, Eq. (2.5–4) can be solved to obtain

$$dx_i = c_{ij} d\bar{x}_j \tag{2.5–6}$$

where c_{ij} are constants. Hence, a small neighborhood of the known point, an inverse transformation [an approximation of Eq. (2.5–2)] can be found in a small neighborhood of the known point. Thus, conditions 1 and 2 stated earlier are sufficient conditions for the existence of an inverse in a small region around the known point. By repeated application of this argument to new known points away from the initial known point, one can extend and find the region R in which a one-to-one inverse transformation given by Eq. (2.5–2) exists.

PROBLEM

2.25 (a) Review the methods of solving linear simultaneous equations. One of the methods uses determinants. Use that method to solve Eq. (2.5–4) for dx_1, dx_2, dx_3. Use the permutation symbol ϵ_{rst}, defined in Eq. (2.3–16), to express the final result.
(b) R is a region in and on a circle of unit radius on a plane. The equation of the circle is $r = 1$ in polar coordinates and $x^2 + y^2 = 1$ in rectangular Cartesian coordinates. Show that the Jacobian J is equal to r and that the area of the circle is

$$\iint_R J\, dr\, d\theta = \iint_R dx\, dy,$$

or

$$\int_0^1 \int_0^{2\pi} r\, dr\, d\theta = \int_0^1 \int_0^{\sqrt{1-x^2}} dx\, dy.$$

Here, an integration of the Jacobian multiplied by the product of the differentials $dr\, d\theta$ gives the area.

2.6 ANALYTICAL DEFINITIONS OF SCALARS, VECTORS, AND CARTESIAN TENSORS

Let (x_1, x_2, x_3) and $(\bar{x}_1, \bar{x}_2, \bar{x}_3)$ be two fixed sets of rectangular Cartesian frames of reference related by the transformation law

$$\bar{x}_i = \beta_{ij} x_j \tag{2.6–1}$$

where β_{ij} is the direction cosine of the angle between unit vectors along the coordinate axes $\bar{x}_i$ and x_j. Thus,

$$\beta_{21} = \cos(\bar{x}_2, x_1), \tag{2.6–2}$$

and so forth. The inverse transform is

$$x_i = \beta_{ji} \bar{x}_j. \tag{2.6–3}$$

A system of quantities is called a *scalar*, a *vector*, or a *tensor*, depending upon how the components of the system are defined in the variables x_1, x_2, x_3 and how they are transformed when the variables x_1, x_2, x_3 are changed to $\bar{x}_1, \bar{x}_2, \bar{x}_3$.

A system is called a *scalar* if it has only a single component Φ in the variables x_i and a single component $\bar{\Phi}$ in the variables $\bar{x}_i$ and if Φ and $\bar{\Phi}$ are numerically equal at the corresponding points,

$$\Phi(x_1, x_2, x_3) = \bar{\Phi}(\bar{x}_1, \bar{x}_2, \bar{x}_3). \tag{2.6–4}$$

A system is called a *vector field* or a *tensor field of rank 1* if it has three components ξ_i in the variables x_i and three components $\bar{\xi}_i$ in the variables $\bar{x}_i$ and if the components are related by the characteristic law

$$\begin{aligned} \bar{\xi}_i(\bar{x}_1, \bar{x}_2, \bar{x}_3) &= \xi_k(x_1, x_2, x_3)\beta_{ik}, \\ \xi_i(x_1, x_2, x_3) &= \bar{\xi}_k(\bar{x}_1, \bar{x}_2, \bar{x}_3)\beta_{ki}. \end{aligned} \tag{2.6–5}$$

Generalizing these definitions to a system that has nine components when i and j range over 1, 2, 3, we define a *tensor field of rank 2* if it is a system that has nine components t_{ij} in the variables x_1, x_2, x_3 and nine components $\bar{t}_{ij}$ in the variables $\bar{x}_1, \bar{x}_2, \bar{x}_3$ and if the components are related by the characteristic law

$$\begin{aligned} \bar{t}_{ij}(\bar{x}_1, \bar{x}_2, \bar{x}_3) &= t_{mn}(x_1, x_2, x_3)\beta_{im}\beta_{jn}, \\ t_{ij}(x_1, x_2, x_3) &= \bar{t}_{mn}(\bar{x}_1, \bar{x}_2, \bar{x}_3)\beta_{mi}\beta_{nj}. \end{aligned} \tag{2.6–6}$$

Further generalization to tensor fields of higher ranks is immediate. These definitions can obviously be modified to two dimensions if the indices range over 1, 2, or to n dimensions if the range of the indices is $1, 2, \ldots, n$. Since our definitions are based on transformations from one rectangular Cartesian frame of reference to another, the systems so defined are called *Cartesian tensors*. For simplicity, only Cartesian tensor equations will be used in this book.

Elaboration on Why Vectors and Tensors Are Defined in This Manner

The analytical definition of vectors is designed to follow the idea of a radius vector. We all know that the radius vector, a vector joining the origin (0, 0, 0) to a point (x_1, x_2, x_3), embodies our idea of a vector and expresses it numerically in terms of the components $(x_1 - 0, x_2 - 0, x_3 - 0)$, i.e., (x_1, x_2, x_3). When this vector is viewed from another frame of reference, the components referred to the new frame can be computed from the old according to Eq. (2.6–1), which is the *law of transformation of the components of a radius vector. Our generalization of Eq. (2.6–1) into Eq. (2.6–5), which defines all vectors, is equivalent to saying that we can call an entity a vector if it behaves like a radius vector, namely, if it has a fixed direction and a fixed magnitude.*

These remarks are intended to differentiate a matrix from a vector. We can list the components of a vector in the form of a column matrix; but not all column matrices are vectors. For example, to identify myself, I can list my age, social security number, street address, and zip code in a column matrix. What can you say about this matrix? Nothing very interesting! It is certainly not a vector.

The mathematical steps we took in generalizing the definition given in Eq. (2.6–5) for a vector to Eq. (2.6–6) for a tensor are natural enough. These equations are so similar that if we call a vector a tensor of rank 1, we cannot help but call the others tensors of rank 2 or 3, etc. What is the physical significance of these higher order tensors? The most effective way to answer this question is to consider some concrete examples, such as the stress tensor. However, before we turn our attention to specific examples to discuss the significance of tensor equations, consider the following problems:

PROBLEMS

2.26 Show that, *if all components of a Cartesian tensor vanish in one coordinate system, then they vanish in all other Cartesian coordinate systems*. This is perhaps the most important property of tensor fields.

Proof. The property follows immediately from Eq. (2.6–6). If every component of t_{mn} vanishes, then the right-hand side vanishes and $\bar{t}_{ij} = 0$ for all i, j.

2.27 Prove the following theorem: *The sum or difference of two Cartesian tensors of the same rank is again a tensor of the same rank*. Thus, any linear combination of tensors of the same rank is again a tensor of the same rank.

Proof. Let A_{ij}, B_{ij} be two tensors. Under the coordinate transformation given by Eq. (2.6–1), we have the new components

$$\overline{A}_{ij} = A_{mn}\beta_{im}\beta_{jn}, \qquad \overline{B}_{ij} = B_{mn}\beta_{im}\beta_{jn}.$$

Adding or subtracting, we obtain

$$\overline{A}_{ij} \pm \overline{B}_{ij} = \beta_{im}\beta_{jn}(A_{mn} \pm B_{mn})$$

and the theorem is proved.

2.28 Prove the following theorem: *Let $A_{\alpha_1\ldots\alpha_r}$, $B_{\alpha_1\ldots\alpha_r}$ be tensors. Then the equation*

$$A_{\alpha_1\ldots\alpha_r}(x_1, x_2, \ldots, x_n) = B_{\alpha_1\ldots\alpha_r}(x_1, x_2, \ldots, x_n)$$

is a tensor equation; i.e., if this equation is true in one Cartesian coordinate system, then it is true in all Cartesian coordinate systems.

Proof. Multiplying both sides of the equation by

$$\beta_{i\alpha_1}\beta_{j\alpha_2}\cdots\beta_{k\alpha_r}$$

and summing over the repeated indices yields the equation

$$\overline{A}_{ij\ldots k}(\bar{x}_1, \bar{x}_2, \ldots, \bar{x}_n) = \overline{B}_{ij\ldots k}(\bar{x}_1, \bar{x}_2, \ldots, \bar{x}_n).$$

Alternatively, write the equation as $\mathbf{A} - \mathbf{B} = 0$. Then every component of $\mathbf{A} - \mathbf{B}$ vanishes. Then apply the results of Probs. 2.27 and 2.26, in turn.

2.7 THE SIGNIFICANCE OF TENSOR EQUATIONS

The theorems stated in the problems at the end of the previous section contain the most important property of tensor fields: *If all the components of a tensor field vanish in one coordinate system, they vanish likewise in all coordinate systems that can be obtained by admissible transformations*. Since the sum and difference of tensor fields of a given type are tensors of the same type, we deduce that *if a tensor equation can be established in one coordinate system, then it must hold for all coordinate systems obtained by admissible transformations*.

Thus, the importance of tensor analysis may be summarized by the following statement: The form of an equation can have general validity with respect to any frame of reference only if every term in the equation has the same tensor characteristics. If this condition is not satisfied, a simple change of the system of reference will destroy the form of the relationship, and the form would, therefore, be merely fortuitous.

We see that tensor analysis is as important as dimensional analysis in any formulation of physical relations. In dimensional analysis, we study the changes a physical quantity undergoes with particular choices of fundamental units. Two physical quantities cannot be equal unless they have the same dimensions. An equation describing a physical relation cannot be correct unless it is invariant with respect to a change of fundamental units.

Because of the design of the tensor transformation laws, the tensorial equations are in harmony with physics.

2.8 NOTATIONS FOR VECTORS AND TENSORS: BOLDFACE OR INDICES?

In continuum mechanics we are concerned with vectors describing displacements, velocities, forces, etc., and with tensors describing stress, strain, constitutive equations, etc. For vectors, the usual notation of boldface letters or an arrow, such as $\mathbf{u}$ or $\vec{u}$, is agreeable to all; but for tensors, there are differences of opinion. A tensor of rank 2 may be printed as a boldface letter or with a double arrow or with a pair of braces. Thus, if T is a tensor of rank 2, it may be printed as $\mathbf{T}$, $\overset{\rightrightarrows}{T}$ or $\{T\}$. The first notation is the simplest, but then you have to remember what the symbol represents; it may be a vector or it may be a tensor. The other notations are cumbersome. More important objections to the simple notation arise when several vectors and tensors are associated together. In vector analysis, we have to distinguish scalar products from vector products. How about tensors? Shall we define many kinds of tensor products? We have to, because there is a variety of ways tensors can be associated. The matter becomes complicated. For this reason, in most theoretical works that require extensive use of tensors, an index notation is used. In this notation, vectors and tensors are resolved into their components with respect to a frame of reference and denoted by symbols such as u_i, u_{ij}, etc. These components are real numbers. Mathematical operations on them follow the usual rules of arithmetic. No special rules of combination need to be introduced. Thus, we gain a measure of simplicity. Furthermore, the index notation exhibits the rank and the range of a tensor clearly. It displays the role of the frame of reference explicitly.

The last-mentioned advantage of the index notation, however, is also a weakness: It draws the attention of the reader away from the physical entity. Hence, one has to be adaptive and familiarize oneself with both systems.

2.9 QUOTIENT RULE

Consider a set of n^3 functions $A(1, 1, 1)$, $A(1, 1, 2)$, $A(1, 2, 3)$, etc., or $A(i, j, k)$ for short, with each of the indices i, j, k ranging over $1, 2, \ldots, n$. Although the set of functions $A(i, j, k)$ has the right number of components, we do not know whether it is a tensor. Now suppose we know something about the nature of the product of $A(i, j, k)$ with an arbitrary tensor. Then there is a method that enables us to establish whether $A(i, j, k)$ is a tensor without going to the trouble of determining the law of transformation directly.

For example, let $\xi_i(x)$ be a vector. Let us suppose that the product $A(i, j, k)\xi_i$ (summation convention used over i) is known to yield a tensor of the type $A_{jk}(x)$, i.e.,

$$A(i, j, k)\xi_i = A_{jk}. \tag{2.9–1}$$

Then we can prove that $A(i, j, k)$ is a tensor of the type $A_{ijk}(x)$.

The proof is very simple. Since $A(i, j, k)\xi_i$ is of the type A_{jk}, it is transformed into $\bar{x}$-coordinates as

$$\bar{A}(i, j, k)\bar{\xi}_i = \bar{A}_{jk} = \beta_{jr}\beta_{ks}A_{rs} = \beta_{jr}\beta_{ks}[A(m, r, s)\xi_m]. \tag{2.9–2}$$

But $\xi_m = \beta_{im}\bar{\xi}_i$. Inserting this in the right-hand side of Eq. (2.9–2) and transposing all terms to one side of the equation, we obtain

$$[\bar{A}(i, j, k) - \beta_{jr}\beta_{ks}\beta_{im}A(m, r, s)]\bar{\xi}_i = 0. \tag{2.9–3}$$

Now $\bar{\xi}_i$ is an arbitrary vector. Hence, the quantity within the brackets must vanish, and we have

$$\bar{A}(i, j, k) = \beta_{im}\beta_{jr}\beta_{ks}A(m, r, s), \tag{2.9–4}$$

which is precisely the law of transformation of the tensor of the type A_{ijk}.

The pattern of the preceding example can be generalized to higher order tensors.

2.10 PARTIAL DERIVATIVES

When only Cartesian coordinates are considered, the partial derivatives of any tensor field behave like the components of a Cartesian tensor. To show this, let us consider two sets of Cartesian coordinates (x_1, x_2, x_3) and $(\bar{x}_1, \bar{x}_2, \bar{x}_3)$ related by

$$\bar{x}_i = \beta_{ij} x_j + \alpha_i, \tag{2.10-1}$$

where β_{ij} and α_i are constants.

Now, if $\xi_i(x_1, x_2, x_3)$ is a tensor, so that

$$\bar{\xi}_i(\bar{x}_1, \bar{x}_2, \bar{x}_3) = \xi_k(x_1, x_2, x_3)\beta_{ik}, \tag{2.10-2}$$

then, on differentiating both sides of this equation, one obtains

$$\frac{\partial \bar{\xi}_i}{\partial \bar{x}_j} = \beta_{ik} \frac{\partial \xi_k}{\partial x_m} \frac{\partial x_m}{\partial \bar{x}_j} = \beta_{ik} \beta_{jm} \frac{\partial \xi_k}{\partial x_m} \tag{2.10-3}$$

which verifies the statement.

It is a common practice to *use a comma to denote partial differentiation*. Thus,

$$\xi_{i,j} \equiv \frac{\partial \xi_i}{\partial x_j}, \qquad \Phi_{,i} \equiv \frac{\partial \Phi}{\partial x_i}, \qquad \sigma_{ij,k} \equiv \frac{\partial \sigma_{ij}}{\partial x_k}.$$

When we restrict ourselves to Cartesian coordinates, $\Phi_{,i}$, $\xi_{i,j}$, and $\sigma_{ij,k}$ are tensors of rank 1, 2, and 3, respectively, provided that Φ, ξ, and σ_{ij} are tensors.

PROBLEMS

2.29 In any tensor $A_{ijk\ldots m}$, equating two indices and summing over that index is called a *contraction*. Thus, for a tensor A_{ijk}, a contraction over i and $j (i, j = 1, 2, 3)$ results in a vector $A_{iik} = A_{11k} + A_{22k} + A_{33k}$. Prove that the contraction of any two indices in a Cartesian tensor of rank n results in a tensor of rank $n - 2$.

Solution. The only significant part of the statement is that the result of contraction is a *tensor*. Let $A_{ijk\ldots n}$ be a tensor of rank n. Then $A_{iik\ldots n}$ has only $(n - 2)$ indices. To show that it is a tensor, consider the definition

$$\bar{A}_{ijk\ldots n} = A_{\alpha_1\alpha_2\alpha_3\ldots\alpha_n} \beta_{i\alpha_1} \beta_{j\alpha_2} \beta_{k\alpha_3} \cdots \beta_{n\alpha_n}.$$

A contraction over i and j yields

$$\bar{A}_{iik\ldots n} = A_{\alpha_1\alpha_2\alpha_3\ldots\alpha_n} \beta_{i\alpha_1} \beta_{i\alpha_2} \beta_{k\alpha_3} \cdots \beta_{n\alpha_n}.$$

But we know from Eq. (2.4–10) that

$$\beta_{i\alpha_1} \beta_{i\alpha_2} = \delta_{\alpha_1\alpha_2}.$$

Hence,

$$\begin{aligned} \bar{A}_{iik\ldots n} &= A_{\alpha_1\alpha_2\alpha_3\ldots\alpha_n} \delta_{\alpha_1\alpha_2} \beta_{k\alpha_3} \cdots \beta_{n\alpha_n} \\ &= A_{\alpha_1\alpha_1\alpha_3\ldots\alpha_n} \beta_{k\alpha_3} \cdots \beta_{n\alpha_n}. \end{aligned}$$

Thus, $A_{\alpha_1\alpha_1\alpha_3\ldots\alpha_n}$ obeys the transformation law for a tensor of rank $(n - 2)$, and we have proved the statement.

2.30 If A_{ij} is a Cartesian tensor of rank 2, show that A_{ii} is a scalar.

Solution. From Prob. 2.29, A_{ii} is a tensor of rank 0 and hence is scalar. More directly, we have

$$\bar{A}_{ij} = A_{mn}\beta_{im}\beta_{jn}$$

$$\bar{A}_{ii} = A_{mn}\beta_{im}\beta_{in} = \delta_{mn}A_{mn} = A_{mm},$$

which obeys the definition of a scalar, Eq. (2.6–4).

2.31 Use the index notation and summation convention to prove the following relations (see the table of notations below):

(a) $\mathbf{u} \times \mathbf{v} = -\mathbf{v} \times \mathbf{u}$

(b) $(\mathbf{s} \times \mathbf{t})\cdot(\mathbf{u} \times \mathbf{v}) = (\mathbf{s}\cdot\mathbf{u})(\mathbf{t}\cdot\mathbf{v}) - (\mathbf{s}\cdot\mathbf{v})(\mathbf{t}\cdot\mathbf{u})$

(c) curl curl $\mathbf{v}$ = grad div $\mathbf{v} - \Delta\mathbf{v}$

Example of solution.

(c) curl curl $\mathbf{v} = \epsilon_{ijk}\dfrac{\partial}{\partial x_j}\left(\epsilon_{klm}\dfrac{\partial v_m}{\partial x_l}\right)$

$$= \epsilon_{ijk}\epsilon_{lmk}\frac{\partial^2 v_m}{\partial x_j \partial x_l}$$

$$= (\delta_{il}\delta_{jm} - \delta_{im}\delta_{jl})\frac{\partial^2 v_m}{\partial x_j \partial x_l}$$

$$= \frac{\partial^2 v_j}{\partial x_j \partial x_i} - \frac{\partial^2 v_i}{\partial x_j \partial x_j} = \frac{\partial}{\partial x_i}\left(\frac{\partial v_j}{\partial x_j}\right) - \frac{\partial}{\partial x_j}\left(\frac{\partial v_i}{\partial x_j}\right)$$

$$= \nabla(\nabla\cdot\mathbf{v}) - \nabla\cdot\nabla\mathbf{v} = \text{grad div } \mathbf{v} - \Delta\mathbf{v}.$$

2.32 Let $\mathbf{r}$ be the radius vector of a typical point in a field and r be the magnitude of $\mathbf{r}$. Prove that, with the notations defined in the following table,

Vector Notation	Index Notation	Rank of Tensor
$\mathbf{v}$ (vector)	v_i	1
$\lambda = \mathbf{u}\cdot\mathbf{v}$ (dot, scalar, or inner product)	$\lambda = u_i v_i$	0
$\mathbf{w} = \mathbf{u} \times \mathbf{v}$ (cross or vector product)	$w_i = \epsilon_{ijk}u_j v_k$	1
grad $\phi = \nabla\phi$ (gradient of scalar field)	$\dfrac{\partial\phi}{\partial x_i}$	1
grad $\mathbf{v} = \nabla\mathbf{v}$ (vector gradient)	$\dfrac{\partial v_i}{\partial x_j}$	2
div $\mathbf{v} = \nabla\cdot\mathbf{v}$ (divergence)	$\dfrac{\partial v_i}{\partial x_i}$	0
curl $\mathbf{v} = \nabla \times \mathbf{v}$ (curl)	$\epsilon_{ijk}\dfrac{\partial v_k}{\partial x_j}$	1
$\nabla^2\mathbf{v} = \nabla\cdot\nabla\mathbf{v} = \Delta\mathbf{v}$ (Laplacian)	$\dfrac{\partial}{\partial x_i}\left(\dfrac{\partial v_j}{\partial x_i}\right) = \dfrac{\partial^2 v_j}{\partial x_i \partial x_i}$	1

(a) div $(r^n\mathbf{r}) = (n + 3)r^n$
(b) curl $(r^n\mathbf{r}) = 0$
(c) $\Delta(r^n) = n(n + 1)r^{n-2}$

Example of solution.
(a) Let the components of $\mathbf{r}$ be x_i $(i = 1, 2, 3)$.

$$\text{div}\,\mathbf{r} = \nabla\cdot\mathbf{r} = \frac{\partial x_i}{\partial x_i} = 3$$

$$r^2 = x_i x_i; \qquad r\frac{\partial r}{\partial x_i} = x_i; \qquad \therefore \frac{\partial r}{\partial x_i} = \frac{x_i}{r}$$

$$\text{div}\,(r^n\mathbf{r}) = \nabla\cdot(r^n\mathbf{r}) = \frac{\partial}{\partial x_i}(r^n x_i) = r^n\frac{\partial x_i}{\partial x_i} + x_i\frac{\partial r^n}{\partial x_i}$$

$$= 3r^n + n_i\left(nr^{n-1}\frac{\partial r}{\partial x_i}\right) = 3r^n + nr^{n-2}x_i x_i = (n + 3)r^n.$$

2.33 A matrix-valued quantity $a_{ij}(i, j = 1, 2, 3)$ is given as follows:

$$\begin{pmatrix} a_{11} & a_{12} & a_{13} \\ a_{21} & a_{22} & a_{23} \\ a_{31} & a_{32} & a_{33} \end{pmatrix} = \begin{pmatrix} 1 & 1 & 0 \\ 1 & 2 & 2 \\ 0 & 2 & 3 \end{pmatrix}$$

What are the values of (a) a_{ii}, (b) $a_{ij}a_{ij}$, (c) $a_{ij}a_{jk}$ when $i = 1$; $k = 1$ and $i = 1$; and $k = 2$.

Answer. 6, 24, 2, 3.

2.34 It is well known that rigid-body rotation is noncommutative. For example, take a book, and fix a frame of reference with x-, y-, z-axes directed along the edges of the book. First rotate the book 90° about y; then rotate it 90° about z. We obtain a certain configuration. But a reversal of the order of rotation yields a different result.

The rotation of coordinates is also noncommutative; i.e., the transformation matrices (β_{ij}) are noncommutative. Demonstrate this in a special case that is analogous to the rigid-body rotation of the book just considered. First transform x, y, z to x', y', z' by a rotation of 90° about the y-axis. Then transform x', y', z' to x'', y'', z'' by a 90° rotation about z'. Thus,

$$\begin{pmatrix} x' \\ y' \\ z' \end{pmatrix} = \begin{pmatrix} 0 & 0 & 1 \\ 0 & 1 & 0 \\ -1 & 0 & 0 \end{pmatrix}\begin{pmatrix} x \\ y \\ z \end{pmatrix}, \qquad \begin{pmatrix} x'' \\ y'' \\ z'' \end{pmatrix} = \begin{pmatrix} 0 & 1 & 0 \\ -1 & 0 & 0 \\ 0 & 0 & 1 \end{pmatrix}\begin{pmatrix} x' \\ y' \\ z' \end{pmatrix}.$$

Derive the transformation matrix from x, y, z to x'', y'', z''. Now, reverse the order of rotation. Show that a different result is obtained.

2.35 Infinitesimal rotations, however, are commutative. Demonstrate this by considering an infinitesimal rotation by an angle θ about y, followed by another infinitesimal rotation ψ about z. Compare the results with the case in which the order of rotations is reversed.

2.36 Express the following set of equations in a single equation using index notation:

$$\epsilon_{xx} = \frac{1}{E}[\sigma_{xx} - \nu(\sigma_{yy} + \sigma_{zz})], \qquad \epsilon_{xy} = \frac{1+\nu}{E}\sigma_{xy}$$

$$\epsilon_{yy} = \frac{1}{E}[\sigma_{yy} - \nu(\sigma_{xx} + \sigma_{zz})], \qquad \epsilon_{yz} = \frac{1+\nu}{E}\sigma_{yz}$$

$$\epsilon_{zz} = \frac{1}{E}[\sigma_{zz} - \nu(\sigma_{xx} + \sigma_{yy})], \qquad \epsilon_{xz} = \frac{1+\nu}{E}\sigma_{xz}$$

2.37 Write out in longhand, in unabridged form, the following equation:

$$G\left(u_{i,kk} + \frac{1}{1-2\nu}u_{k,ki}\right) + X_i = \rho\frac{\partial^2 u_i}{\partial t^2}.$$

Let

$$x_1 = x,\ x_2 = y,\ x_3 = z; \qquad u_1 = u,\ u_2 = v,\ u_3 = w.$$

2.38 Show that $\epsilon_{ijk}\sigma_{jk} = 0$, where ϵ_{ijk} is the permutation symbol and σ_{jk} is a symmetric tensor, i.e., $\sigma_{jk} = \sigma_{kj}$.

2.39 Write down a full set of basic laws of physics in tensor notation, using the indicial system. Take a good physics book and go through it from beginning to end.

3 STRESS

In Chapter 1 we introduced the concept of stress. In Chapter 2 we defined and analyzed the Cartesian tensors. In this chapter we discuss the properties of the stress tensor.

3.1 NOTATIONS OF STRESS

The concept of *stress* has been discussed in Sec. 1.6. Consider a continuum in a rectangular parallelepiped, as shown in Fig. 3.1. Let a rectangular Cartesian frame

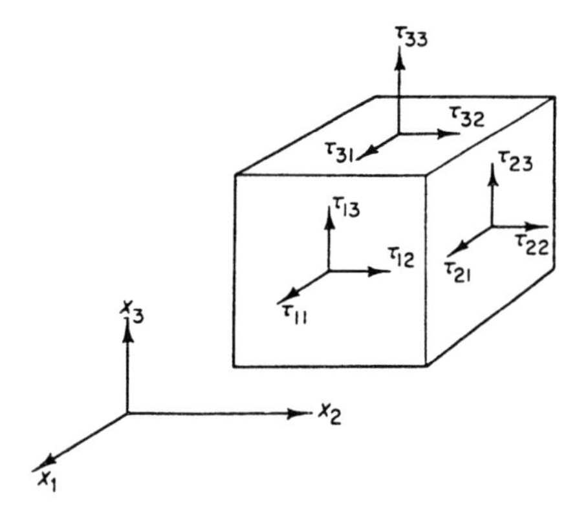

Figure 3.1 Notations of stress components.

of reference with coordinate axes x_1, x_2, x_3 parallel to the edges of the parallelepiped be used. Let the surface ΔS_1 be a surface of the parallelepiped with an outer normal vector pointing in the positive direction of the x_1-axis. Let the stress vector acting on ΔS_1 be denoted by $\overset{1}{\mathbf{T}}$, with three components $\overset{1}{T}_1$, $\overset{1}{T}_2$, $\overset{1}{T}_3$ in the directions of the coordinate axes x_1, x_2, x_3, respectively. In this special case, we introduce a new set of symbols for these stress components:

$$\overset{1}{T}_1 = \tau_{11}, \; \overset{1}{T}_2 = \tau_{12}, \; \overset{1}{T}_3 = \tau_{13}. \tag{3.1–1}$$

Similarly, let ΔS_2 be the surface with an outer normal pointing in the direction of the x_2 axis. The stress vector acting on ΔS_2, $\overset{2}{\mathbf{T}}$, has three components in the directions of x_1, x_2, x_3. These stress components shall be denoted by

$$\overset{2}{T}_1 = \tau_{21}, \quad \overset{2}{T}_2 = \tau_{22}, \quad \overset{2}{T}_3 = \tau_{23}. \tag{3.1-2}$$

The situation is similar for ΔS_3. If we arrange the components of the stresses, or tractions, acting on the three surfaces in a square matrix, we obtain

	Components of Stresses		
	1	2	3
Surface normal to x_1	τ_{11}	τ_{12}	τ_{13}
Surface normal to x_2	τ_{21}	τ_{22}	τ_{23}
Surface normal to x_3	τ_{31}	τ_{32}	τ_{33}

(3.1-3)

This is illustrated in Fig. 3.1. The components τ_{11}, τ_{22}, τ_{33} are called *normal stresses*, and the remaining components τ_{12}, τ_{13}, etc., are called *shear stresses*. Each of these components has the dimension of force per unit area, or M/LT^2.

A great diversity in notations for stress components exists in the literature. The most widely used notations in American literature are, in reference to a system of rectangular Cartesian coordinates x, y, z,

$$\begin{pmatrix} \sigma_x & \tau_{xy} & \tau_{xz} \\ \tau_{yx} & \sigma_y & \tau_{yz} \\ \tau_{zx} & \tau_{zy} & \sigma_z \end{pmatrix} \tag{3.1-4}$$

or

$$\begin{pmatrix} \sigma_{xx} & \sigma_{xy} & \sigma_{xz} \\ \sigma_{yx} & \sigma_{yy} & \sigma_{yz} \\ \sigma_{zx} & \sigma_{zy} & \sigma_{zz} \end{pmatrix} \tag{3.1-5}$$

Love* writes X_x, Y_x for σ_x and τ_{xy}, and Todhunter and Pearson† use $\widehat{xx}$, $\widehat{xy}$. Since the reader is likely to encounter all these notations in the literature, we shall not insist on uniformity and would use whichever notation that happens to be convenient. There should be no confusion.

It is important to emphasize again that a stress will always be understood to be the force (per unit area) that the part lying on the positive side of a surface element (the side on the positive side of the outer normal) exerts on the part lying on the negative side. Thus, if the outer normal of a surface element points in the

*A. E. H. Love. *A Treatise on the Mathematical Theory of Elasticity*. Cambridge: University Press. 1st ed., 1892. 4th ed. 1927.

†I. Todhunter and K. Pearson. *A History of the Theory of Elasticity and of the Strength of Materials*. Cambridge: University Press. Vol. 1, 1886. Vol. 2, 1893.

positive direction of the x_2-axis and τ_{22} is positive, the vector representing the component of normal stress acting on the surface element will point in the positive x_2-direction. But if τ_{22} is positive while the outer normal points in the negative x_2-axis direction, then the stress vector acting on the element also points to the negative x_2-axis direction. (See Fig. 3.2).

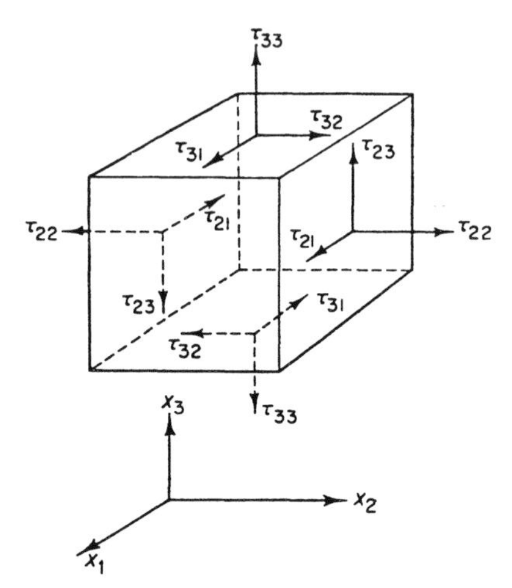

Figure 3.2 Directions of positive stress components.

Similarly, positive values of τ_{21}, τ_{23} will imply shearing stress vectors pointing to the positive x_1-, x_3-axes if the outer normal agrees in sense with x_2-axis, whereas the stress vectors point to the negative x_1-, x_3-directions if the outer normal disagrees in sense with the x_2-axis, as illustrated in Fig. 3.2. A careful study of the figure is essential. Naturally, these rules agree with the usual notation of tension, compression, and shear.

3.2 THE LAWS OF MOTION

Continuum mechanics is founded on Newton's laws of motion. Let the coordinate system x_1, x_2, x_3 be a rectangular Cartesian inertial frame of reference. Let the space occupied by a material body at any time t be denoted by $B(t)$. See Fig. 3.3. Let $\mathbf{r}$ be the position vector of a point with respect to the origin of the coordinate system. Now, consider an infinitesimal element of volume dv enclosing the point at $\mathbf{r}$. Let ρ be the density of the material, and $\mathbf{V}$ be the velocity at $\mathbf{r}$. Then the mass of the infinitesimal element is $\rho\, dv$, and the linear momentum is $(\rho\, dv)\,\mathbf{V}$. An integration of the momentum over the domain $B(t)$, i.e.,

$$\mathcal{P} = \int_{B(t)} \mathbf{V}\,\rho\, dv \tag{3.2–1}$$

is the *linear momentum* of the body in the configuration $B(t)$. The integral of the

$$\int_B \mathbf{X}\, dv.$$

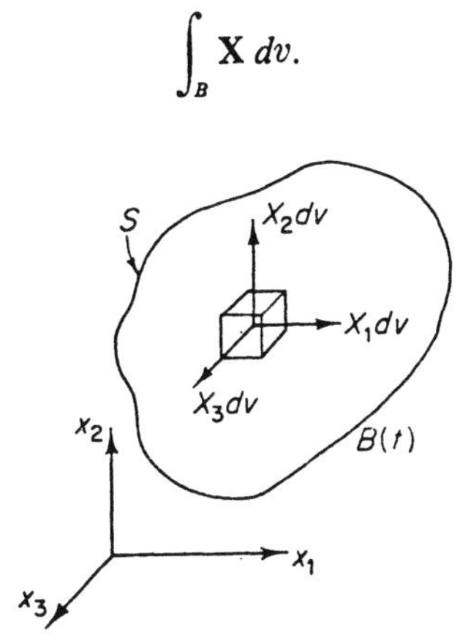

Figure 3.3 Body forces.

moment of momentum of the element about the origin, $\mathbf{r} \times \mathbf{V}\, \rho\, dv$, over the domain $B(t)$, i.e.,

$$\mathcal{H} = \int_{B(t)} \mathbf{r} \times \mathbf{V}\, \rho\, dv \tag{3.2–2}$$

is the *moment of momentum* of the body. *Newton's laws*, as stated by Euler for a continuum, assert that *the rate of change of the linear momentum is equal to the total applied force* $\mathcal{F}$ *acting on the body*, i.e.,

$$\dot{\mathcal{P}} = \mathcal{F}, \tag{3.2–3}$$

and that the rate of change of moment of momentum is equal to the total applied torque $\mathcal{L}$ *about the origin*, i.e.,

$$\dot{\mathcal{H}} = \mathcal{L}. \tag{3.2–4}$$

It is easy to verify that if Eq. (3.2–3) holds, then when Eq. (3.2–4) is valid for one choice of origin, it is valid for all choices of origin.*

As we have mentioned before, there are two types of external forces acting on material bodies in the mechanics of continuous media:

(1) Body forces, acting on elements of volume of the body.

(2) Surface forces, or stresses, acting on surface elements.

Examples of body forces are gravitational forces and electromagnetic forces. Examples of surface forces are aerodynamic pressure acting on a body and stress due to mechanical contact between two bodies, or between one part of a body on another.

To specify a body force, we consider a volume bounded by an arbitrary surface S (Fig. 3.3). The resultant force vector contributed by the body force is assumed

*The derivatives $\dot{\mathcal{P}}$ and $\dot{\mathcal{H}}$ refer to the time rate of change of $\mathcal{P}$ and $\mathcal{H}$ of a fixed set of material particles. Later we shall denote them by $D\mathcal{P}/Dt$ and $D\mathcal{H}/Dt$, respectively. (See Sec. 10.3.)

to be representable in the form of a volume integral taken over the domain B enclosed in S, viz.,

$$\int_B \mathbf{X}\, dv.$$

The vector $\mathbf{X}$, with three components X_1, X_2, X_3, all of the dimensions of force per unit volume, i.e., $M(LT)^{-2}$, is called the body force per unit volume. For example, in a gravitational field,

$$X_i = \rho g_i,$$

where g_i are components of a gravitational acceleration field and ρ is the density (mass per unit volume) of the material.

The surface force acting on an imagined surface in the interior of a body is the stress vector conceived in Euler and Cauchy's stress principle. The surface force acting on an external surface of a body can also be expressed in terms of a stress vector. According to this concept, the total force acting upon the material occupying the region B interior to a closed surface S is

$$\mathscr{F} = \oint_S \overset{\nu}{\mathbf{T}}\, dS + \int_B \mathbf{X}\, dv, \tag{3.2–5}$$

where $\overset{\nu}{\mathbf{T}}$ is the stress vector acting on dS whose outer normal vector is $\boldsymbol{\nu}$. Similarly, the torque about the origin is given by the expression

$$\mathscr{L} = \oint_S \mathbf{r} \times \overset{\nu}{\mathbf{T}}\, ds + \int_B \mathbf{r} \times \mathbf{X}\, dv. \tag{3.2–6}$$

Combining these equations, we have the equations of motion,

$$\oint_S \overset{\nu}{\mathbf{T}}\, dS + \int_B \mathbf{X}\, dv = \frac{D}{Dt}\int_B \mathbf{V}\rho\, dv, \qquad \blacktriangle \quad (3.2\text{–}7)$$

$$\oint_S \mathbf{r} \times \overset{\nu}{\mathbf{T}}\, dS + \int_B \mathbf{r} \times \mathbf{X}\, dv = \frac{D}{Dt}\int_B \mathbf{r} \times \mathbf{V}\rho\, dv. \qquad \blacktriangle \quad (3.2\text{–}8)$$

No demand was made on the domain $B(t)$ other than that it must consist of the same material particles at all times. No special rule was made about the choice of the particles, other than that of continuity, i.e., that they form a continuum. Equations (3.2–7) and (3.2–8) are applicable to any material bodies. They can be applied to an ocean, but they are also applicable to a spoonful of water. The boundary surface of $B(t)$ may coincide with the external boundary of an elastic solid, but it may also include only a small portion thereof.

3.3 CAUCHY'S FORMULA

From the equations of motion, we shall first derive a simple result which states that *the stress vector* $\mathbf{T}^{(+)}$ *representing the action of material exterior to a surface element on the interior is equal in magnitude and opposite in direction to the stress vector* $\mathbf{T}^{(-)}$ *representing the action of the interior material on the exterior across the same surface element*:

$$\mathbf{T}^{(-)} = -\mathbf{T}^{(+)}. \qquad \blacktriangle \quad (3.3\text{–}1)$$

To prove this, we consider a small "pillbox" with two parallel surfaces of area ΔS and thickness δ, as shown in Fig. 3.4. When δ shrinks to zero, while ΔS remains small but finite, the volume forces and the linear momentum and its rate of change

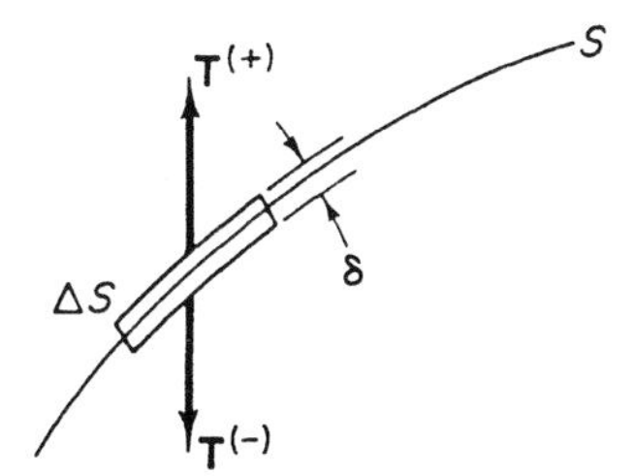

Figure 3.4 Equilibrium of a "pillbox" across a surface S.

with time vanish, as does the contribution of surface forces on the sides of the pillbox. The equation of motion (3.2–3) implies, therefore, that for small ΔS,

$$\mathbf{T}^{(+)}\,\Delta S + \mathbf{T}^{(-)}\,\Delta S = 0.$$

Equation (3.3–1) then follows.

Another way of stating this result is to say that the stress vector is a function of the normal vector to a surface. When the sense of direction of the normal vector reverses, that of the stress vector reverses also.

Now we shall show that *knowing the components* τ_{ij}, *we can write down at once the stress vector acting on any surface with unit outer normal vector* $\boldsymbol{\nu}$ *whose components are* ν_1, ν_2, ν_3. *This stress vector is denoted by* $\overset{\nu}{\mathbf{T}}$, *with components* $\overset{\nu}{T}_1$, $\overset{\nu}{T}_2$, $\overset{\nu}{T}_3$ *given by Cauchy's formula*:

$$\overset{\nu}{T}_i = \nu_j \tau_{ji}. \qquad \blacktriangle \quad (3.3\text{–}2)$$

Cauchy's formula can be derived in several ways. We shall give an elementary derivation.

Let us consider an infinitesimal tetrahedron formed by three surfaces parallel to the coordinate planes and one normal to the unit vector $\boldsymbol{\nu}$. (See Fig. 3.5.) Let

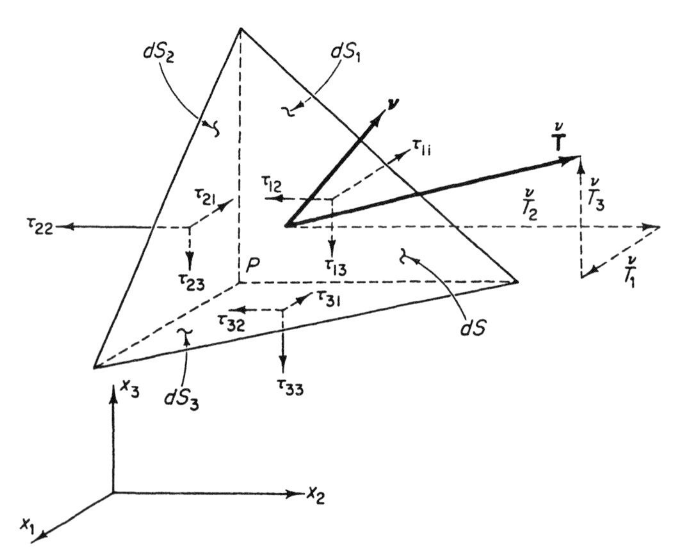

Figure 3.5 Surface tractions on a tetrahedron.

the area of the surface normal to $\boldsymbol{\nu}$ be dS. Then the areas of the other three surfaces are

$$\begin{aligned} dS_1 &= dS \cos(\boldsymbol{\nu}, \mathbf{x}_1) \\ &= \nu_1\, dS = \text{area of surface parallel to the } x_2x_3\text{-plane}, \\ dS_2 &= \nu_2\, dS = \text{area of surface parallel to the } x_3x_1\text{-plane}, \\ dS_3 &= \nu_3\, dS = \text{area of surface parallel to the } x_1x_2\text{-plane}, \end{aligned}$$

and the volume of the tetrahedron is

$$dv = \tfrac{1}{3} h\, dS,$$

where h is the height of the vertex P from the base dS. The forces in the positive direction of x_1, acting on the three coordinate surfaces, can be written as

$$(-\tau_{11} + \epsilon_1)dS_1, \qquad (-\tau_{21} + \epsilon_2)dS_2, \qquad (-\tau_{31} + \epsilon_3)dS_3,$$

where τ_{11}, τ_{21}, τ_{31} are the stresses at the vertex P opposite to dS. The negative sign is obtained because the outer normals to the three surfaces are opposite in sense with respect to the coordinate axes, and the ϵ's are inserted because the tractions act at points slightly different from P. If we assume that the stress field is continuous, then ϵ_1, ϵ_2, ϵ_3 are infinitesimal quantities. On the other hand, the force acting on the triangle normal to ν has a component $(\overset{\nu}{T}_1 + \epsilon)dS$ in the positive x_1-axis direction, the body force has an x_1-component equal to $(X_1 + \epsilon')dv$, and the rate of change of linear momentum has a component $\rho\dot{V}_1\, dv$, where $\dot{V}_1$ is the component of acceleration in the direction of x_1. Here, $\overset{\nu}{T}_1$ and X_1 refer to the point P, and ϵ and ϵ' are again infinitesimal. The first equation of motion is thus

$$(-\tau_{11} + \epsilon_1)\nu_1\, dS + (-\tau_{21} + \epsilon_2)\nu_2\, dS$$

$$+(-\tau_{31} + \epsilon_3)\nu_3\, dS + (\overset{\nu}{T}_1 + \epsilon)dS + (X_1 + \epsilon')\tfrac{1}{3}\, h\, dS = \rho\dot{V}_1\, \tfrac{1}{3}\, h\, dS. \quad (3.3\text{–}3)$$

Dividing through by dS, taking the limit as $h \to 0$, and noting that ϵ_1, ϵ_2, ϵ_3, ϵ, ϵ' vanish with h and dS, one obtains

$$\overset{\nu}{T}_1 = \tau_{11}\nu_1 + \tau_{21}\nu_2 + \tau_{31}\nu_3, \quad (3.3\text{–}4)$$

which is the first component of Eq. (3.3–2). Other components follow similarly.

Cauchy's formula assures us that the nine components of stresses τ_{ij} are necessary and sufficient to define the traction across any surface element in a body. Hence, the stress state in a body is characterized completely by the set of quantities τ_{ij}. Since $\overset{\nu}{T}_i$ is a vector and Eq. (3.3–2) is valid for an arbitrary vector ν_j, it follows that τ_{ij} is a tensor. Henceforth, τ_{ij} will be called a stress tensor.

Checking Acceptable Errors

In Sec. 1.5, we defined continua on the basis of acceptable variability and a limiting approach that has a lower bound of dimensions. In Sec. 1.6, the concept was applied to the definition of stress. In Section 1.7, we adopted an abstract copy of the real material as a way of idealization. In the proof of Cauchy's formula, Eq. (3.3–4), we have used the abstract copy and followed the usual method of calculus to throw away a number of terms in Eq. (3.3–3) and reach Eq. (3.3–4). We claimed that the sum of the terms

$$\epsilon_1\nu_1 + \epsilon_2\nu_2 + \epsilon_3\nu_3 + \epsilon + \tfrac{1}{3}h(\epsilon' - \rho\dot{V}_1) \quad (3.3\text{–}5)$$

is small, compared with the terms that are retained; i.e.,

$$\overset{\nu}{T}_1,\ \tau_{11}\nu_1,\ \nu_{21}\nu_2,\ \tau_{31}\nu_3, \quad (3.3\text{–}6)$$

when we take Eq. (3.3–3) to the limit as $h \to 0$ and $\Delta S \to 0$. Now, if we are not allowed to take the limit as $h \to 0$ and $\Delta S \to 0$, but instead we are restricted to accept h no smaller than a constant h^* and ΔS no smaller than a constant multiplied by $(h^*)^2$, then the quantity listed in line (3.3–5) must be evaluated for $h = h^*$ and $\Delta S = \text{const.}\cdot(h^*)^2$ and must be compared with the quantities listed in line (3.3–6). A standard of how small is negligible must be defined, and the comparison be made under that definition. If we find the quantity in line (3.3–5) negligible compared with those listed in line (3.3–6), then we can say that Eq. (3.3–3) or Eq. (3.3–2) is valid. This tedious step should be done, in principle, to apply the continuum theory to objects of the real world.

3.4 EQUATIONS OF EQUILIBRIUM

We shall now transform the equations of motion, Eqs. (3.2–7) and (3.2–8), into differential equations. This can be done elegantly by means of Gauss's theorem and Cauchy's formula, as is shown in Chapter 10, but here we shall pursue an elementary course to assure physical clarity.

Consider the static equilibrium state of an infinitesimal parallelepiped with surfaces parallel to the coordinate planes. The stresses acting on the various surfaces are shown in Fig. 3.6. The force $\tau_{11}\, dx_2\, dx_3$ acts on the left-hand side, the force $[\tau_{11} + (\partial\tau_{11}/\partial x_1)\, dx_1]\, dx_2\, dx_3$ acts on the right-hand side, etc. As it will be explained below, these expressions are based on the assumption of continuity of the stresses. The body force is $X_i\, dx_1\, dx_2\, dx_3$.

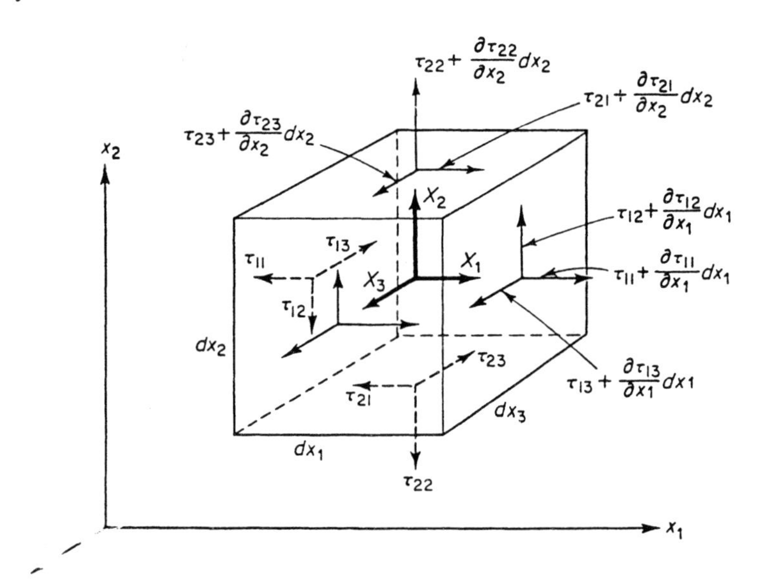

Figure 3.6 Equilibrating stress components on an infinitesimal parallelepiped.

The stresses indicated in the figure may be explained as follows. We are concerned with a nonuniform stress field. Every stress component is a function of position. Thus, the stress component τ_{11} is a function of x_1, x_2, x_3: $\tau_{11}(x_1, x_2, x_3)$. At a point slightly to the right of the point (x_1, x_2, x_3), namely, at $(x_1 + dx_1, x_2, x_3)$, the value of the stress τ_{11} is $\tau_{11}(x_1 + dx_1, x_2, x_3)$. But if τ_{11} is a continuously differentiable function of x_1, x_2, x_3, then, according to Taylor's theorem with a remainder, we have

$$\tau_{11}(x_1 + dx_1, x_2, x_3) = \tau_{11}(x_1, x_2, x_3) + dx_1 \frac{\partial\tau_{11}}{\partial x_1}(x_1, x_2, x_3) + dx_1^2 \frac{1}{2}\frac{\partial^2\tau_{11}}{\partial x_1^2}(x_1 + \alpha\, dx_1, x_2, x_3)$$

where $0 \le \alpha \le 1$. If $\partial^2\tau_{11}/\partial x_1^2$ is finite, then the last term can be made arbitrarily small compared with the other terms by choosing dx_1 sufficiently small. With such a choice, we have

$$\tau_{11}(x_1 + dx_1, x_2, x_3) = \tau_{11}(x_1, x_2, x_3) + \frac{\partial \tau_{11}}{\partial x_1}(x_1, x_2, x_3)dx_1.$$

In Fig. 3.6 we write, for short, τ_{11} and $\tau_{11} + (\partial\tau_{11}/\partial x_1)dx_1$ on the surfaces where the stresses act. The left, bottom, and rear surfaces are located at x_1, x_2, x_3. The edges of the element have lengths dx_1, dx_2, dx_3.

All stresses and their derivatives are evaluated at (x_1, x_2, x_3). Equilibrium of the body demands that the resultant forces vanish. Consider the forces in the x_1-direction. As shown in Fig. 3.7, we have six components of surface force and one component of body force. The sum is

$$\left(\tau_{11} + \frac{\partial \tau_{11}}{\partial x_1} dx_1\right) dx_2\, dx_3 - \tau_{11} dx_2\, dx_3 + \left(\tau_{21} + \frac{\partial \tau_{21}}{\partial x_2} dx_2\right) dx_3\, dx_1$$
$$- \tau_{21}\, dx_3\, dx_1 + \left(\tau_{31} + \frac{\partial \tau_{31}}{\partial x_3} dx_3\right) dx_1\, dx_2$$
$$- \tau_{31}\, dx_1\, dx_2 + X_1\, dx_1\, dx_2\, dx_3 = 0. \qquad (3.4\text{–}1)$$

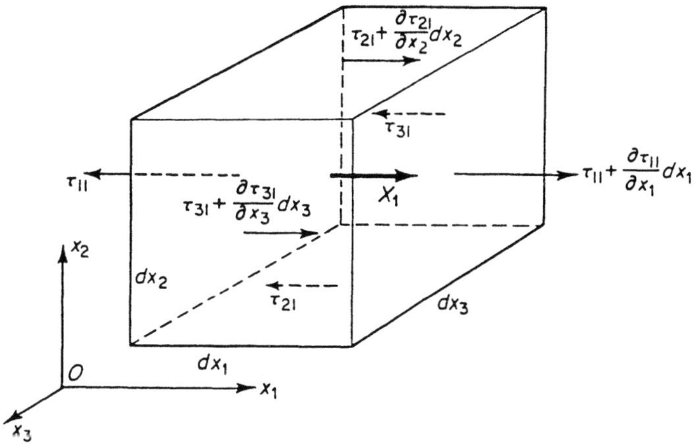

Figure 3.7 Components of tractions in x_1-direction.

Dividing by $dx_1\, dx_2\, dx_3$, we obtain

$$\frac{\partial \tau_{11}}{\partial x_1} + \frac{\partial \tau_{21}}{\partial x_2} + \frac{\partial \tau_{31}}{\partial x_3} + X_1 = 0. \qquad (3.4\text{–}2)$$

A cyclic permutation of the indices leads to similar equations of equilibrium of forces in the x_2- and x_3-directions. The whole set, written concisely, is

$$\frac{\partial \tau_{ij}}{\partial x_j} + X_i = 0. \qquad \blacktriangle \quad (3.4\text{–}3)$$

This is an important result. A shorter derivation will be given later, in Sec. 10.6.

The equilibrium of an element requires also that the resultant moment vanish. If there do not exist external moments proportional to a volume, the consideration of moments will lead to the important conclusion that *the stress tensor is symmetric*, i.e.,

$$\tau_{ij} = \tau_{ji}. \qquad ▲ \quad (3.4\text{–}4)$$

This is demonstrated as follows. Referring to Fig. 3.6 and considering the moment of all the forces about the x_3-axis, we see that those components of forces parallel to Ox_3 or lying in planes containing Ox_3 do not contribute any moment. The

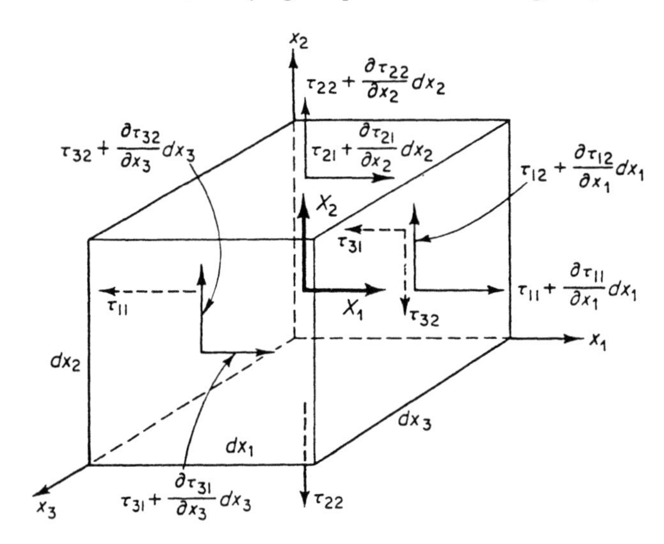

Figure 3.8 Components of tractions that contribute moment about the Ox_3-axis.

components that do contribute a moment about the x_3-axis are shown in Fig. 3.8. Therefore, taking care of the moment arm properly, we have

$$\begin{aligned}
&-\left(\tau_{11} + \frac{\partial \tau_{11}}{\partial x_1} dx_1\right) dx_2\, dx_3 \frac{dx_2}{2} + \tau_{11}\, dx_2\, dx_3 \frac{dx_2}{2} \\
&+ \left(\tau_{12} + \frac{\partial \tau_{12}}{\partial x_1} dx_1\right) dx_2\, dx_3\, dx_1 - \left(\tau_{21} + \frac{\partial \tau_{21}}{\partial x_2} dx_2\right) dx_1\, dx_3\, dx_2 \\
&+ \left(\tau_{22} + \frac{\partial \tau_{22}}{\partial x_2} dx_2\right) dx_1\, dx_3 \frac{dx_1}{2} - \tau_{22}\, dx_1\, dx_3 \frac{dx_1}{2} \\
&+ \left(\tau_{32} + \frac{\partial \tau_{32}}{\partial x_3} dx_3\right) dx_1\, dx_2 \frac{dx_1}{2} - \tau_{32}\, dx_1\, dx_2 \frac{dx_1}{2} \\
&- \left(\tau_{31} + \frac{\partial \tau_{31}}{\partial x_3} dx_3\right) dx_1\, dx_2 \frac{dx_2}{2} + \tau_{31}\, dx_1\, dx_2 \frac{dx_2}{2} \\
&- X_1\, dx_1\, dx_2\, dx_3 \frac{dx_2}{2} + X_2\, dx_1\, dx_2\, dx_3 \frac{dx_1}{2} = 0.
\end{aligned}$$

On dividing through by $dx_1\, dx_2\, dx_3$ and passing to the limit as $dx_1 \to 0$, $dx_2 \to 0$, and $dx_3 \to 0$, we obtain

$$\tau_{12} = \tau_{21}. \qquad (3.4\text{–}5)$$

Similar considerations of resultant moments about Ox_2 and Ox_1 lead to the general result given by Eq. (3.4–3). A shorter derivation will be given later, in Sec. 10.7.

So far, we have considered the condition of equilibrium. If it is desired to derive the equation of motion instead of that of equilibrium, it is necessary only to apply the D'Alembert principle to our cubical element. According to the D'Alembert principle, a particle in motion may be considered in equilibrium if the negative of the product of the mass and the acceleration of the particle is applied as an external force on the particle. This is the inertial force. For a system of particles, D'Alembert's principle applies if the resultant of the inertial forces on all particles is applied to the center of mass of the system.

For the element considered in this section, if **a** (with components a_1, a_2, a_3) represents the acceleration vector of the particle referred to an inertial frame of reference, then since the mass of the element is $\rho\, dx_1\, dx_2\, dx_3$, the inertial force is $-\rho a_i\, dx_1\, dx_2\, dx_3$. An addition of this to Eq. (3.4–1) and dividing through by $dx_1\, dx_2\, dx_3$ leads to the equation of motion,

$$\rho a_1 = \frac{\partial \tau_{11}}{\partial x_1} + \frac{\partial \tau_{21}}{\partial x_2} + \frac{\partial \tau_{31}}{\partial x_3} + X_1, \text{ etc.,} \qquad (3.4\text{–}6)$$

i.e.,

$$\rho a_i = \frac{\partial \tau_{ij}}{\partial x_j} + X_i. \qquad (3.4\text{–}7)$$

3.5 CHANGE OF STRESS COMPONENTS IN TRANSFORMATION OF COORDINATES

In the previous section, the components of stress τ_{ij} are defined with respect to a rectangular Cartesian system x_1, x_2, x_3. Let us now take a second set of rectangular Cartesian coordinates x'_1, x'_2, x'_3, with the same origin but oriented differently, and consider the stress components in the new reference system (Fig. 3.9). Let these coordinates be connected by the linear relations

$$x'_k = \beta_{ki} x_i, \qquad (k = 1, 2, 3) \qquad (3.5\text{–}1)$$

where β_{ki} are the direction cosines of the x'_k-axis with respect to the x_i-axis. Since τ_{ij} is a tensor (Sec. 3.3) we can write down the transformation law at once. However, to emphasize the importance of the result, we shall insert an elementary derivation based on Cauchy's formula (derived in Sec. 3.3), which states that if dS is a surface element whose unit outer normal vector $\boldsymbol{\nu}$ has components ν_i, then the force per unit area acting on dS is a vector $\overset{\nu}{\mathbf{T}}$ with components

$$\overset{\nu}{T}_i = \tau_{ji} \nu_j. \qquad (3.5\text{–}2)$$

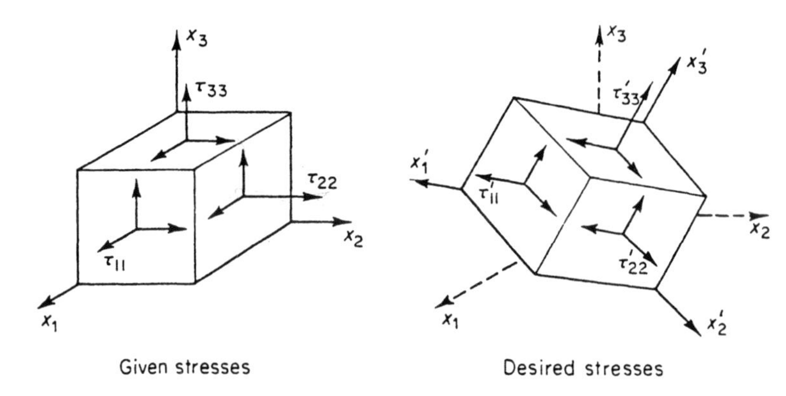

Figure 3.9 Transformation of stress components under rotation of coordinate systems.

If the normal $\boldsymbol{\nu}$ is chosen to be parallel to the axis x_k', so that

$$\nu_1 = \beta_{k1}, \qquad \nu_2 = \beta_{k2}, \qquad \nu_3 = \beta_{k3},$$

then the stress vector $\overset{k}{\mathbf{T}}{}'$ has components

$$\overset{k}{T}{}_i' = \tau_{ji}\beta_{kj}.$$

The component of the vector $\overset{k}{\mathbf{T}}{}'$ in the direction of the axis x_m' is given by the product of $\overset{k}{T}{}_i'$ and β_{mi}. Hence, the stress component

$$\begin{aligned}\tau_{km}' &= \text{projection of } \overset{k}{T}{}' \text{ on the } x_m'\text{-axis} \\ &= \overset{k}{T}{}_1'\beta_{m1} + \overset{k}{T}{}_2'\beta_{m2} + \overset{k}{T}{}_3'\beta_{m3} \\ &= \tau_{j1}\beta_{kj}\beta_{m1} + \tau_{j2}\beta_{kj}\beta_{m2} + \tau_{j3}\beta_{kj}\beta_{m3};\end{aligned}$$

i.e.,

$$\tau_{km}' = \tau_{ji}\beta_{kj}\beta_{mi}. \tag{3.5–3}$$

If we compare Eq. (3.5–3) with Eq. (2.5–6), we see that the stress components transform like a Cartesian tensor of rank 2. Thus, the physical concept of stress described by τ_{ij} agrees with the mathematical definition of a tensor of rank 2 in a Euclidean space.

3.6 STRESS COMPONENTS IN ORTHOGONAL CURVILINEAR COORDINATES

Orthogonal curvilinear coordinates are often introduced in continuum mechanics if the boundary conditions are simplified by such a frame of reference. For example, if we want to study the flow in a circular cylindrical tube or the torsion of a circular shaft, it is natural to use cylindrical coordinates. If we wish to study the stress distribution in a sphere, it is natural to use spherical coordinates. In fact, if we want to study the explosive forming of a flat sheet of metal into a spherical cap, it may be useful to use a rectangular frame of reference for the original state of the plate and a spherical-polar frame of reference for the deformed state.

It is appropriate to resolve the components of stress in the directions of the curvilinear coordinates and denote them by corresponding subscripts. For example, in a set of cylindrical coordinates, r, θ, z, which are related to the rectangular Cartesian coordinates x, y, z by

$$\begin{cases} x = r\cos\theta, \\ y = r\sin\theta, \\ z = z, \end{cases} \qquad \begin{cases} \theta = \tan^{-1}\dfrac{y}{x}, \\ r^2 = x^2 + y^2, \\ z = z, \end{cases} \tag{3.6–1}$$

it is natural to denote the components of the stress tensor at a point (r, θ, z) by

$$\begin{pmatrix} \tau_{rr} & \tau_{r\theta} & \tau_{rz} \\ \tau_{\theta r} & \tau_{\theta\theta} & \tau_{\theta r} \\ \tau_{zr} & \tau_{z\theta} & \tau_{rr} \end{pmatrix} \text{ or } \begin{pmatrix} \sigma_r & \tau_{r\theta} & \tau_{rz} \\ \tau_{\theta r} & \sigma_\theta & \tau_{\theta z} \\ \tau_{zr} & \tau_{z\theta} & \sigma_z \end{pmatrix}. \tag{3.6–2}$$

To relate these stress components to σ_x, τ_{xy}, etc., let us erect a local rectangular Cartesian frame of reference $x'y'z'$ at the point (r, θ, z), with the origin located at the point (r, θ, z), the x'-axis in the direction of increasing r, the y'-axis in the direction of increasing θ, and the z'-axis parallel to z (see Fig. 3.10). Then, in

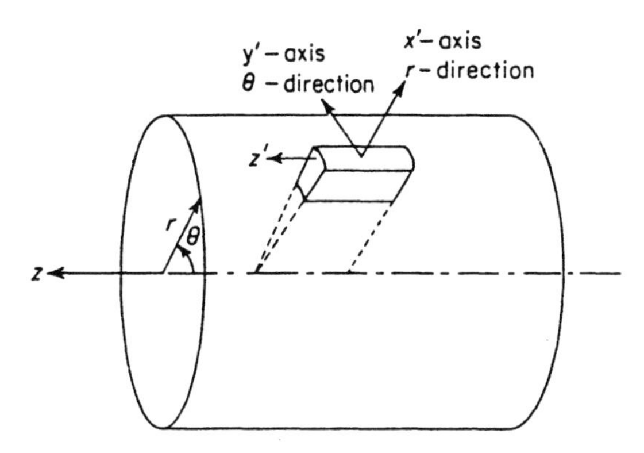

Figure 3.10 Stress components in cylindrical polar coordinates.

conventional notation, the stresses $\tau_{x'x'}$, $\tau_{y'y'}$, . . . are well defined. Now we can define the stress components listed in Eq. (3.6–2) by identifying r, θ, z with x', y', z':

$$\tau_{rr} = \tau_{x'x'}, \qquad \tau_{r\theta} = \tau_{x'y'}, \qquad \tau_{\theta\theta} = \tau_{y'y'}, \tag{3.6–3}$$

etc. Since the coordinate systems x', y', z' and x, y, z are both Cartesian, we can apply the transformation law of Eq. (3.5–3). The direction cosines of the axes x', y', z' relative to x, y, z are (see Fig. 2.2 and Eq. 2.4–3)

$$(\beta_{ij}) = \begin{pmatrix} \cos\theta & \sin\theta & 0 \\ -\sin\theta & \cos\theta & 0 \\ 0 & 0 & 1 \end{pmatrix}. \tag{3.6–4}$$

Hence, by virtue of Eqs. (3.5–3) and (3.6–3), we have

$$\begin{aligned}
\sigma_x &= \sigma_r \cos^2\theta + \sigma_\theta \sin^2\theta - \tau_{r\theta} \sin 2\theta, \\
\sigma_y &= \sigma_r \sin^2\theta + \sigma_\theta \cos^2\theta + \tau_{r\theta} \sin 2\theta, \\
\sigma_z &= \sigma_z, \\
\tau_{xy} &= (\sigma_r - \sigma_\theta)\sin\theta\cos\theta + \tau_{r\theta}(\cos^2\theta - \sin^2\theta), \\
\tau_{zx} &= \tau_{zr}\cos\theta - \tau_{z\theta}\sin\theta, \\
\tau_{zy} &= \tau_{zr}\sin\theta + \tau_{z\theta}\cos\theta.
\end{aligned} \tag{3.6–5}$$

Spherical or other orthogonal curvilinear coordinates can be treated in a similar manner.

3.7 STRESS BOUNDARY CONDITIONS

Problems in mechanics usually appear this way: We know something about the forces or velocities or displacements on the surface of a solid or fluid body and inquire into what happens inside the body. For example, the wind blows on a building whose foundation we known is firm. What are the stresses acting in the columns and beams? Are they safe? To resolve such questions, we set down the known facts concerning the external world in the form of boundary conditions and then use the differential equations (field equations) to extend the information to the interior of the body. If a solution is found to satisfy all the field equations and boundary conditions, then complete information is obtained for the entire interior of the body.

On the surface of a body or at an interface between two bodies, the traction (force per unit area) acting on the surface must be the same on both sides of the surface. This, indeed, is the basic concept of stress that defines the interaction of one part of a body on another.

Consider a cube composed of a hard material joined to a soft material, as shown in Fig. 3.11(a). Let the block be compressed between two plane walls. Both the soft material and the hard material will be stressed. At a point P on the interface AB, the situation may be illustrated by a sequence of free-body diagrams as shown in the figure. For the hard material, on the positive side of the interface at P, there acts a surface traction $\overset{\nu}{\mathbf{T}}{}^{(1)}$, Fig. 3.11(b). With x_1, x_2, x_3 identified with the coordinates

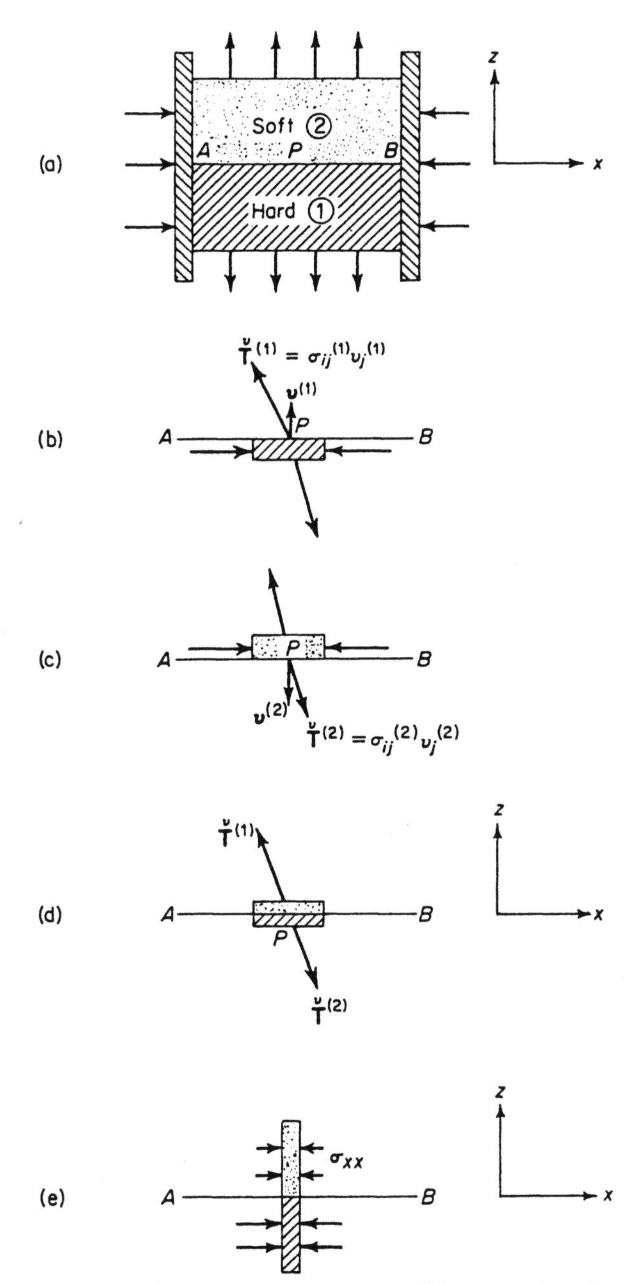

Figure 3.11 Derivation of the stress boundary condition at an interface between two materials. (a) An interface *AB* between two continuous media 1 and 2. (b) Free-body diagram of a small element of material No. 1 at a point *P* on the interface. The stress vector $\overset{\nu}{\mathbf{T}}^{(1)}$ acts on the surface *AB* of this element. (c) Free-body diagram of a small element of material No. 2 at *P*. (d) Free-body diagram of a small flat element including both materials. (e) Free-body diagram of a small vertical element, showing that σ_{xx} can be discontinuous at the interface.

x, y, z, the unit normal vector $\boldsymbol{\nu}^{(1)}$ has three components (0, 0, 1), and the traction vector $\overset{\nu}{\mathbf{T}}{}^{(1)}$ has three components $\sigma_{ij}^{(1)} \nu_j^{(1)}$, $(i = 1, 2, 3)$, where $\sigma_{ij}^{(1)}$ is the stress tensor in the hard material. For the soft material, there must exist a similar traction $\overset{\nu}{\mathbf{T}}{}^{(2)}$, with components $\sigma_{ij}^{(2)} \nu_j^{(2)}$, Fig. 3.11(c). The equilibrium of an infinitesimally thin pillbox, as shown in the free-body diagram [Fig. 3.11(d)], requires that

$$\overset{\nu}{\mathbf{T}}{}^{(1)} = \overset{\nu}{\mathbf{T}}{}^{(2)}. \tag{3.7–1}$$

This is the condition of equality of surface traction on the two sides of an interface. More explicitly, let the interface be the xy-plane, and let the z-axis be normal to xy. Then, the vector equation (3.7–1) implies the three equations

$$\sigma_{zz}^{(1)} = \sigma_{zz}^{(2)}, \qquad \sigma_{xz}^{(1)} = \sigma_{xz}^{(2)}, \qquad \sigma_{yz}^{(1)} = \sigma_{yz}^{(2)}, \tag{3.7–2}$$

which are the boundary conditions on the stresses in the media 1 and 2 at their interface.

Note that these interface conditions indicate nothing about the stress components σ_{xx}, σ_{yy}, σ_{xy}. These components are not required to be continuous across the boundary. Indeed, if the elastic moduli of materials 1 and 2 are unequal and the compressive strain is uniform, then, in general,

$$\sigma_{xx}^{(1)} \neq \sigma_{xx}^{(2)}, \qquad \sigma_{yy}^{(1)} \neq \sigma_{yy}^{(2)}, \qquad \sigma_{xy}^{(1)} \neq \sigma_{xy}^{(2)}. \tag{3.7–3}$$

That these discontinuities are not in conflict with any conditions of equilibrium can be seen in Fig. 3.11(e).

A special case is one in which medium 2 is so soft that its stresses are completely negligible compared with those in medium 1 (for example, air vs. steel). Then the surface is said to be *free*, and the boundary conditions are

$$\sigma_{zz} = 0, \qquad \sigma_{xz} = 0, \qquad \sigma_{yz} = 0. \tag{3.7–4}$$

On the other hand, if the traction in medium 2 is known, then it can be considered as the "external" load acting on medium 1. Thus, the stress boundary conditions on a solid body usually take the form

$$\sigma_{nn} = p_1, \qquad \sigma_{nt_1} = p_2, \qquad \sigma_{nt_2} = p_3, \tag{3.7–5}$$

where p_1, p_2, p_3 are specific functions of location and time and n, t_1, t_2 are a set of local orthogonal axes with n pointing in the direction of the outer normal.

Although every surface is an interface between two spaces, it is a general practice to confine one's attention to one side of the surface and call the other side "external." For example, structural engineers speak of the wind load on a building as the "external load" applied to the building. Reciprocally, to the fluid dynamicist, the building is merely a rigid border to the flow of air. The same interface presents two different kinds of boundary conditions to the two media. The basic justification for such a divergence of attitude is that the small elastic deformation of the structure is unimportant to the aerodynamicist who computes the aerodynamic pressure acting on the structure, whereas the elastic deformation is all important to the structural analyst who determines the safety of the building. Hence, for the aerodynamicist the building is rigid, whereas for the elastician it is not. In other words, both boundary conditions are approximations.

PROBLEMS

3.1 Consider a long string. If you pull on it with a force T, it is clear that the same total tension T acts on every cross section of the string. If we consider the strength of the string, it is intuitively clear that the larger the cross section, the stronger it would be. Thus, if we have several strings and wish to compare the strengths of their materials, the comparison should be based on the *stress* (which in this case is equal to the tension T divided by the cross-sectional area), rather than on the total force. It is not too much to hope that all strings of the same material will break at the same ultimate stress. In fact, the problem would be very interesting if an experiment were done and one discovered that all strings made of the same material did not break at the same stress. Could you conceive of such a contingency? What if the strings were extremely small? For concreteness, consider nylon threads of diameters 1 cm, 0.1 cm, 10^{-2} cm, 10^{-3} cm, . . . , 10^{-6} cm. When would you begin to feel a little uncertain that some other factors might enter the picture in defining the strength of the threads? What are the factors?

3.2 Take a piece of chalk and break it (a) by bending, and (b) by twisting. The way the piece of chalk breaks will be different in these two cases. Why? Can we predict the mode of failure? The cleavage surface?

3.3 A gentle breeze blowing over an expanse of water generates ripples, Fig. P3.3. Describe the stress vector acting on the water surface. Write down the boundary conditions at the water surface.

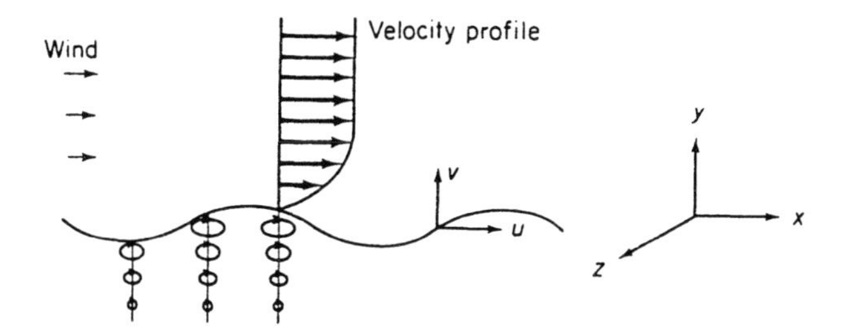

Figure P3.3 Dynamic boundary conditions at the water-air interface.

3.4 In Fig. P3.4, water is shown in a reservoir. At a point P, let us consider surfaces A-A, B-B, etc. Draw stress vectors acting on these surfaces. Consider all possible surfaces passing through P. What is the locus of all the stress vectors?

Answer: A sphere.

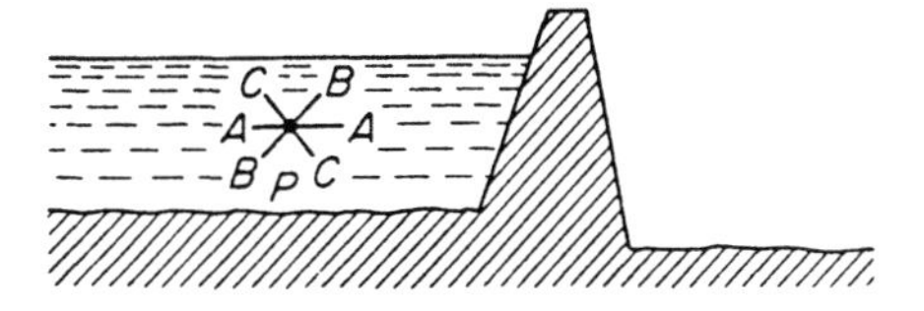

Figure P3.4 Water in a reservoir.

3.5 Water in a reservoir is pouring over a dam (Fig. P3.5). Consider a point close to the top of the dam, say, 10 cm above it. Again (as in Prob. 3.4), consider all surfaces passing

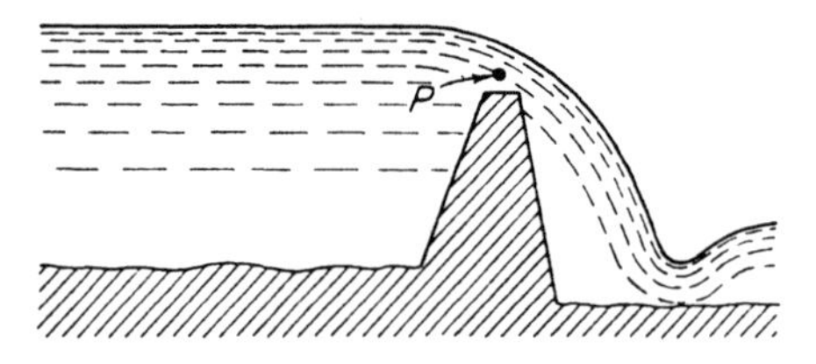

Figure P3.5 Water pouring over a dam.

through this point, and describe the stress vectors acting on these surfaces. Is the locus of all the stress vectors a sphere?

Now consider a sequence of points closer and closer to the solid surface on the top of the dam, say, at distances 1 cm, 10^{-1} cm, 10^{-2} cm, 10^{-3} cm, and 10^{-4} cm. Would you expect the stress-vector locus to change as the distance becomes very small? Pay particular attention to the viscosity of water.

3.6 Label the stresses shown in Fig. P3.6.

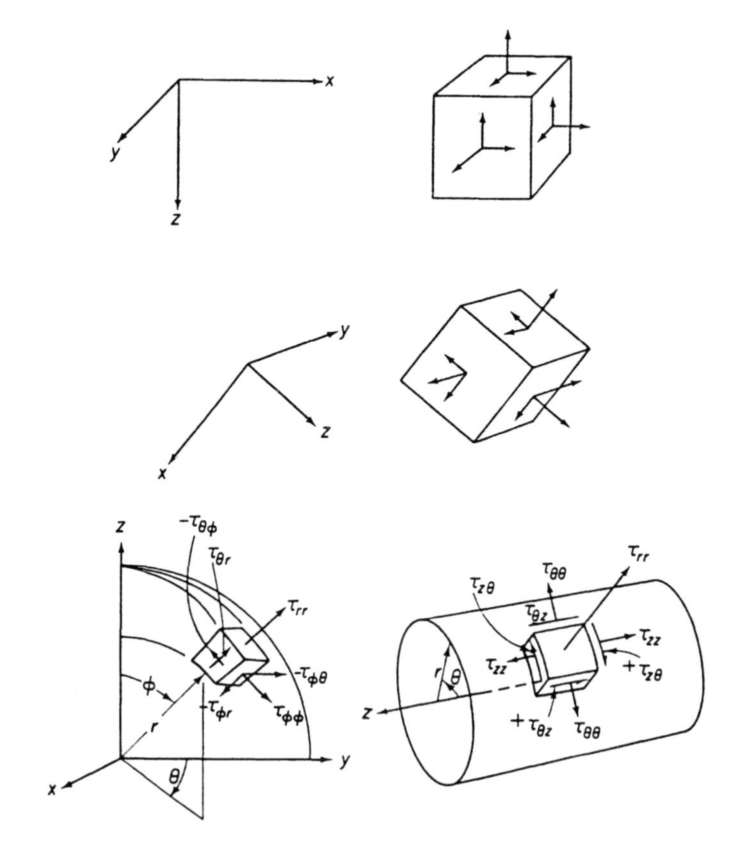

Figure P3.6 Stresses.

3.7 The components of a stress tensor at a certain place in a body may be presented as a matrix:

$$\begin{array}{c|ccc} & x & y & z \\ x & 0 & 1 & 2 \\ y & 1 & 2 & 0 \\ z & 2 & 0 & 1 \end{array}.$$

What is the stress vector acting on the outer side (the side away from the origin) of the following plane, which passes through the place in question?

$$x + 3y + z = 1.$$

What are the normal and tangential components of the stress vector to this plane?

Answer: $\overset{\nu}{\mathbf{T}}_i = (5, 7, 3)/\sqrt{11}$; $T^{(n)} = \frac{29}{11}\cdot$shear $= 0.771$.

Solution: The plane has a normal vector with direction cosines $(1, 3, 1)/\sqrt{11}$. Hence,

$$\overset{\nu}{T}_1 = \frac{(0\cdot1) + (1\cdot3) + (2\cdot1)}{\sqrt{11}} = \frac{5}{\sqrt{11}}, \qquad \overset{\nu}{T}_2 = \frac{7}{\sqrt{11}}, \qquad \overset{\nu}{T}_3 = \frac{3}{\sqrt{11}}.$$

If we use $\mathbf{i}$, $\mathbf{j}$, $\mathbf{k}$ to denote unit vectors in the directions of the x-, y-, z-axes, respectively, we have $\overset{\nu}{\mathbf{T}} = (5\mathbf{i} + 7\mathbf{j} + 3\mathbf{k})/\sqrt{11}$. The normal component is $\overset{\nu}{T}_i \nu_i = \frac{29}{11}$. The shear (tangential component) can be obtained by several methods:

(1) Let the shear $= s$ and the normal component $= n$. Then

$$s^2 + n^2 = |\overset{\nu}{T}_i|^2 = \frac{25 + 49 + 9}{11} = \frac{83}{11}.$$

$$s^2 = \frac{83}{11} - \left(\frac{29}{11}\right)^2;$$

hence

$$s = \frac{6\sqrt{2}}{11}.$$

(2) The vector of the normal component plus the vector of the shear component equals the vector $\overset{\nu}{\mathbf{T}}$. The normal component lies in the direction of the unit normal $(1\mathbf{i} + 3\mathbf{j} + 1\mathbf{k})29/(11\sqrt{11})$. Let the shear component vector be $x\mathbf{i} + y\mathbf{j} + z\mathbf{k}$; then $29/(11\sqrt{11}) + x = 5/\sqrt{11}$, implying that $x = (55 - 29)/36.5 = 0.712$. Similarly, $y = -0.274$, $z = 0.109$, and the shear $= (x^2 + y^2 + z^2)^{1/2} = 0.771$.

3.8 With reference to the x-, y-, and z-coordinates, the state of stress at a certain point of a body is given by the following matrix:

$$(\sigma_{ij}) = \begin{pmatrix} 200 & 400 & 300 \\ 400 & 0 & 0 \\ 300 & 0 & -100 \end{pmatrix} \text{kPa}$$

Find the stress vector acting on a plane passing through the point and parallel to the plane $x + 2y + 2z - 6 = 0$.

Answer: $\overset{\nu}{\mathbf{T}} = 533\mathbf{i} + 133\mathbf{j} + 33\mathbf{k}$.

3.9 Does equilibrium exist for the following stress distribution in the absence of body force?

$$\sigma_x = 3x^2 + 4xy - 8y^2, \qquad \tau_{xy} = \tfrac{1}{2}x^2 - 6xy - 2y^2,$$

$$\sigma_y = 2x^2 + xy + 3y^2, \qquad \sigma_z = \tau_{xz} = \tau_{yz} = 0.$$

Answer: Yes, according to Eq. (3.4–2).

3.10 The stress at a point is $\sigma_x = 5{,}000$ kPa, $\sigma_y = 5{,}000$ kPa, $\tau_{xy} = \sigma_z = \tau_{xz} = \tau_{yz} = 0$. Consider all planes passing through this point. On each plane, there acts a stress vector that can be resolved into two components: a *normal* stress and a *shear stress*. Consider planes oriented in all directions. Show that the maximum shear stress in the material at the point is 2,500 kPa.

Solution: Let a coordinate system be so chosen that the point in question is located at the origin. A plane passing through this point may be represented by the equation.

$$lx + my + nz = 0 \tag{1}$$

where (l, m, n) is the direction cosine of the normal to the plane. Hence, $(\nu_1, \nu_2, \nu_3) = (l, m, n)$ and $l^2 + m^2 + n^2 = 1$. By permitting the normal vector to assume all possible directions, we obtain all the planes named in the problem. Now, the stress vector acting on the plane [Eq. (1)] is $(\overset{\nu}{T}_1, \overset{\nu}{T}_2, \overset{\nu}{T}_3) = (5{,}000l, 5{,}000m, 0)$. The normal component of surface traction is the component of $(\overset{\nu}{T}_i)$ in the direction of (ν_i), i.e., the scalar product of these vectors. That is, the normal stress $= 5{,}000(l^2 + m^2)$. Hence, (shear stress)$^2 = (\overset{\nu}{T}_i)^2 -$ (normal stress)$^2 = (5{,}000)^2(l^2 + m^2) - (5{,}000)^2(l^2 + m^2)^2 = (5{,}000)^2[l^2 + m^2 - (l^2 + m^2)^2]$. But $l^2 + m^2 + n^2 = 1$; hence,

$$(\text{shear stress})^2 = (5{,}000)^2[1 - n^2 - (1 - n^2)^2] \tag{2}$$

To find the value of n that is less than 1 and that renders the shear stress a relative maximum, we set

$$0 = \frac{\partial}{\partial n}(\text{shear stress})^2 = (5{,}000)^2[-2n + 2(1 - n^2)\cdot 2n].$$

The solution is $n^2 = \frac{1}{2}$. Hence, from Eq. (2), we obtain the maximum shear stress squared, $(5{,}000)^2/4$, and the final result that the maximum shear stress is 2,500 kPa.

3.11 If the state of stress at a point (x_0, y_0, z_0) is

$$(\sigma_{ij}) = \begin{pmatrix} 100 & 0 & 0 \\ 0 & 50 & 0 \\ 0 & 0 & -100 \end{pmatrix} \text{kPa},$$

find the stress vector and the magnitude of the normal and shearing stress acting on the plane $x - x_0 + y - y_0 + z - z_0 = 0$.

Answer: $\overset{\nu}{\mathbf{T}} = \frac{1}{\sqrt{3}}(100, 50, -100)$ kPa, $\sigma^{(n)} = 16.7$ kPa, $\tau = 81.7$ kPa.

3.12 For the keyed shaft shown in Fig. P3.12, determine the maximum permissible value of the load P if the stress in the shear key is not to exceed 70 MPa.

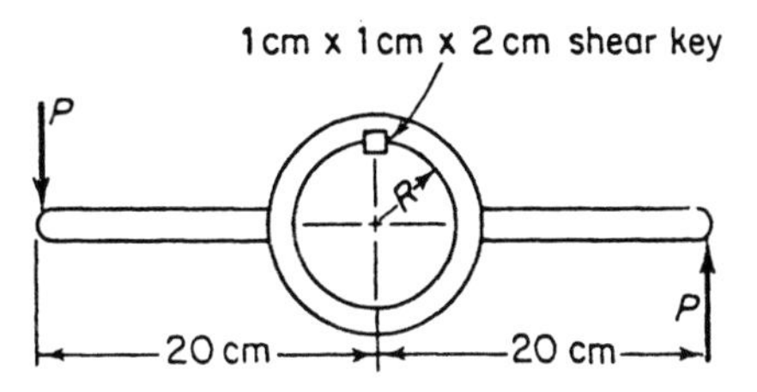

Figure P3.12 Key on a shaft.

3.13 Prove that if $\sigma_z = \sigma_{zx} = \sigma_{zy} = 0$, then under the coordinate transformation given by Eq. (2.4–3), we have $\sigma_x + \sigma_y = \sigma_{x'} + \sigma_{y'}$; that is, in a planar stress distribution, the sum of the two normal stresses is invariant.

3.14 Two sheets of plywood are spliced together as shown in Fig. P3.14. If the allowable shear stress in the glue is 1.4 MPa, what must be the minimum length L of the splice pieces if a 40 kN load is to be carried?

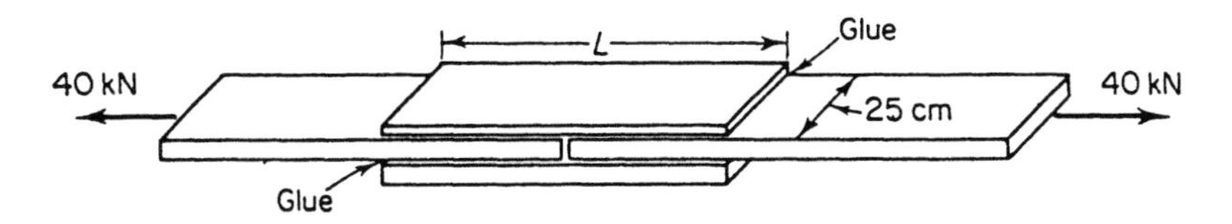

Figure P3.14 Glued seam.

3.15 A windmill propeller has been safely operated for a long time. Suppose that you want to scale it up for a factor of R_L in size and to rotate R_N times faster, using a geometrically similar design, but perhaps a different material. How would the tensile stress due to the centrifugal force vary with R_L and R_N? The aerodynamic force varies with the square of the relative wind speed. How would the bending stress in the propeller blade vary with R_L and R_N?

3.16 The set of eight planes with direction numbers $(\pm 1, \pm 1, \pm 1)$, with one of the $\pm$ signs chosen in each case [e.g., $(1, 1, -1)$, corresponding to the plane $x + y - z = 0$] is called the *octahedral* planes. Let the state of stress be specified by τ_{ij}, for which $\tau_{ij} = 0$ whenever $i \neq j$. Determine the stress vector and the shear stress acting on each of the octahedral planes.

3.17 Have you seen tree branches broken by wind and observed how they split? What does this tell us about the strength characteristics of the wood?

3.18 Experiment by attempting to break various materials such as macaroni, celery, carrots, high-carbon steel tools like drills and files, strips of aluminum and magnesium, and silicone Silly Putty. Discuss the strength characteristics of these materials.

3.19 A circular cylindrical rod is twisted. Describe the stress state in the rod. Use the notations shown in Fig. P3.6 or Fig. 3.10. Discuss, in particular, the stress components at a point on the outer, free surface of the rod.

3.20 A water tower that consists of a big raindrop-shaped tank supported on top of a column is hit by an earthquake. The tower shakes. The maximum lateral acceleration (in a direction perpendicular to the column) is estimated to be 0.2 times the gravitational acceleration. The maximum lateral inertial force induced by the earthquake is therefore equal to 20% of the weight of the tank and water and acts in a horizontal direction.

The maximum vertical acceleration is about the same magnitude. Discuss the stress state in the column.

3.21 Discuss the state of stress distribution in an airplane wing during flight and during landing.

3.22 *Couette Flow*. The space between two concentric cylinders is filled with a fluid (Fig. P3.22). Let the inner cylinder be stationary, while the outer cylinder is rotated at an angular speed ω radians/sec. If the torque measured on the inner cylinder is T, what is the torque acting on the outer cylinder? Why?

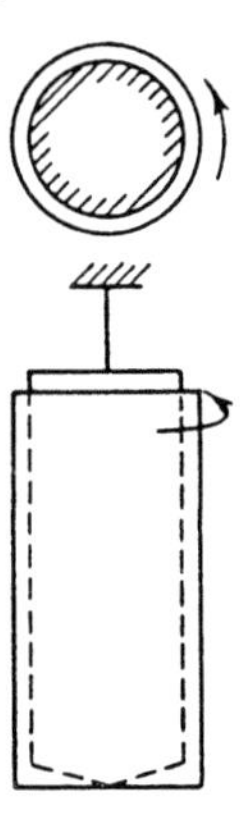

Figure P3.22 Couette flow meter.

3.23 In designing a tie rod, it is decided that the maximum shear stress must not exceed 20,000 kPa (because of possible yielding). What is the maximum tension the rod can take? Use steel.

Answer: 40,000 kPa.

3.24 Take a thin strip of steel of rectangular cross section (say, 0.5 cm × 1 cm × 100 cm). Using a handbook, find the ultimate strength of the steel. Let the strip be used to sustain a compressive load in the direction of the longest edge. On the basis of the ultimate strength alone, how large a force should the strip be able to resist?

Now try to compress the strip. The strip buckles at a load far smaller than is expected. Explain this phenomenon of elastic buckling.

3.25 Roll a sheet of paper into a circular cylinder of radius about 3 or 4 cm. Such a tube can sustain a fairly sizable end compression.

Stand the tube on the table and compress it with the palm of your hand. The cylinder will fail by buckling. Describe the buckling pattern. How large is the buckling load compared with the strength of the paper in compression if buckling can be avoided?

Since the paper does not tear after buckling, nor does it stretch, the *metric* of the deformed surface is identicai with that of the original one. Hence, the transformation from the cylinder to the buckled surface is an *isometric* transformation.

It is known in differential geometry that if one surface can be transformed isometrically into another, their total curvature must be the same at corresponding points. Now, the total curvature of a surface is the product of the principal curvatures. For a flat sheet of paper, the total curvature is zero; so is that of the cylinder, and so also must be the postbuckling surface. In this way, we expect the postbuckling surface to be composed of areas with zero total curvature, namely, flat triangular portions that are assembled together into a diamond pattern. Compare this with the experimental findings.

Note. The subject in this problem is of great interest to aeronautical and astronautical engineering. Thin-walled structures are used extensively when light weight is mandatory. Elastic stability is a main concern in designing these structures.

3.26 A rope is hung from the ceiling. Let the density of the rope be 2 g/cm^3. Find the stress in the rope.

Solution: Let the x-axis be chosen in the direction of the rope. The only stress of concern is σ_x. We shall assume that $\tau_{xy} = \tau_{xz} = 0$. Then the equation of equilibrium is

$$\frac{\partial \sigma_x}{\partial x} + \rho g = 0$$

where g is the gravitational acceleration. The solution is

$$\sigma_x = -\rho g x + \text{const.}$$

But $\sigma_x = 0$ when $x = L$, the length of the rope. Hence, the constant is $\rho g L$. Thus, $\sigma_x = \rho g(L - x)$. The maximum tension is at the ceiling, where $\sigma_x = \rho g L$.

3.27 Consider a vertical column of an isothermal atmosphere that obeys the gas law $p/\rho = RT$, or $\rho = p/RT$, where ρ is the density of the gas, p is the pressure, R is the gas constant, and T is the absolute temperature. This gas is subjected to a gravitational acceleration g so that the body force is ρg per unit volume, pointing to the ground. If the pressure at the ground level $z = 0$ is p_0, determine the relation between the pressure and the height z above the ground.

Answer: $p = p_0 \exp[-(g/RT)z]$.

3.28 Discuss why the solution given in Prob. 3.27 is unrealistic for the earth's atmosphere. If the temperature T is a known function of the height z, what would the solution be?

3.29 Consider a two-dimensional state of stress in a thin plate in which $\tau_{zz} = \tau_{zx} = \tau_{zy} = 0$. The equations of equilibrium acting in the plate in the absence of body force are

$$\frac{\partial \sigma_x}{\partial x} + \frac{\partial \tau_{xy}}{\partial y} = 0, \qquad \frac{\partial \tau_{xy}}{\partial x} + \frac{\partial \sigma_y}{\partial y} = 0.$$

Show that these equations are satisfied identically if σ_x, σ_y, σ_{xy} are derived from an arbitrary function $\Phi(x, y)$ such that

$$\sigma_x - \frac{\partial^2 \Phi}{\partial y^2}, \qquad \sigma_y = \frac{\partial^2 \Phi}{\partial x^2}, \qquad \tau_{xy} = -\frac{\partial^2 \Phi}{\partial x \partial y}$$

Thus, the equations of equilibrium can be satisfied by infinitely many solutions.

3.30 In a square plate in the region $-1 \le x, y \le 1$, the following stresses hold:

$$\sigma_{xx} = cx + dy,$$

$$\sigma_{yy} = ax + by, \ (a, b, c, d \text{ are constants})$$

$$\sigma_{zz} = \sigma_{zx} = \sigma_{zy} = 0.$$

What must the shear stress σ_{xy} be in order to assure equilibrium? Assume that the material the plate is made of is isotropic and obeys Hooke's law. What are the strains? Are they compatible? What kind of boundary conditions are satisfied if the constant $a = b = 0$, but c and d do not vanish?

4 PRINCIPAL STRESSES AND PRINCIPAL AXES

Principal stresses, stress invariants, stress deviations, and the maximum shear are important concepts. They tell us the state of stress in the simplest numerical way. They are directly related to the strength of materials. One has to evaluate them frequently; therefore, we devote a chapter to them.

4.1 INTRODUCTION

We have seen that nine components of stress, of which six are independent, are necessary to specify the state of material interaction at any given point in a body. These nine components of stress form a symmetric matrix

$$\sigma = \begin{pmatrix} \sigma_{11} & \sigma_{12} & \sigma_{13} \\ \sigma_{21} & \sigma_{22} & \sigma_{23} \\ \sigma_{31} & \sigma_{32} & \sigma_{33} \end{pmatrix}, \qquad (\sigma_{ij} = \sigma_{ji}),$$

the components of which transform as the components of a tensor under rotation of coordinates. Later we shall show that because the stress tensor is symmetric, a set of coordinates can be found with respect to which the matrix of stress components can be reduced to a diagonal matrix of the form

$$\sigma = \begin{pmatrix} \sigma_1 & 0 & 0 \\ 0 & \sigma_2 & 0 \\ 0 & 0 & \sigma_3 \end{pmatrix}.$$

The particular set of coordinates axes with respect to which the stress matrix is diagonal are called the *principal axes*, and the corresponding stress components are called the *principal stresses*. The coordinate planes determined by the principal axes are called the *principal planes*. Physically, each of the principal stresses is a normal stress acting on a principal plane. On a principal plane, the stress vector is normal to the plane, and there is no shear component.

To know the principal axes and principal stresses is obviously useful, because they help us visualize the state of stress at any point. In fact, the matter is so important, that in solving problems in continuum mechanics, we rarely stop before the final answer is reduced to principal values. We need to know, therefore, not only that the principal stresses exist and can be found *in principle*, but also the practical means of finding them. We shall show that the *symmetry of the stress tensor* is the basic reason for the existence of principal axes. Other symmetric tensors, such as the strain tensor, by analogy and an identical mathematical process, must also have principal axes and principal values. Indeed, the proof we shall give for the possibility of reducing a real-valued symmetric matrix into a principal one is not limited to three dimensions, but can be extended to n dimensions. We shall find that such an extension is of great importance when we consider mechanical vibrations of elastic bodies, or acoustics in general. In the vibration theory, the principal values correspond to the *vibration frequencies*, and the principal coordinates describe the *normal modes* of vibration. We shall not discuss these subjects now; rather, we merely point out that the subject we are going to study has much broader applications than to stress alone.

On the other hand, if a tensor is not symmetric, then neither the existence of real-valued principal values nor the possibility of reduction to diagonal form by rotation of coordinates can be assured. Symmetry is thus a great asset.

As an introduction, we shall consider the two-dimensional case in greater detail. Then we shall proceed to the three-dimensional case in abridged notation. Finally, we shall use the principal stresses to discuss some geometric representations of the stress state, as well as to introduce some additional definitions.

4.2 PLANE STATE OF STRESS

Let us consider a simplified physical situation in which a thin membrane is stretched by forces acting on its edges and lying in its plane. An example is shown in Fig. 4.1. We shall leave the faces $z = h$ and $z = -h$ free (unstressed). In this case, we can safely say that since the stress components σ_{zz}, σ_{zy}, σ_{zx} are zero on the

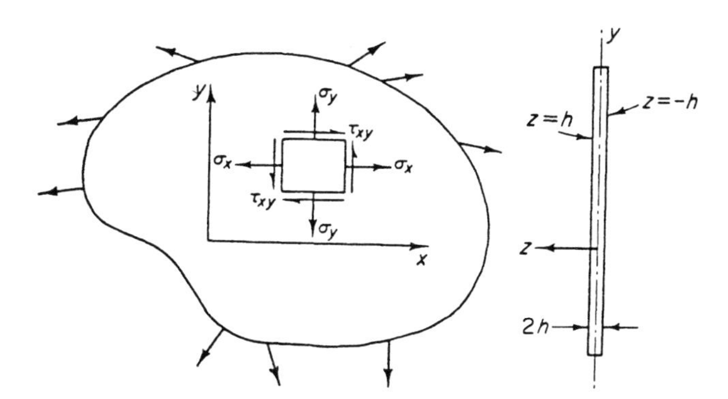

Figure 4.1 An approximate plane state of stress.

surface, they are approximately zero throughout the membrane because the membrane is very thin; that is,

$$\sigma_{zz} = \sigma_{zx} = \sigma_{zy} = 0. \tag{4.2–1}$$

The state of stress in which these equations hold is called a *plane state of stress* in the xy-plane. Obviously, in plane stress we are concerned only with the stress components in the symmetric matrix

$$\begin{pmatrix} \sigma_x & \tau_{xy} & 0 \\ \tau_{xy} & \sigma_y & 0 \\ 0 & 0 & 0 \end{pmatrix}.$$

Here, for clarity, we write σ_x for σ_{xx}, σ_y for σ_{yy}, and τ_{xy} for σ_{xy}.

We shall now consider a rotation of coordinates from xy to $x'y'$ and apply the results of Sec. 3.5 to find the stress components in the new frame of reference:

$$\begin{pmatrix} \sigma_{x'} & \tau_{x'y'} & 0 \\ \tau_{x'y'} & \sigma_{y'} & 0 \\ 0 & 0 & 0 \end{pmatrix}.$$

In this case, the direction cosines between the two systems of rectangular Cartesian coordinates can be expressed in terms of a single angle θ. (See Fig. 4.2.) The matrix of direction cosines is

$$\begin{pmatrix} \beta_{11} & \beta_{12} & \beta_{13} \\ \beta_{21} & \beta_{22} & \beta_{23} \\ \beta_{31} & \beta_{32} & \beta_{33} \end{pmatrix} = \begin{pmatrix} \cos\theta & \sin\theta & 0 \\ -\sin\theta & \cos\theta & 0 \\ 0 & 0 & 1 \end{pmatrix}. \tag{4.2–2}$$

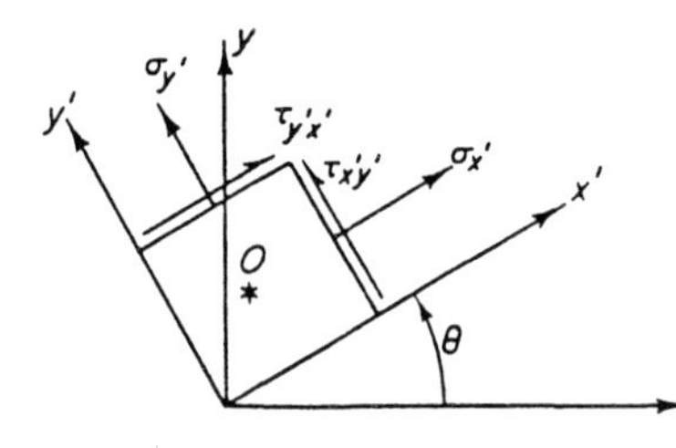

Figure 4.2 Change of coordinates in plane state of stress.

Writing x, y, and x', y' in place of x_1, x_2 and x_1', x_2'; σ_x for τ_{11}; τ_{xy} for τ_{12}, etc., and identifying direction cosines β_{ij} according to Eq. (4.2–2), we obtain, on substituting into Eq. (3.5–3), the new components:

$$\sigma_{x'} = \sigma_x \cos^2\theta + \sigma_y \sin^2\theta + 2\tau_{xy} \sin\theta\cos\theta, \tag{4.2–3}$$

$$\sigma_{y'} = \sigma_x \sin^2\theta + \sigma_y \cos^2\theta - 2\tau_{xy} \sin\theta\cos\theta, \tag{4.2–4}$$

$$\tau_{x'y'} = (-\sigma_x + \sigma_y)\sin\theta\cos\theta + \tau_{xy}(\cos^2\theta - \sin^2\theta). \tag{4.2–5}$$

Since

$$\sin^2 \theta = \tfrac{1}{2}(1 - \cos 2\theta), \qquad \cos^2 \theta = \tfrac{1}{2}(1 + \cos 2\theta),$$

we may write the preceding equations as

$$\sigma_{x'} = \frac{\sigma_x + \sigma_y}{2} + \frac{\sigma_x - \sigma_y}{2} \cos 2\theta + \tau_{xy} \sin 2\theta, \tag{4.2–6}$$

$$\sigma_{y'} = \frac{\sigma_x + \sigma_y}{2} - \frac{\sigma_x - \sigma_y}{2} \cos 2\theta - \tau_{xy} \sin 2\theta, \tag{4.2–7}$$

$$\tau_{x'y'} = -\frac{\sigma_x - \sigma_y}{2} \sin 2\theta + \tau_{xy} \cos 2\theta. \tag{4.2–8}$$

From these equations, it follows that

$$\sigma_{x'} + \sigma_{y'} = \sigma_x + \sigma_y, \tag{4.2–9}$$

$$\frac{\partial \sigma_{x'}}{\partial \theta} = 2\tau_{x'y'}, \qquad \frac{\partial \sigma_{y'}}{\partial \theta} = -2\tau_{x'y'}, \tag{4.2–10}$$

$$\tau_{x'y'} = 0 \text{ when } \tan 2\theta = \frac{2\tau_{xy}}{\sigma_x - \sigma_y}. \tag{4.2–11}$$

The directions of the x'-, y'-axes corresponding to the particular values of θ given by Eq. (4.2–11) are called the *principal directions*; the axes x' and y' are then called the *principal axes*, and $\sigma_{x'}$, $\sigma_{y'}$ are called the *principal stresses*.

If x', y' are principal axes, then $\tau_{x'y'} = 0$, and Eqs. (4.2–10) show that $\sigma_{x'}$ is either a maximum or a minimum with respect to all choices of θ. Similarly, so is $\sigma_{y'}$. On substituting θ from Eq. (4.2–11) into Eqs. (4.2–6) and (4.2–7), we obtain the result

$$\begin{matrix}\sigma_{\max} \\ \sigma_{\min}\end{matrix} = \frac{\sigma_x + \sigma_y}{2} \pm \sqrt{\left(\frac{\sigma_x - \sigma_y}{2}\right)^2 + \tau_{xy}^2}. \quad ▲ \tag{4.2–12}$$

On the other hand, on differentiating $\tau_{x'y'}$ from Eq. (4.2–8) with respect to θ and setting the derivative equal to zero, we can find an angle θ at which $\tau_{x'y'}$ reaches its extreme value. It can be shown that this angle is $\pm 45°$ from the principal directions given by Eq. (4.2–11) and that the maximum value of $\tau_{x'y'}$ is

$$\tau_{\max} = \frac{\sigma_{\max} - \sigma_{\min}}{2} = \sqrt{\left(\frac{\sigma_x - \sigma_y}{2}\right)^2 + \tau_{xy}^2}. \quad ▲ \tag{4.2–13}$$

This is the maximum of the shear stress acting on all planes parallel to the z-axis. When planes inclined to the z-axis are also considered, some other planes may have a shear higher than this. (See Sec. 4.8.)

4.3 MOHR'S CIRCLE FOR PLANE STRESS

A geometric representation of Eqs. (4.2–6)–(4.2–13) was given by Otto Mohr (*Zivilingenieur*, 1882, p. 113). An example is shown in Fig. 4.3. The normal and shear stresses acting on a surface are plotted on a *stress plane* in which the abscissa represents the normal stress and the ordinate the shear. For the normal stress, a tension is plotted as positive and a compression as negative. For the shear, a special rule is needed. We specify (for Mohr's circle construction only) that a shear is taken as positive on a face of an element when it yields a clockwise moment about the center point O of the element. (See Fig. 4.2.) A shear stress yielding a counterclockwise moment about the center O is taken to be negative. Thus, $\tau_{x'y'}$ in Fig. 4.2 is considered negative, and $\tau_{y'x'}$ is considered positive. Following this special rule, we plot, in Fig. 4.3, the point A whose abscissa is σ_x and whose ordinate is $-\tau_{xy}$ and the point B with abscissa σ_y and ordinate τ_{yx}. Then we join the line AB, which intersects the σ-axis at C. Next, with C as a center, we draw a circle through A and B. This is Mohr's circle.

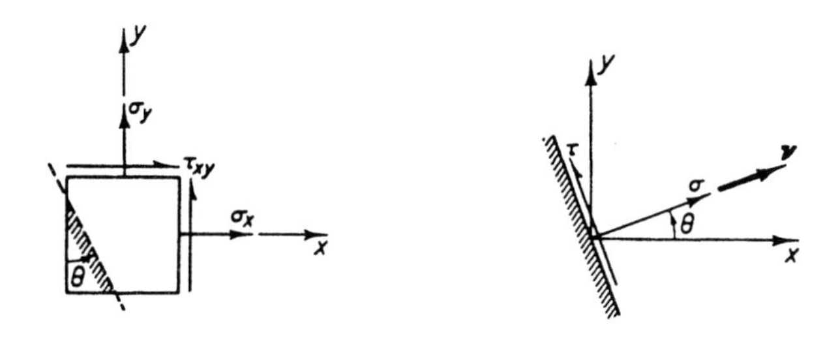

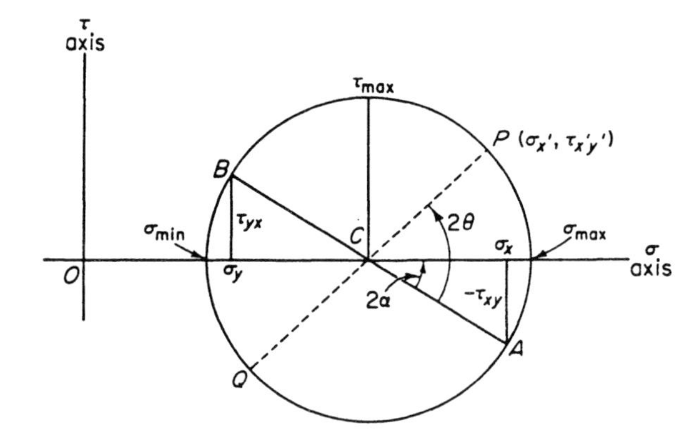

Figure 4.3 Mohr's circle for plane stress.

To obtain the stresses acting on a surface whose normal makes an angle θ with the x-axis in the counterclockwise direction, we draw a radius CP that makes an angle 2θ with the line CA, as shown in Fig. 4.3. Then, the abscissa of P gives the normal stress on that surface and the ordinate the shear. The point located on the other end of the diameter PQ represents the stress acting on a surface whose normal makes an angle $\theta + (\pi/2)$ with the x-axis.

To prove that this construction is valid, we note that Mohr's circle has a center located at C, where

$$\overline{OC} = \frac{\sigma_x + \sigma_y}{2}, \tag{4.3–1}$$

and a radius

$$\overline{AC} = \overline{CP} = \sqrt{\left(\frac{\sigma_x - \sigma_y}{2}\right)^2 + \tau_{xy}^2}. \tag{4.3–2}$$

From Fig. 4.3, we see that the abscissa of P is

$$\begin{aligned} \sigma_{x'} &= \overline{OC} + \overline{CP}\cos(2\theta - 2\alpha) \\ &= \overline{OC} + \overline{CP}(\cos 2\theta \cos 2\alpha + \sin 2\theta \sin 2\alpha). \end{aligned} \tag{4.3–3}$$

But we see also from the diagram that

$$\cos 2\alpha = \frac{\sigma_x - \sigma_y}{2\overline{CP}}, \qquad \sin 2\alpha = \frac{\tau_{xy}}{\overline{CP}}. \tag{4.3–4}$$

Substituting these results into Eq. (4.3–3), we get

$$\sigma_{x'} = \frac{\sigma_x + \sigma_y}{2} + \frac{\sigma_x - \sigma_y}{2}\cos 2\theta + \tau_{xy}\sin 2\theta, \tag{4.3–5}$$

which is exactly the same as Eq. (4.2–6).

Similarly, we have the ordinate of P:

$$\begin{aligned} \tau_{x'y'} &= \overline{CP}\sin(2\theta - 2\alpha) = \overline{CP}(\sin 2\theta \cos 2\alpha - \cos 2\theta \sin 2\alpha) \\ &= \frac{\sigma_x - \sigma_y}{2}\sin 2\theta - \tau_{xy}\cos 2\theta, \end{aligned} \tag{4.3–6}$$

which agrees with Eq. (4.2–8) in magnitude but differs in sign. The sign is fixed by the convention adopted here for Mohr's circle. A positive-valued $\tau_{x'y'}$, according to Eq. (4.2–8), would be a counterclockwise moment and would have been plotted with negative ordinate on Mohr's circle. Hence, everything agrees, and the validity of Mohr's circle is proved.

Mohr's circle gives a visual picture of how the stress varies with the orientation of the surface. It tells us how to locate the principal axes. It shows us that the planes on which the maximum shear occurs are oriented at 45° from the principal planes. In practice, however, a direct calculation by solving Eq. (4.5–3) in Sec. 4.5 is a faster way to compute the principal stresses.

4.4 MOHR'S CIRCLES FOR THREE-DIMENSIONAL STRESS STATES

Let σ_1, σ_2, σ_3 be the principal stresses at a point. The components of the stress vector acting on any section can be obtained by the tensor transformation law, Eq. (3.5–3). Otto Mohr has shown the interesting result that if the normal stress $\sigma_{(n)}$ and the shearing stress τ acting on any section are plotted on a plane, with σ and τ as coordinates as shown in Fig. 4.4, they will necessarily fall in a closed domain represented by the shaded area bounded by the three circles with centers on the σ-axis.

This result is very instructive in showing that, indeed, if $\sigma_1 \geqslant \sigma_2 \geqslant \sigma_3$, *then* σ_1 *is the largest stress and* $(\sigma_1 - \sigma_3)/2$ *is the largest shear for all possible surfaces. The plane on which the largest shear acts is inclined at* 45° *from the principal planes on which* σ_1 *and* σ_3 *act.*

The meaning of the three bounding circles in Fig. 4.4 can be explained easily. Let x-, y-, z-axes be chosen in the directions of the principal axes. On a plane perpendicular to the x-axis, there acts a normal stress, say, σ_1, and no shear. On a plane normal to y, there acts a normal stress, say, σ_2, and no shear. Now consider all planes parallel to the z-axis. For these planes, the normal and shear stresses acting on them are given exactly by Eqs. (4.2–3) through (4.2–5) or Eqs. (4.2–6) through (4.2–8). Hence, the Mohr-circle construction described in Sec. 4.3 applies, and the circle passing through σ_1, σ_2 represents the totality of all stress states on these planes. Similarly, the other two circles (one passing through σ_2, σ_3 and the other through σ_3, σ_1) represent the totality of all stress states acting on all planes parallel to either the x-axis or the y-axis. It remains only to show that the stresses on all other planes lie in the shaded area. The proof is given in earlier editions of this book. It is quite lengthy and is omitted here.

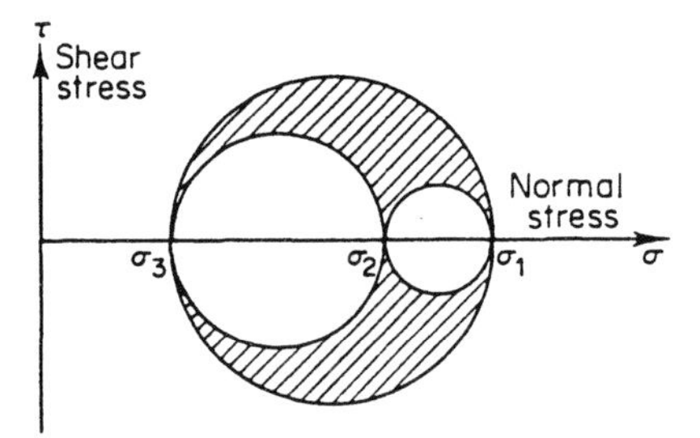

Figure 4.4 Mohr's circles.

4.5 PRINCIPAL STRESSES

In a general state of stress, the stress vector acting on a surface with normal $\boldsymbol{\nu}$ depends on the direction of $\boldsymbol{\nu}$. At a given point in a body, the angle between the stress vector and the normal $\boldsymbol{\nu}$ varies with the orientation of the surface. We shall show that we can always find a surface so oriented that the stress vector is exactly normal to it. In fact, we shall show that there are at least three mutually orthogonal surfaces that fulfill this requirement at any point in the body. Such a surface is called a *principal plane*, its normal a *principal axis*, and the value of normal stress acting on the principal plane a *principal stress*.

Let $\boldsymbol{\nu}$ be a unit vector in the direction of a principal axis, and let σ be the corresponding principal stress. Then the stress vector acting on the surface normal to $\boldsymbol{\nu}$ has components $\sigma\nu_i$, referred to a set of rectangular cartesian coordinates x_1, x_2, x_3. On the other hand, this same vector is given by the expression $\tau_{ji}\nu_j$, where τ_{ij} is the stress tensor. Hence, writing $\nu_i = \delta_{ji}\nu_j$, we have, on equating these two expressions and transposing to the same side,

$$(\tau_{ji} - \sigma\delta_{ji})\nu_j = 0, \qquad (i = 1, 2, 3). \tag{4.5–1}$$

These three equations, with $i = 1, 2, 3$, are to be solved for ν_1, ν_2, and ν_3. Since $\boldsymbol{\nu}$ is a unit vector, we must find a set of nontrivial solutions for which $\nu_1^2 + \nu_1^2 + \nu_3^2 = 1$. Thus, Eqs. (4.5–1) pose an eigenvalue problem. Since τ_{ij} as a matrix is real valued and symmetric, we need only to recall a result in the theory of matrices to assert that *there exist three real-valued principal stresses and a set of orthonormal principal axes. Whether the principal stresses are all positive, all negative, or mixed depends on whether the quadratic form $\tau_{ij}x_ix_j$ is positive definite, negative definite, or uncertain, respectively*. However, because of the importance of these results, we shall derive them anew below.

The system of Eqs. (4.5–1) has a set of nonvanishing solutions ν_1, ν_2, ν_3 if and only if the determinant of the coefficients vanishes, i.e.,

$$|\tau_{ij} - \sigma\delta_{ij}| = 0. \tag{4.5–2}$$

Equation (4.5–2) is a cubic equation in σ; its roots are the principal stresses. For each value of the principal stress, a unit normal vector $\boldsymbol{\nu}$ can be determined.

On expanding Eq. (4.5–2), we have

$$\begin{aligned}|\tau_{ij} - \sigma\delta_{ij}| &= \begin{vmatrix} \tau_{11} - \sigma & \tau_{12} & \tau_{13} \\ \tau_{21} & \tau_{22} - \sigma & \tau_{23} \\ \tau_{31} & \tau_{32} & \tau_{33} - \sigma \end{vmatrix} \\ &= -\sigma^3 + I_1\sigma^2 - I_2\sigma + I_3 = 0,\end{aligned} \tag{4.5–3}$$

where

$$I_1 = \tau_{11} + \tau_{22} + \tau_{33}, \tag{4.5–4}$$

$$I_2 = \begin{vmatrix} \tau_{22} & \tau_{23} \\ \tau_{32} & \tau_{33} \end{vmatrix} + \begin{vmatrix} \tau_{33} & \tau_{31} \\ \tau_{13} & \tau_{11} \end{vmatrix} + \begin{vmatrix} \tau_{11} & \tau_{12} \\ \tau_{21} & \tau_{22} \end{vmatrix}, \tag{4.5–5}$$

$$I_3 = \begin{vmatrix} \tau_{11} & \tau_{12} & \tau_{13} \\ \tau_{21} & \tau_{22} & \tau_{23} \\ \tau_{31} & \tau_{32} & \tau_{33} \end{vmatrix}. \tag{4.5–6}$$

On the other hand, if σ_1, σ_2, σ_3 are the roots of Eq. (4.5–3), which can then be written as

$$(\sigma - \sigma_1)(\sigma - \sigma_2)(\sigma - \sigma_3) = 0, \tag{4.5–7}$$

it can be seen that the following relations between the roots and the coefficients must hold:

$$I_1 = \sigma_1 + \sigma_2 + \sigma_3, \tag{4.5–8}$$

$$I_2 = \sigma_1\sigma_2 + \sigma_2\sigma_3 + \sigma_3\sigma_1, \tag{4.5–9}$$

$$I_3 = \sigma_1\sigma_2\sigma_3. \tag{4.5–10}$$

Since the principal stresses characterize the physical state of stress at a point, they are independent of any coordinates of reference. Hence, Eq. (4.5–7) is independent of the orientation of the coordinates of reference. But Eq. (4.5–7) is exactly the same as Eq. (4.5–3). Therefore, Eq. (4.5–3) and the coefficients I_1, I_2, I_3 are invarient with respect to the rotation of coordinates. I_1, I_2, I_3 are called the *invariants* of the stress tensor with respect to rotation of coordinates.

We shall show now that, for a symmetric stress tensor, the three principal stresses are all real and the three principal planes are mutually orthogonal. These important properties can be established when the stress tensor is symmetric, i.e.,

$$\tau_{ij} = \tau_{ji}. \tag{4.5–11}$$

The proof is as follows. Let $\overset{1}{\boldsymbol{\nu}}$, $\overset{2}{\boldsymbol{\nu}}$, $\overset{3}{\boldsymbol{\nu}}$ be unit vector in the direction of the principal axes, with components $\overset{1}{\nu}_i$, $\overset{2}{\nu}_i$, $\overset{3}{\nu}_i$ $(i = 1, 2, 3)$ that are the solutions of Eq. (4.5–1) corresponding to the roots, σ_1, σ_2, σ_3, respectively; then

$$\begin{aligned}(\tau_{ij} - \sigma_1\delta_{ij})\overset{1}{\nu}_j &= 0,\\ (\tau_{ij} - \sigma_2\delta_{ij})\overset{2}{\nu}_j &= 0,\\ (\tau_{ij} - \sigma_3\delta_{ij})\overset{3}{\nu}_j &= 0.\end{aligned} \tag{4.5–12}$$

Multiplying the first equation by $\overset{2}{\nu}_i$ and the second by $\overset{1}{\nu}_i$, summing over i, and subtracting the resulting equations, we obtain

$$(\sigma_2 - \sigma_1)\overset{1}{\nu}_i\overset{2}{\nu}_i = 0 \tag{4.5–13}$$

on account of the symmetry condition (4.5–11), which implies that

$$\tau_{ij}\overset{1}{\nu}_j\overset{2}{\nu}_i = \tau_{ji}\overset{1}{\nu}_j\overset{2}{\nu}_i = \tau_{ij}\overset{2}{\nu}_j\overset{1}{\nu}_i, \tag{4.5–14}$$

the last equality being obtained by interchanging the dummy indices i and j.

Now, if we assume tentatively that Eq. (4.5–3) has a complex root, then, since the coefficients in that equation are real valued, a complex conjugate root must also exist, and the set of roots may be written as

$$\sigma_1 = \alpha + i\beta, \qquad \sigma_2 = \alpha - i\beta, \qquad \sigma_3$$

where α, β, σ_3 are real numbers and i stands for the imaginary number $\sqrt{-1}$. In this case, Eqs. (4.5–12) show that $\overset{1}{\nu}_j$ and $\overset{2}{\nu}_j$ are conjugate to each other and can be written as

$$\overset{1}{\nu}_j \equiv a_j + ib_j, \qquad \overset{2}{\nu}_j \equiv a_j - ib_j.$$

Therefore,

$$\begin{aligned}\overset{1}{\nu}_j\overset{2}{\nu}_j &= (a_j + ib_j)(a_j - ib_j) \\ &= a_1^2 + a_2^2 + a_3^2 + b_1^2 + b_2^2 + b_3^2 \neq 0.\end{aligned}$$

It follows from Eq. (4.5–13) that $\sigma_1 - \sigma_2 = 2i\beta = 0$. So B $= 0$. But this contradicts the original assumption that the roots are complex. Thus, the assumption of the existence of complex roots is untenable, and the roots σ_1, σ_2, σ_3 are all real.

When $\sigma_1 \neq \sigma_2 \neq \sigma_3$, Eq. (4.5–13) and similar equations imply that

$$\overset{1}{\nu}_i\overset{2}{\nu}_i = 0, \qquad \overset{2}{\nu}_i\overset{3}{\nu}_i = 0, \qquad \overset{3}{\nu}_i\overset{1}{\nu}_i = 0; \tag{4.5–15}$$

i.e., the principal vectors are mutually orthogonal to each other. If $\sigma_1 = \sigma_2 \neq \sigma_3$, $\overset{3}{\nu}_i$ will be fixed, but we can determine an infinite number of pairs of vectors $\overset{1}{\nu}_i$ and $\overset{2}{\nu}_i$ orthogonal to $\overset{3}{\nu}_i$. If $\sigma_1 = \sigma_2 = \sigma_3$, then any set of orthogonal axes may be taken as the principal axes.

If the reference axes x_1, x_2, x_3 are chosen to coincide with the principal axes, then the matrix of stress components becomes

$$(\tau_{ij}) = \begin{pmatrix} \sigma_1 & 0 & 0 \\ 0 & \sigma_2 & 0 \\ 0 & 0 & \sigma_3 \end{pmatrix}. \tag{4.5–16}$$

4.6 SHEARING STRESSES

We have seen that on an element of surface with a unit outer normal $\boldsymbol{\nu}$ (with components ν_i), there acts a traction $\overset{\nu}{\mathbf{T}}$ (with components $\overset{\nu}{T}_i = \tau_{ji}\nu_j$). The component of $\overset{\nu}{\mathbf{T}}$ in the direction of $\boldsymbol{\nu}$ is the normal stress acting on the surface element. Let this normal stress be denoted by $\sigma_{(n)}$. Since the component of a vector in the direction of a unit vector is given by the scalar product of the two vectors, we obtain

$$\sigma_{(n)} = \overset{\nu}{T}_i\nu_i = \tau_{ij}\nu_i\nu_j. \tag{4.6–1}$$

On the other hand, since the vector $\overset{\nu}{\mathbf{T}}$ can be decomposed into two orthogonal components $\sigma_{(n)}$ and τ, where τ denotes the shearing stress tangent to the surface (see Fig. 4.5), we see that the magnitude of the shearing stress on a surface having the normal $\boldsymbol{\nu}$ is given by the equation

$$\tau^2 = |\overset{\nu}{T}_i|^2 - \sigma_{(n)}^2. \tag{4.6–2}$$

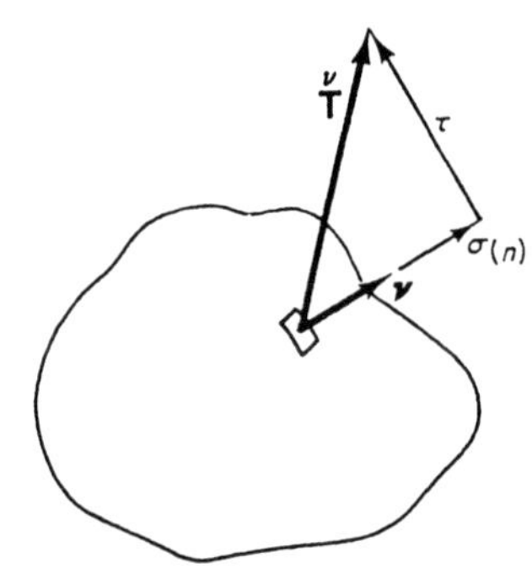

Figure 4.5 Notations.

Let the principal axes be chosen as the coordinate axes, and let σ_1, σ_2, σ_3 be the principal stresses. Then

$$\begin{aligned} \overset{\nu}{T}_1 &= \sigma_1 \nu_1, \qquad \overset{\nu}{T}_2 = \sigma_2 \nu_2, \qquad \overset{\nu}{T}_3 = \sigma_3 \nu_3, \\ |\overset{\nu}{T}_i|^2 &= (\sigma_1 \nu_1)^2 + (\sigma_2 \nu_2)^2 + (\sigma_3 \nu_3)^2, \end{aligned} \tag{4.6–3}$$

and, from Eq. (4.6–1),

$$\sigma_{(n)} = \sigma_1 \nu_1^2 + \sigma_2 \nu_2^2 + \sigma_3 \nu_3^2, \tag{4.6–4}$$

$$\sigma_{(n)}^2 = [\sigma_1 \nu_1^2 + \sigma_2 \nu_2^2 + \sigma_3 \nu_3^2]^2. \tag{4.6–5}$$

On substituting into Eq. (4.6–2) and noting that

$$(\nu_1)^2 - (\nu_1)^4 = (\nu_1)^2[1 - (\nu_1)^2] = (\nu_1)^2[(\nu_2)^2 + (\nu_3)^2], \tag{4.6–6}$$

we see that

$$\begin{aligned} \tau^2 = {} & (\nu_1)^2(\nu_2)^2(\sigma_1 - \sigma_2)^2 + (\nu_2)^2(\nu_3)^2(\sigma_2 - \sigma_3)^2 \\ & + (\nu_3)^2(\nu_1)^2(\sigma_3 - \sigma_1)^2. \end{aligned} \tag{4.6–7}$$

For example, if $\nu_1 = \nu_2 = 1/\sqrt{2}$ and $\nu_3 = 0$, then $\tau = \pm\frac{1}{2}(\sigma_1 - \sigma_2)$.

PROBLEMS

4.1 Show that $\tau_{\max} = \frac{1}{2}(\sigma_{\max} - \sigma_{\min})$ and that the plane on which $\tau_{\max}$ acts makes an angle of 45° with the direction of the largest and the smallest principal stresses.

Solution. The problem is to find the maximum or minimum of τ. Now, τ^2 is given by Eq. (4.6–7). We must find the extremum of τ^2 as a function of ν_1, ν_2, ν_3 under the restriction that $\nu_1^2 + \nu_2^2 + \nu_3^2 = 1$. Using the method of the Lagrangian multiplier, we seek to minimize the function

$$f \equiv \nu_1^2\nu_2^2(\sigma_1 - \sigma_2)^2 + \nu_2^2\nu_3^3(\sigma_2 - \sigma_3)^2 + \nu_3^2\nu_1^2(\sigma_3 - \sigma_1)^2 + \lambda(\nu_1^2 + \nu_2^2 + \nu_3^3 - 1).$$

By the usual method, we compute the partial derivatives $\partial f/\partial \nu_i$, $\partial f/\partial \lambda$, equate them to zero, and solve for ν_1, ν_2, ν_3, λ. This leads to the following equations:

$$\frac{\partial f}{\partial \nu_1} = 0: \qquad 2\nu_1\nu_2^2(\sigma_1 - \sigma_2)^2 + 2\nu_1\nu_3^2(\sigma_3 - \sigma_1)^2 + 2\lambda\nu_1 = 0, \tag{1}$$

$$\frac{\partial f}{\partial \nu_2} = 0: \qquad 2\nu_2\nu_1^2(\sigma_1 - \sigma_2)^2 + 2\nu_2\nu_3^2(\sigma_2 - \sigma_3)^2 + 2\lambda\nu_2 = 0, \tag{2}$$

$$\frac{\partial f}{\partial \nu_3} = 0: \qquad 2\nu_3\nu_2^2(\sigma_2 - \sigma_3)^2 + 2\nu_3\nu_1^2(\sigma_3 - \sigma_1)^2 + 2\lambda\nu_3 = 0, \tag{3}$$

$$\frac{\partial f}{\partial \lambda} = 0: \qquad \nu_1^2 + \nu_2^2 + \nu_3^2 = 1. \tag{4}$$

One of the solutions of Eq. (1) is obviously $\nu_1 = 0$. On setting $\nu_1 = 0$, Eqs. (2) and (3) become

$$\nu_3^2(\sigma_2 - \sigma_3)^2 + \lambda = 0, \qquad \nu_2^2(\sigma_2 - \sigma_3)^2 + \lambda = 0.$$

These equations are consistent only if $\nu_2 = \nu_3$. On setting $\nu_2 = \nu_3$, Eq. (4) becomes $0 + \nu_2^2 + \nu_2^2 = 1$, or $\nu_2 = 1/\sqrt{2}$. Hence, the first set of solutions is

$$\nu_1 = 0, \qquad \nu_2 = \nu_3 = \frac{1}{\sqrt{2}}, \qquad \lambda = -\frac{(\sigma_2 - \sigma_3)^2}{2}.$$

Substituting this back into f, or Eq. (4.6–7), we find the extremum of τ^2:

$$\tau_{\text{ext}}^2 = \frac{(\sigma_2 - \sigma_3)^2}{4}, \quad \text{or} \quad \tau_{\max} \quad \text{or} \quad \tau_{\min} = \frac{\sigma_2 - \sigma_3}{2}.$$

Other sets of solutions of Eqs. (1), (2), (3), and (4) can be obtained by setting, in turn, $\nu_2 = 0$ and $\nu_3 = 0$. We have then the relative maxima or minima

$$\frac{\sigma_2 - \sigma_3}{2}, \qquad \frac{\sigma_3 - \sigma_1}{2}, \qquad \frac{\sigma_1 - \sigma_2}{2}.$$

The largest of the three is the absolute maximum of τ.

The direction of the normal to the plane on which the absolute maximum shear occurs is given by the appropriate ν's. Whichever the solution is, we have

$$\nu_i = \nu_j = \frac{1}{\sqrt{2}} \qquad (i \neq j),$$

which implies a 45° inclination the x_i-, x_j-axes.

4.7 STRESS-DEVIATION TENSOR

The tensor

$$\tau'_{ij} = \tau_{ij} - \sigma_0\delta_{ij} \tag{4.7–1}$$

is called the *stress-deviation tensor*, in which δ_{ij} is the Kronecker delta and σ_0 is the mean *stress*:

$$\sigma_0 = \tfrac{1}{3}(\sigma_1 + \sigma_2 + \sigma_3) = \tfrac{1}{3}(\tau_{11} + \tau_{22} + \tau_{33}) = \tfrac{1}{3}I_1, \tag{4.7–2}$$

where I_1 is the first invariant of Sec. 4.5. The separation of τ_{ij} into a hydrostatic part $\sigma_0 \delta_{ij}$ and the deviation τ'_{ij} is very important in describing the plastic behavior of metals.

The first invariant of the stress-deviation tensor always vanishes:

$$I'_1 = \tau'_{11} + \tau'_{22} + \tau'_{33} = 0. \tag{4.7–3}$$

To determine the principal stress deviations, the procedure of Sec. 4.5 may be followed. The determinantal equation

$$|\tau'_{ij} - \sigma' \delta_{ij}| = 0 \tag{4.7–4}$$

may be expanded in the form

$$\sigma'^3 - J_2 \sigma' - J_3 = 0. \tag{4.7–5}$$

It is easy to verify the following equations relating J_2, J_3 to the invariants I_2, I_3 defined in Sec. 4.5:

$$J_2 = 3\sigma_0^2 - I_2, \tag{4.7–6}$$

$$J_3 = I_3 - I_2\sigma_0 + 2\sigma_0^3 = I_3 + J_2\sigma_0 - \sigma_0^3, \tag{4.7–7}$$

It is also easy to verify the following alternative expressions on account of Eq. (4.7–3):

$$\begin{aligned} J_2 &= -\tau'_{11}\tau'_{22} - \tau'_{22}\tau'_{33} - \tau'_{33}\tau'_{11} + (\tau_{12})^2 + (\tau_{23})^2 + (\tau_{31})^2 \\ &= \tfrac{1}{6}[(\tau_{11} - \tau_{22})^2 + (\tau_{22} - \tau_{33})^2 + (\tau_{33} - \tau_{11})^2] + (\tau_{12})^2 + (\tau_{23})^2 + (\tau_{31})^2 \\ &= \tfrac{1}{2}[(\tau'_{11})^2 + (\tau'_{22})^2 + (\tau'_{33})^2] + (\tau_{12})^2 + (\tau_{23})^2 + (\tau_{31})^2 \end{aligned} \tag{4.7–8}$$

Hence,

$$J_2 = \tfrac{1}{2}\tau'_{ij}\tau'_{ij}. \tag{4.7–9}$$

To verify all four equations, we note first that since J_2, J_3 and I_2, I_3 are all invariants, it is sufficient to verify Eqs. (4.7–6) and (4.7–7) with a particular choice of frame of reference. We observe that the principal axes of the stress tensor and the stress-deviation tensor coincide. We choose x_1, x_2, x_3 in the direction of the principal axes. Then if σ'_1, σ'_2, σ'_3, are the principal stress deviations, we have

$$\sigma'_1 = \sigma_1 - \sigma_0, \qquad \sigma'_2 = \sigma_2 - \sigma_0, \qquad \sigma'_3 = \sigma_3 - \sigma_0, \tag{4.7–10}$$

$$J_2 = -(\sigma'_1\sigma'_2 + \sigma'_2\sigma'_3 + \sigma'_3\sigma'_1), \tag{4.7–11}$$

$$J_3 = \sigma'_1\sigma'_2\sigma'_3. \tag{4.7–12}$$

Note the negative sign in Eq. (4.7–11) because of our choice of signs in Eq. (4.7–5). The reason for this choice will become evident if we observe, from the last two lines of Eq. (4.7–8), that J_2 so defined is indeed positive definite. From Eq. (4.7–11), by direct substitution, we have

$$\begin{aligned} J_2 &= -(\sigma_1 - \sigma_0)(\sigma_2 - \sigma_0) - (\sigma_2 - \sigma_0)(\sigma_3 - \sigma_0) - (\sigma_3 - \sigma_0)(\sigma_1 - \sigma_0) \\ &= -(\sigma_1\sigma_2 + \sigma_2\sigma_3 + \sigma_3\sigma_1) + 2\sigma_0(\sigma_1 + \sigma_2 + \sigma_3) - 3\sigma_0^2 \\ &= -I_2 + 6\sigma_0^2 - 3\sigma_0^2 = 3\sigma_0^2 - I_2, \end{aligned}$$

which verifies Eq. (4.7–6). A similar substitution of Eq. (4.7–10) into Eq. (4.7–12) verifies Eq. (4.7–7). Now we revert to an arbitrary orientation of frame of reference. A direct identification of the coefficients in Eq. (4.7–5) with those of Eq. (4.7–4) yields, as in Eq. (4.5–5),

$$J_2 = -\begin{vmatrix} \tau'_{22} & \tau'_{23} \\ \tau'_{32} & \tau'_{33} \end{vmatrix} - \begin{vmatrix} \tau'_{33} & \tau'_{31} \\ \tau'_{13} & \tau'_{11} \end{vmatrix} - \begin{vmatrix} \tau'_{11} & \tau'_{12} \\ \tau'_{21} & \tau'_{22} \end{vmatrix}. \tag{4.7–13}$$

Expansion of the determinants yields the first line of Eq. (4.7–8). The primes in τ'_{12}, τ'_{23}, τ'_{31}, can be omitted because these quantities are equal to τ_{12}, τ_{23}, τ_{31}, respectively. The third line of Eq. (4.7–8) is obtained if we add the null quantity $\frac{1}{2}(\tau'_{11} + \tau'_{22} + \tau'_{33})^2$ to the first line and simplify the results. To obtain the second line of Eq. (4.7–8), we note first that

$$\tau_{11} - \tau_{22} = (\tau_{11} - \sigma_0) - (\tau_{22} - \sigma_0) = \tau'_{11} - \tau'_{22}.$$

Hence,

$$\begin{aligned} &(\tau_{11} - \tau_{22})^2 + (\tau_{22} - \tau_{33})^2 + (\tau_{31} - \tau_{11})^2 \\ &\qquad = 2(\tau'^2_{11} + \tau'^2_{22} + \tau'^2_{33}) - 2(\tau'_{11}\tau'_{22} + \tau'_{22}\tau'_{33} + \tau'_{33}\tau'_{11}). \end{aligned}$$

Adding the null quantity $(\tau'_{11} + \tau'_{22} + \tau'_{33})^2$ to the right-hand side reduces the sum to $3(\tau'^2_{11} + \tau'^2_{22} + \tau'^2_{33})$. The equality of the third line of Eq. (4.7–8) with the second line is then evident. The last equation, Eq. (4.7–9), is nothing but a restatement of the third line of Eq. (4.7–8). Thus, every equation is verified.

Example of Application: Testing of Material in a Pressurized Chamber

If we test a simply supported steel beam [Fig. 4.6(a)] in the laboratory by a lateral load P at the center, the relationship between P and the deflection δ under the load will be a curve such as that shown in Fig. 4.6(b). The spot at which the P–δ curve deviates by a specified amount from a straight line through the origin is the *yield point*. If the beam is designed to support an engineering structure, it should not be loaded beyond the yield point, because beyond this point the deflection increases rapidly and irreversibly, and "permanent set" occurs.

(a)

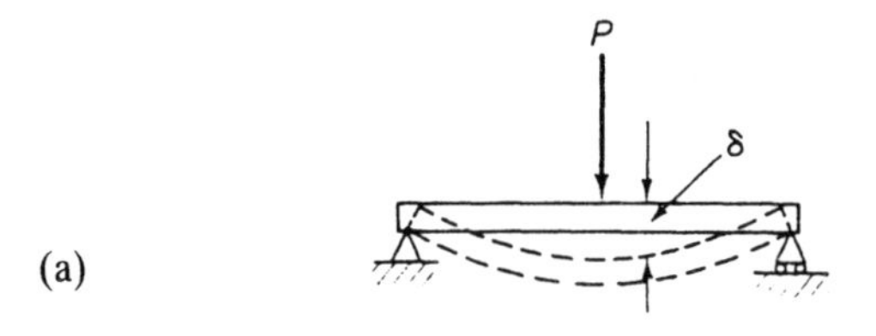

(b)

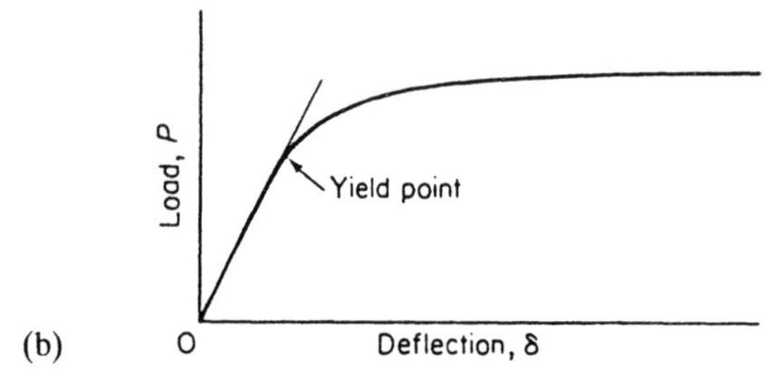

Figure 4.6 Testing of a steel beam in a pressurized chamber. (a) The beam. (b) The load-deflection curve.

Now, let us ask these questions: If we are going to build a beam to be used as an instrument in the Mariana trench in the Pacific Ocean, 10,911 m. under the sea, what would the load-deflection curve be? Would the hydrostatic pressure of the ocean depth change the load-deflection curve of the beam?

Questions like these are of great interest to seismologists, geologists, engineers, and material scientists. Although nobody has performed such a test in the ocean depths, a simulated test was done by Percy Williams Bridgman (1882–1961) at Harvard. He built a test chamber in which high pressures that approach those in the ocean depths was achieved. The test results indicate that the P–δ curve of steel beam is virtually unaffected by the hydrostatic pressure.

The yielding of steel, then, is unaffected by the hydrostatic pressure. In other words, yielding is related to stress or strain, but only to that part of the stress tensor that is independent of hydrostatic pressure. This leads to the consideration of the stress-deviation tensor τ'_{ij} defined in Eq. (4.7–1), for which the hydrostatic part $\tau'_{\alpha\alpha}$ is zero. Yielding of most materials is related not to τ_{ij}, but to τ'_{ij}.

4.8 LAMÉ'S STRESS ELLIPSOID

On any surface element with a unit outer normal vector $\boldsymbol{\nu}$ (with components ν_i), there acts a traction vector $\overset{\nu}{\mathbf{T}}$ with components given by

$$\overset{\nu}{T}_i = \tau_{ji}\nu_j.$$

Let the principal axes of the stress tensor be chosen as the coordinate axes x_1, x_2, x_3, and let the principal stresses be written as σ_1, σ_2, σ_3. Then,

$$\tau_{ij} = 0, \qquad \text{if } i \neq j,$$

and

$$\overset{\nu}{T}_1 = \sigma_1 \nu_1, \qquad \overset{\nu}{T}_2 = \sigma_2 \nu_2, \qquad \overset{\nu}{T}_3 = \sigma_3 \nu_3. \tag{4.8–1}$$

Since $\boldsymbol{\nu}$ is a unit vector, we have

$$(\nu_1)^2 + (\nu_2)^2 + (\nu_3)^2 = 1. \tag{4.8–2}$$

On solving Eq. (4.8–1) for ν_i and substituting into Eq. (4.8–2), we see that the components of $\overset{\nu}{T}_i$ satisfy the equation

$$\frac{(\overset{\nu}{T}_1)^2}{(\sigma_1)^2} + \frac{(\overset{\nu}{T}_2)^2}{(\sigma_2)^2} + \frac{(\overset{\nu}{T}_3)^2}{(\sigma_3)^2} = 1, \tag{4.8–3}$$

which is the equation of an ellipsoid with reference to a system of rectangular coordinates with axes labeled $\overset{\nu}{T}_1$, $\overset{\nu}{T}_2$, $\overset{\nu}{T}_3$. This ellipsoid is the locus of the end points of vectors $\overset{\nu}{\mathbf{T}}$ issuing from a common center. (See Fig. 4.7.)

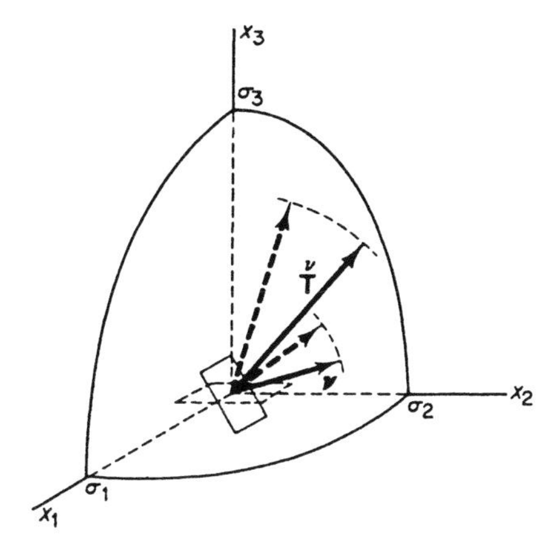

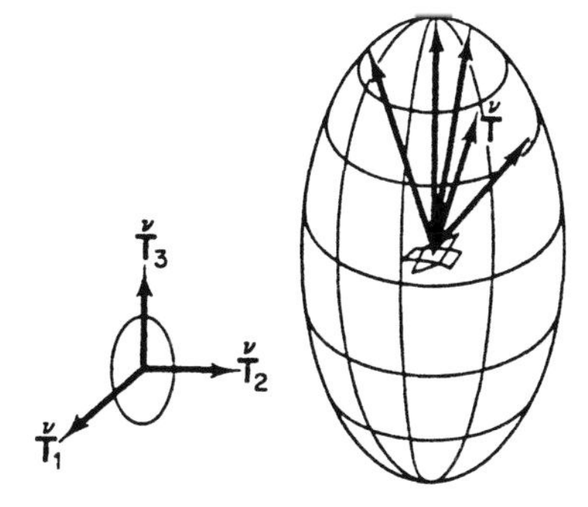

Figure 4.7 Stress ellipsoid as the locus of the end of the vector $\overset{\nu}{\mathbf{T}}$ as $\boldsymbol{\nu}$ varies.

PROBLEMS

4.2 Let $\tau_{xx} = 1{,}000$ kPa, $\tau_{yy} = -1{,}000$ kPa, $\tau_{zz} = 0$, $\tau_{xy} = 500$ kPa, $\tau_{yz} = -200$ kPa, $\tau_{zx} = 0$. What is the magnitude of the traction acting on a surface whose normal vector is

$$\boldsymbol{\nu} = 0.10\,\mathbf{i} + 0.30\,\mathbf{j} + \sqrt{0.90}\,\mathbf{k}?$$

What are the three components (in the x-, y-, and z-directions) of the stress vector acting on the surface? What is the normal stress acting on the surface? What is the resultant shear stress acting on the surface?

Answer. $(\overset{\nu}{T}_i) = (250, -440, -60)$, traction $= 509$ kPa, normal stress $= -164$ kPa, shear $= 481$ kPa.

4.3 In 1850, George Stokes gave the solution to the problem of a sphere of radius a (Fig. P4.3) moving in a (Newtonian) viscous fluid at a constant velocity U in the direction of the x-axis. On the surface of the sphere, the three components of the stress vector are

$$\overset{r}{T}_x = -\frac{x}{a}p_0 + \frac{3}{2}\mu\frac{U}{a}, \qquad \overset{r}{T}_y = -\frac{y}{a}p_0, \qquad \overset{r}{T}_z = -\frac{z}{a}p_0 \tag{1}$$

where p_0 is the pressure at a large distance from the sphere. What is the resultant force acting on the sphere?

Solution. The total surface force acting on the sphere is

$$F_x = \oint \overset{\nu}{T}_x\,dS, \qquad F_y = \oint \overset{\nu}{T}_y\,dS, \qquad F_z = \oint \overset{\nu}{T}_z\,dS. \tag{2}$$

By symmetry,

$$\oint x\,dS = \oint y\,dS = \oint z\,dS = 0. \tag{3}$$

Hence, the only nonvanishing component of the resultant force is

$$F_x = \oint \frac{3}{2}\mu\frac{U}{a}dS = 4\pi a^2\frac{3}{2}\mu\frac{U}{a} = 6\pi\mu a U. \tag{4}$$

Note. Robert Andrew Millikan, who won the Nobel Prize in 1923, measured the charge of an electron with a *cloud chamber*. A chamber was sparsely filled with tiny, approximately spherical oil drops. Two parallel condenser plates in the chamber were charged, creating an electric field. Uncharged drops fell in the direction of gravity. Any drops with an electron attached also moved in the direction of the electric field. By shining a light on the drops, the tracks of the charged drops were photographed and their velocities measured. Millikan used Stokes's formula, Eq. (4), to compute the resultant force acting on each particle due to fluid friction. This force was balanced by the electric force acting on the electron. Millikan showed that the electron charge came in a definite unit—that it was quantized—and then measured what that unit was. In this way, he obtained a basic physical constant.

Nobel prizes were first given in 1901. Millikan was the second American to win the Nobel Prize in physics. Stoke's formula, however, is not entirely satisfactory from a theoretical point of view. Many improvements have been suggested in the literature.

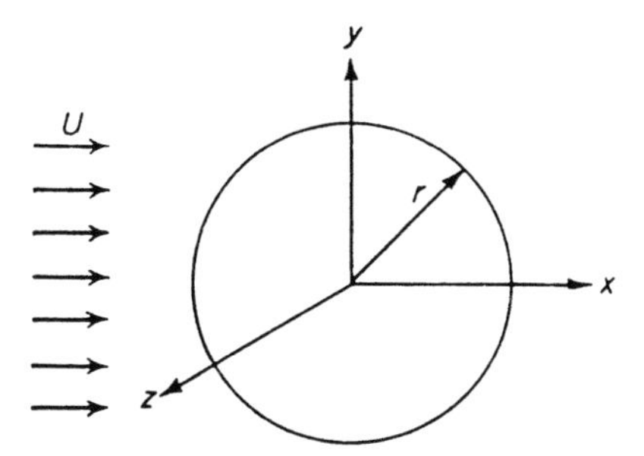

Figure P4.3 A sphere in viscous fluid (Stokes's problem).

4.4 A thick-walled elastic circular cylindrical tube with an inner radius a and an outer radius b is subjected to an internal pressure p_i and an external pressure p_o. (See Fig. P4.4.) The stresses in the cylinder are

$$\sigma_{rr} = \frac{a^2b^2(p_o - p_i)}{r^2(b^2 - a^2)} + \frac{p_i a^2 - p_o b^2}{b^2 - a^2},$$

$$\sigma_{\theta\theta} = -\frac{a^2b^2(p_o - p_i)}{r^2(b^2 - a^2)} + \frac{p_i a^2 - p_o b^2}{b^2 - a^2},$$

$$\sigma_{r\theta} = 0.$$

where r is the radial location of a point in the tube. Find the radius for which the maximum principal stress occurs and the value of the absolute maximum principal stress in the tube wall. What is the average value of $\sigma_{\theta\theta}$ in the wall?

This solution was due to Lamé under the hypotheses of the linearized theory of elasticity (Chapter 12). The nonuniformity of the stress distribution in the wall should be noted. The stress concentration at the inner wall is significant.

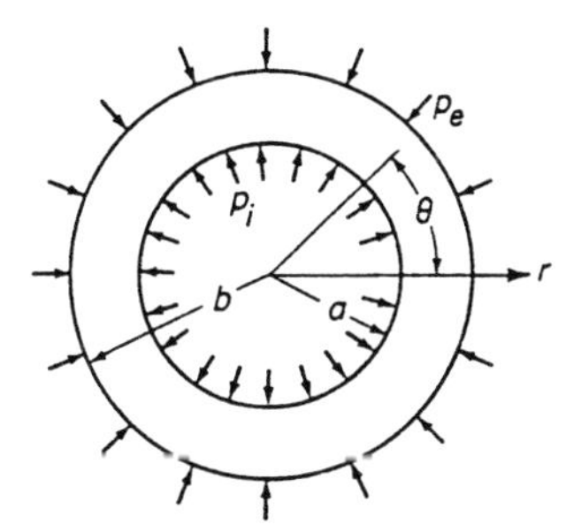

Figure P4.4 Thick-walled cylinder subjected to internal and external pressures.

4.5 Suppose that you are designing a high-pressure chamber, such as a gun barrel. The pressure generated by the explosive is so high that the maximum circumferential stress $\sigma_{\theta\theta}$ exceeds the allowable tensile stress σ_{cr}. To reduce the tensile stress at the inner wall, you may shrink-fit an outer shell on the barrel. The outer shell is put on hot, and then, as it cools, it subjects the outside of the gun barrel to a compressive load of pressure p_o that tends to reduce the tensile stress $\sigma_{\theta\theta}$ at the inside of the gun bore. Suppose $\sigma_{\theta\theta} > \sigma_{cr}$, i.e., the stress exceeds the allowable stress. Present a multi-layered gun barrel design that will make the maximum principal stress $< \sigma_{cr}$ by using the results given in Prob. 4.4.

4.6 A human blood vessel is a remarkable organ. It has considerable residual stress when there is no load acting on it. If a segment is cut in vivo, its length will shorten by 30 to 40%. If the segment is then cut radially, it will open into a sector. The open sector is a good approximation to the zero-stress state of the blood vessel because any further cutting of the specimen yields no further measurable change of strain in the vessel wall. The meaning of the open-sector shaped zero-stress state has been investigated, (see Y.C. Fung, "What are the residual stresses doing in our blood vessels?", *Annals of Biomedical Engineering* **19:**237–249, 1991.) It is shown that because of the existence of residual strains, the distribution of the circumferential strain in the vessel wall is quite uniform throughout the vessel wall at the normal living condition. In the full range of strain from zero stress to physiological condition, the stress-strain relationship of the blood vessel is nonlinear. But, if one considers only small changes of stress and strain from the normal living condition, then the incremental-stress-strain relationship can be linearized. Now, if we assume further that the elastic constants of the linearized stress-strain relationship are constant throughout the vessel wall, then under the restriction to small changes from the normal condition, the Lamé solution given in Prob. 4.4 applies. Now suppose a normal, healthy person suddenly becomes hypertensive, he or she incurs an abnormally higher blood pressure by an increment of Δp_i. Plot the incremental stresses in the blood vessel wall. Are they uniform? Where is the largest incremental stress? In Chapter 13, we shall see that the blood vessel will remodel itself in response to the incremental stresses.

4.7 *Stress concentration*. Describe the boundary conditions for a plate with a circular hole subjected to a static uniform tensile loading with the normal stress σ_x = const. = p acting on the ends. (See Fig. P4.7.)

It is known that if this plate is made of a linear elastic material, the solution is

$$\sigma_r = \frac{p}{2}\left(1 - \frac{a^2}{r^2}\right)\left[1 + \left(1 - 3\frac{a^2}{r^2}\right)\cos 2\theta\right],$$

$$\sigma_\theta = \frac{p}{2}\left[1 + \frac{a^2}{r^2} - \left(1 + 3\frac{a^4}{r^4}\right)\cos 2\theta\right],$$

$$\tau_{r\theta} = -\frac{p}{2}\left(1 - \frac{a^2}{r^2}\right)\left(1 + 3\frac{a^2}{r^2}\right)\sin 2\theta.$$

(a) Check the stress boundary conditions to see whether they are satisfied.
(b) Find the location of the point where the normal stress σ_θ is the maximum.
(c) Find the maximum shearing stress in the entire plate.
(d) Obtain the maximum principal stress in the plate.

Note. You see that the maximum stress is increased around the hole. This is the phenomenon of *stress concentration*.

Answer. (a) The horizontal edges and the circular hole are stress free. On the hole, boundary conditions are

$$\sigma_{rr} = 0, \qquad \tau_{r\theta} = 0 \qquad \text{when } r = a.$$

(b) σ_θ reaches the maximum $3p$ when $\theta = \pi/2$.

(c) The maximum shear equals $3p/2$ and occurs at $r = a$, $\theta = \pi/2$, acting in a plane inclined at 45° from the z-axis.

(d) The maximum principal stress is $3p$.

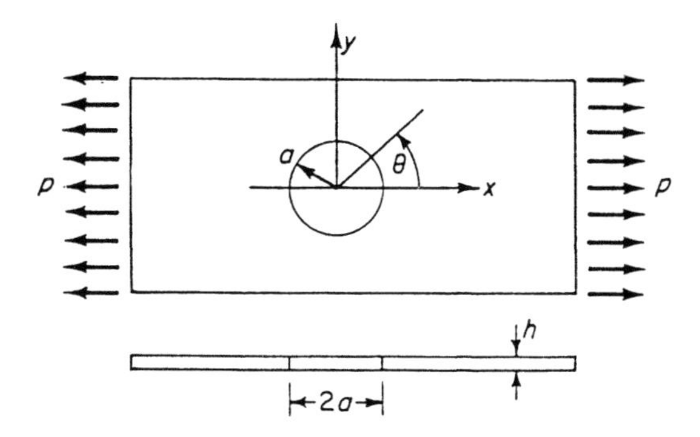

Figure P4.7 Circular hole in a thin plate.

4.8 Windows on the sides of airplanes are stress raisers. Suppose that you were given the job of designing the windows for the passengers to look out. To help you make design decisions, consider an idealized problem of placing an elliptical hole in an infinite plate of aluminium alloy under a tensile stress S, with the minor axis of the ellipse parallel to the tension, Fig. P4.8. This problem has been solved, and the result states that the tensile stress in the wall at the ends of the major axis of the hole is

$$\sigma = S\left(1 + 2\frac{a}{b}\right),$$

where $2a$ is the major axis and $2b$ is the minor axis of the ellipse. (See Fig. P4.8.) What do you learn from this result?

A crack in the wall of the airplane may be simulated as an elongated elliptical hole. Why is it very dangerous if the crack is perpendicular to the direction of tension?

Explain the benefit that can be derived by drilling holes at the ends of the crack. Would the holes help stop the spreading of the crack?

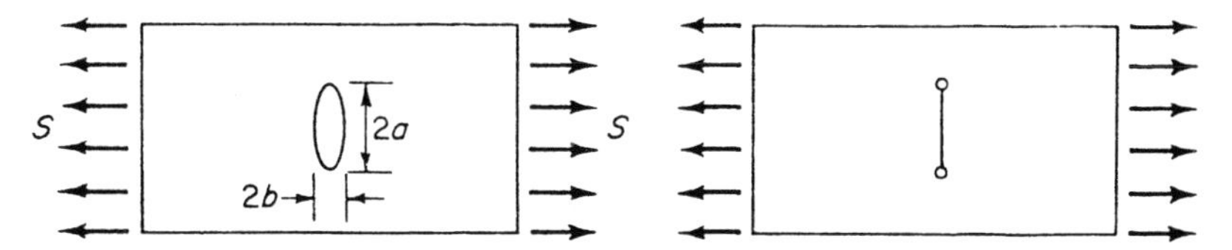

Figure P4.8 Elliptical hole in plate; stress-relieving drilling.

4.9 An earthquake is initiated at time $t = 0$ at an epicenter C. (See Fig. P4.9) Analyzing seismic waves around the world following an earthquake is not simple, but a formulation of the mathematical problem is not difficult. Leaving the equations of motion and

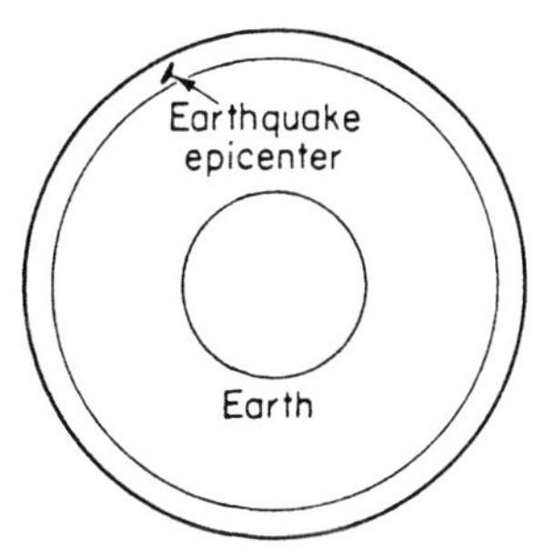

Figure P4.9 Earthquake.

continuity till after Chapter 10, formulate all the boundary conditions involved: at the surface of the earth, at the boundary of the mantle, at the boundary of the core, and at the epicenter. Earthquakes are usually caused by shear fracture at the epicenter, releasing residual strains and an associated strain energy.

4.10 Suppose you are an engineer responsible for an important project, and you must test the strength of the materials you are using. One of the most important tests is to determine the strength of the materials in uniaxial tension. Design a testing machine to do this. Design the shape and construction of the test specimen. Explain in detail the rationale of your design. Fig. P.4.10 shows a test specimen. Is it good? In this discussion, the stress distribution in the specimen must be considered, and the way the force is transmitted to the specimen is important.

Suppose you are designing this equipment for a biologist who is interested in the strength of soft tissues such as muscles, tendons, skin, and blood vessels. To make a test specimen in the shape of Fig. P.4.10 is then impractical. What would you do instead?

Figure P4.10 Tension test specimen.

4.11 Concrete, rocks, and bone are strong in compression and usually function in compressive mode. To test their strength in compression, some considerations quite different from those used in answering Problem 4.10 must be taken into account. Design specimens of these materials to be tested in compression.

4.12 For the impact of a hammer on a semi-infinite elastic body (Fig. P4.12), what boundary conditions apply?

Solution: Initial condition. The deformation equals zero everywhere. When the hammer strikes, the *boundary conditions* are:

(a) On the flat surface, but not under the hammer:

$$\overset{\nu}{T}_i = 0 \qquad (i = 1, 2, 3).$$

(b) For the conditions at infinity in the semiinfinite body, let u_i be the components of displacement caused by the deformation of the body, and let σ_{ij} be the stresses; then

$$(u_i) = 0, \qquad (\sigma_{ij}) = 0 \qquad (i, j = 1, 2, 3).$$

(c) On the surface under the hammer, the traction and the displacement normal to the interface must be consistent. Hence, if we denote the table and the hammer by (T) and (H), respectively, we must have, on the common interface,

$$\overset{\nu}{u}_i^{(T)} = \overset{\nu}{u}_i^{(H)}, \qquad \overset{\nu}{T}_i^{(T)} = \overset{\nu}{T}_i^{(H)}.$$

The normal ν is the normal to the interface, which may be described by the equation $z = f(x, y, t)$. Then $\nu_1 : \nu_2 : \nu_3 = \partial f/\partial x : \partial f/\partial y : -1$. However, we do not know the function $f(x, y, t)$, which can be determined rigorously only by solving the whole problem of stress distribution in the hammer and the table together.

In lieu of an exact solution, one may propose an approximate problem. For example, we may assume that when the hammer strikes the table, the component of the stress vector normal to the table is much larger than the tangential components. Hence, if we ignore the

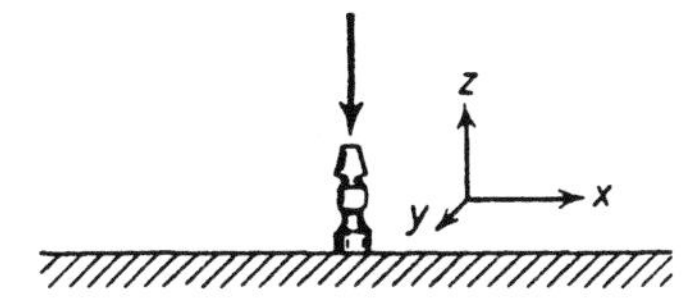

Figure P4.12 Hammer.

latter components (in a case that might be called a "lubricated" hammer), then the boundary condition under the hammer may be written as

$$\overset{z}{T}_x = 0, \qquad \overset{z}{T}_y = 0, \qquad \overset{z}{T}_z = F(x, y)\,\delta(t)$$

where $\delta(t)$ is the Dirac unit-impulse function, which is zero when t is finite, but tends to ∞ when $t \to 0$ in such a way that the integral of $\delta(t)$ for t from $-\epsilon$ to $+\epsilon$ is exactly equal to 1, ϵ being a positive number. $F(x, y)$ is unknown. A simplifying assumption could be $F(x, y) = \text{const}$.

Note. Perhaps there are uneven contacts between the hammer and the tabletop, local failure, slippage, etc. If one is serious about these possibilities, one must specify them precisely and then investigate their consequences.

4.13 Suppose that the semi-infinite body of the preceding problem is a large expanse of water and the load is a package dropped from an airplane (Fig. P4.13). The water will surely splash. What boundary conditions are known in this case?

Figure P4.13 Package drop.

4.14 A palm tree support its own weight. (See Fig. P4.14.)

(a) Assume that the treetop weighs 100 kg, the cross-sectional area at the top is 100 cm^2, the tree trunk has a mass density of 2 grams/cm^3, and the tree is 10 m tall. If the tree trunk is a uniform cylinder, what is the stress in the tree trunk at a distance x from the top due to the weight of the tree alone? Solve this problem by means of the equation of equilibrium $\tau_{ij,j} + X_i = 0$.

(b) The stress due to the weight of the tree can be reduced by increasing the cross sectional area of the trunk toward the base. Consider such a trunk of circular cross section with variable diameter. Compute the average stress by means of a free-body diagram.

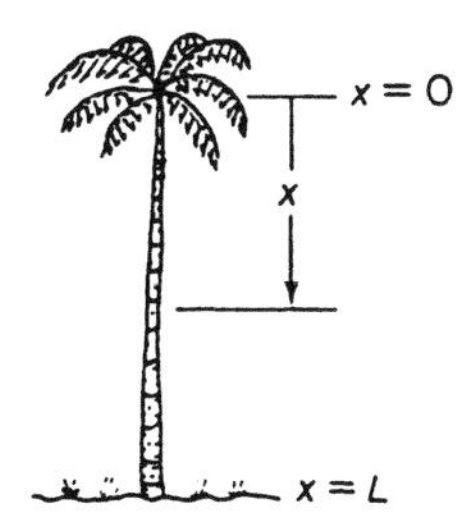

Figure P4.14 A palm tree supporting its own weight.

(c) Solve part (b) by means of the differential equations of equilibrium. What additional consideration is needed?

(d) How should the diameter of the tree trunk vary with x if we want the longitudinal stress due to the weight of the tree to be uniform throughout?

(e) In Prob. 1.19 (Chap. 1) we considered the wind load acting on a palm tree and determined a diameter D as a function of x that will make the bending stress σ_o computed from Eq. (1.11–31) uniform in the tree. Now, consider both the wind load and the weight. Let the strength of the tree trunk material be σ_1 in tension, and σ_2 in compression. When the maximum principal stress equals σ_1 the material breaks. When the compressive stress exceeds σ_2 the material is crushed. If the tree is "designed" to have a uniform maximum tensile stress (independent of x), what should $D(x)$ be? If the "design" is for uniform maximum compressive stress, what $D(x)$ should be?

4.15 What is the physical meaning of the sum $\tau_{\alpha\alpha}$?

Answer. It is the sum of the normal stresses in three orthogonal directions. If we consider a cube of water in static condition, so that all its sides are subjected to a pressure p, and there is no shear stress acting on the surface, then $\tau_{\alpha\alpha} = -3p$, or $p = -\tau_{\alpha\alpha}/3$.

If a uniform tension of equal intensity acts on each side of a cube, then $\tau_{\alpha\alpha}/3$ represents the tension.

If the three stresses τ_{xx}, τ_{yy}, τ_{zz} are not equal, then $\tau_{\alpha\alpha}/3$ represents the mean normal stress.

4.16 The stress at a point in a body has the following components with respect to a set of rectangular Cartesian coordinates x_1, x_2, x_3:

$$(\sigma_{ij}) = \begin{pmatrix} 1 & 0 & -1 \\ 0 & -1 & 0 \\ -1 & 0 & 1 \end{pmatrix}.$$

Find the values of the invariants I_1, I_2, I_3 and the principal stresses.

Answer. $I_1 = 1$, $I_2 = -2$, $I_3 = 0$. $(\sigma_1, \sigma_2, \sigma_3) = (0, 2, -1)$.

4.17 Let τ_{ij} be a stress tensor. Evaluate the products (a) $e_{ijk}\tau_{jk}$ and (b) $e_{ijk}e_{ist}\tau_{kt}$.

4.18 A plate is stretched in the x-direction, compressed in the y-direction, and free in the z-direction. There is a flaw in a plane that is parallel to the z-axis and inclined at 45° to the x-axis. If the shear stress acting on the flaw exceeds a critical stress τ_{cr}, the plate will fail. Determine the critical combinations of σ_x and σ_y at which the plate fails.

4.19 Consider a rod that has a cross-sectional area of 1 cm^2.

(a) Assume that the material has the following strength characteristics, beyond which the rod breaks: maximum shear stress, 400 kPa; maximum tensile stress, 1.0 MPa; maximum compressive stress, 10.0 MPa. Let a tension P act on the rod. At what value of P will the rod break? What is the expected angle of inclination of the broken section?

(b) Answer the same questions as in part (a) if the strength characteristics are maximum shear, 500 kPa; maximum tension, 0.9 MPa; and maximum compression, 10.0 MPa.

4.20 A circular cylindrical rod is stretched by an axial load, bent by a bending moment, and twisted by a torque, so that the stresses in a little element at a point on the surface of the cylinder are

$$\sigma_r = 0, \qquad \tau_{rz} = \tau_{r\theta} = 0, \qquad \sigma_z = 1 \text{ kn/m}^2, \qquad \tau_{z\theta} = 2\text{kn/m}^2, \qquad \sigma_\theta = 0.$$

What are the principal stresses at that point?

4.21 In the earth, there is hydrostatic pressure due to the earth's weight and there is a shear stress due to strain in the earth's crust. At a point in the earth, the hydrostatic pressure is 10 MPa, and the shear stress, evaluated with respect to a chosen frame of reference x_1, x_2, x_3, is $\tau_{12} = 5$ MPa, $\tau_{23} = \tau_{31} = 0$. Find the principal stresses and planes at that point.

Answer. $\sigma_1 = -5$ MPa acts on a plane with a normal vector $\nu_1 = -\nu_2 = \sqrt{2}/2$, $\nu_3 = 0$, which is inclined at 45° to the negative x_1-axis. The principal axis associated with the principal stress $\sigma_2 = -10$ MPa is the x_3-axis, and that associated with $\sigma_3 = -15$ MPa is a vector inclined at 45° to the positive x_1-axis.

4.22 A driver of a moving car that weighs 1,600 kg made a sudden panic stop by slamming on the brakes; this promptly locked the wheels. Assume a maximum coefficient of friction between the tire and the ground of $\frac{1}{4}$, and assume that each of the car wheels is attached to the hub by four bolts.

(a) Compute the shear force that must act in each bolt. The bolt has a diameter of 1 cm, and its axis is 6 cm away from the axis of the wheel, which is 36 cm above the ground.

(b) The allowable shear stress of the bolt material is 150 Mn/m^2. Are the shear stresses in the bolts within the allowable limit? (Assume that the bolts are initially stress free).

(c) The garage mechanic who put on the wheels for the car used a large wrench and tightened the nuts most vigorously, so that a tensile stress of 140 Mn/m^2 was imposed on the bolts. This tensile stress was the initial stress in the bolts. *Now* when the brakes are applied and a shear stress as computed in part (b) is induced, are the bolts still safe? To answer this question, compute the maximum shear stress in the bolts under the combined tension and shear; then compare it with the allowable shear stress.

Answer. (a) 1,470 n. (b) shear stress due to braking = 18.71 Mn/m^2. Shear stress due to car weight = 12.47 Mn/m^2. Shear in most severe configuration = 31.18 Mn/m^2. It is less than the allowable 150 Mn/m^2; hence, the bolts are safe. (c) By Mohr's circle construction, or by Eqs. (4.2–12), $\tau_{max} = 76.63$ Mn/m^2 and $\sigma_{max} = 146.63$ Mn/m^2. The bolts are clearly safe with respect to shear. But better check the handbook with regard to tensile stress to see whether σ_{max} is allowable.

4.23 The foundation of a deep-sea drilling platform is subjected to a hydrostatic pressure p, an additional stress q in the vertical direction due to the weight of the platform, and a shear stress τ due to an earthquake. Determine the three principal stresses and the maximum shear.

5 ANALYSIS OF DEFORMATION

Forces applied to solids cause deformation, and forces applied to liquids cause flow. Often, the major objective of an analysis is to find the deformation or flow. It is our objective in this chapter to analyze the deformation of solid bodies in such a way as to be relevant to the state of stress in those bodies.

5.1 DEFORMATION

If we pull a rubber band, it stretches. If we compress a cylinder, it shortens. If we bend a rod, it bends. If we twist a shaft, it twists. See Fig. 5.1. Tensile stress causes tensile strain. Shear stress causes shear strain. This is common sense. To express these phenomena quantitatively, it is necessary to define measures of strain.

Consider a string of an initial length L_0. If it is stretched to a length L, as shown in Fig. 5.1(a), it is natural to describe the change by dimensionless ratios such as L/L_0, $(L - L_0)/L_0$, and $(L - L_0)/L$. The use of dimensionless ratios eliminates the absolute length from consideration. It is commonly felt that these ratios, and not the lengths L_0 or L, are related to the stress in the string. This expectation can be verified in the laboratory. The ratio L/L_0 is called the *stretch ratio* and is denoted by the symbol λ. The ratios

$$\epsilon = \frac{L - L_0}{L_0}, \qquad \epsilon' = \frac{L - L_0}{L} \tag{5.1–1}$$

are strain measures. Either of them can be used, although numerically, they are different. For example, if $L = 2$, and $L_0 = 1$, we have $\lambda = 2$, $\epsilon = 1$, and $\epsilon' = \frac{1}{2}$. We shall have reasons (to be discussed later) also to introduce the measures

$$e = \frac{L^2 - L_0^2}{2L^2}, \qquad \varepsilon = \frac{L^2 - L_0^2}{2L_0^2}. \tag{5.1–2}$$

If $L = 2$ and $L_0 = 1$, we have $e = \frac{3}{8}$ and $\varepsilon = \frac{3}{2}$. But if $L = 1.01$ and $L_0 = 1.00$, then $e \doteq 0.01$, $\varepsilon \doteq 0.01$, $\epsilon \doteq 0.01$, and $\epsilon' \doteq 0.01$. Hence, in infinitesimal elongations,

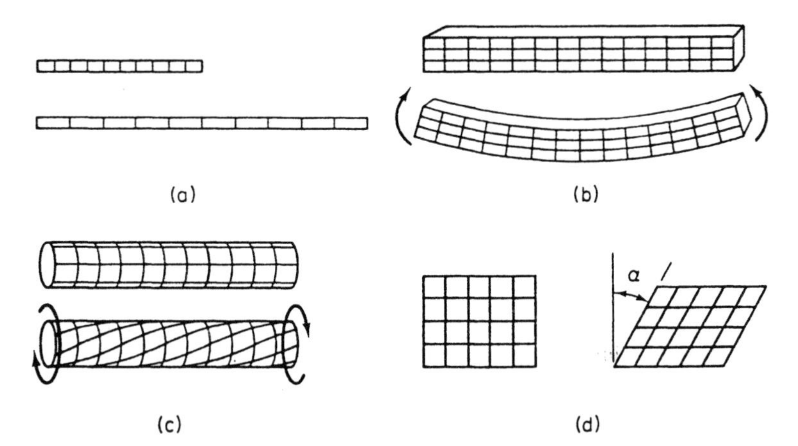

Figure 5.1 Patterns of deformation. (a) Stretching. (b) Bending. (c) Twisting. (d) Simple shear.

all of these strain measures are approximately equal. In finite elongations, however, they are different.

The preceding strain measures can be used to describe more complex deformations. For example, if we bend a rectangular beam by moments acting at the ends, as shown in Fig. 5.1(b), the beam will deflect into an arc. The "fibers" on top will be shortened, and those on the bottom will be elongated. These longitudinal strains are related to the bending moment acting on the beam.

To illustrate shear, consider a circular cylindrical shaft, as shown in Fig. 5.1(c). When the shaft is twisted, the elements in the shaft are distorted in a manner shown in Fig. 5.1(d). In this case, the angle α may be taken as a measure of strain. It is more customary, however, to take $\tan \alpha$ or $\frac{1}{2} \tan \alpha$ as the shear strain; the reasons for this will be elucidated later.

The selection of proper measures of strain is dictated basically by the stress-strain relationship (i.e., the constitutive equation of the material). For example, if we pull on a string, it elongates. The experimental results can be presented as a curve of the tensile stress σ plotted against the stretch ratio λ or strain e. An empirical formula relating σ to e can then be determined. The case of infinitesimal strain is simple because the different measures of strain just presented all coincide. It was found that, for most engineering materials subjected to an infinitesimal strain in uniaxial stretching, a relation like

$$\sigma = Ee \tag{5.1–3}$$

where E is a constant called *Young's modulus*, is valid within a certain range of stresses. Equation (5.1–3) is called *Hooke's law*. A material obeying it is said to be a *Hookean material*. Steel is a Hookean material if σ is less than a certain bound that is called a *yield stress* in tension.

Corresponding to Eq. (5.1–3), the relationship for a Hookean material subjected to an infinitesimal shear strain is

$$\tau = G \tan \alpha \tag{5.1–4}$$

where G is another constant called the *shear modulus* or *modulus of rigidity*. The range of validity of Eq. (5.1–4) is again bounded by a *yield stress*, this time in shear. The yield stresses in tension, in compression, and in shear are different in general.

Equations (5.1–3) and (5.1–4) are the simplest of the constitutive equations. The more general cases will be discussed in Chapters 7, 8, and 9.

Deformations of most things in nature and in engineering are much more complex than those just discussed. We therefore need a general method of treatment. First, however, let us consider the mathematical description of deformation.

Let a body occupy a space S. Referred to a rectangular Cartesian frame of reference, every particle in the body has a set of coordinates. When the body is deformed, every particle takes up a new position, which is described by a new set of coordinates. For example, a particle P, located originally at a place with coordinates (a_1, a_2, a_3), is moved to the place Q with coordinates (x_1, x_2, x_3) when the body moves and deforms. Then the vector $\overrightarrow{PQ}$ is called the *displacement vector* of the particle. (See Fig. 5.2.) The components of the displacement vector are, clearly,

$$x_1 - a_1, \qquad x_2 - a_2, \qquad x_3 - a_3.$$

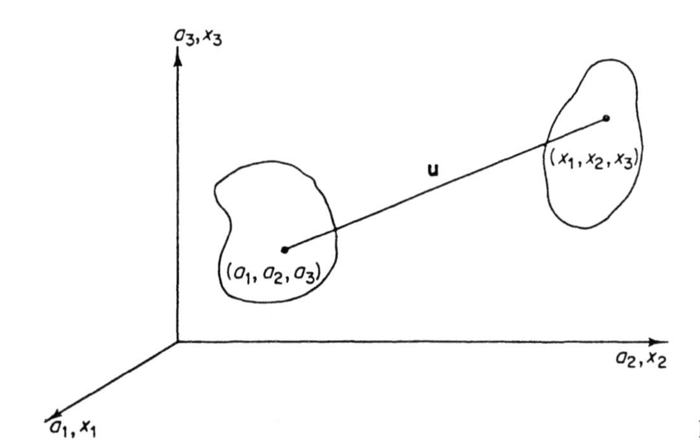

Figure 5.2 Displacement vector.

If the displacement is known for every particle in the body, we can construct the deformed body from the original. Hence, a deformation can be described by a displacement field. Let the variable (a_1, a_2, a_3) refer to any particle in the original configuration of the body, and let (x_1, x_2, x_3) be the coordinates of that particle when the body is deformed. Then the deformation of the body is known if x_1, x_2, x_3 are known functions of a_1, a_2, a_3:

$$x_i = x_i(a_1, a_2, a_3). \tag{5.1–5}$$

This is a transformation (mapping) from a_1, a_2, a_3 to x_1, x_2, x_3. In continuum mechanics, we assume that deformation is continuous. Thus, a neighborhood is transformed into a neighborhood. We also assume that the transformation is one to one; i.e., the functions in Eq. (5.1–5) are single valued, continuous, and have the unique inverse

$$a_i = a_i(x_1, x_2, x_3) \tag{5.1–6}$$

for every point in the body.

The displacement vector $\mathbf{u}$ is then defined by its components

$$u_i = x_i - a_i. \tag{5.1–7}$$

If a displacement vector is associated with every particle in the original position, we may write

$$u_i(a_1, a_2, a_3) = x_i(a_1, a_2, a_3) - a_i. \tag{5.1–8}$$

If that displacement is associated with the particle in the deformed position, we write

$$u_i(x_1, x_2, x_3) = x_i - a_i(x_1, x_2, x_3). \tag{5.1–9}$$

PROBLEM

In order that the transformation (5.1–5) be single valued, continuous, and differentiable, what conditions must be satisfied by the functions $x_i(a_1, a_2, a_3)$?

Note: If the transformation is single valued, continuous, and differentiable, then the functions $x_i(a_1, a_2, a_3)$ must be single valued, continuous, and differentiable, and the Jacobian $|\partial x_i/\partial a_j|$ must not vanish in the space occupied by the body. The last statement is nontrivial. (See Sec. 2.5.)

5.2 THE STRAIN

The idea that the stress in a body is related to the strain was first announced by Robert Hooke (1635–1703) in 1676 in the form of an anagram, *ceiiinosssttuv*. He explained it in 1678 as

Ut tensio sic vis,

or "The power of any springy body is in the same proportion with the extension." The meaning of this statement is clear to anyone who ever handled a spring or pulled a rubber band.

A rigid-body motion induces no stress. Thus, the displacements themselves are not directly related to the stress. To relate deformation with stress, we must consider the stretching and distortion of the body. For this purpose, let us consider three neighboring points P, P', P'' in the body. (See Fig. 5.3.) If they are transformed to the points Q, Q', Q'' in the deformed configuration, the change in area

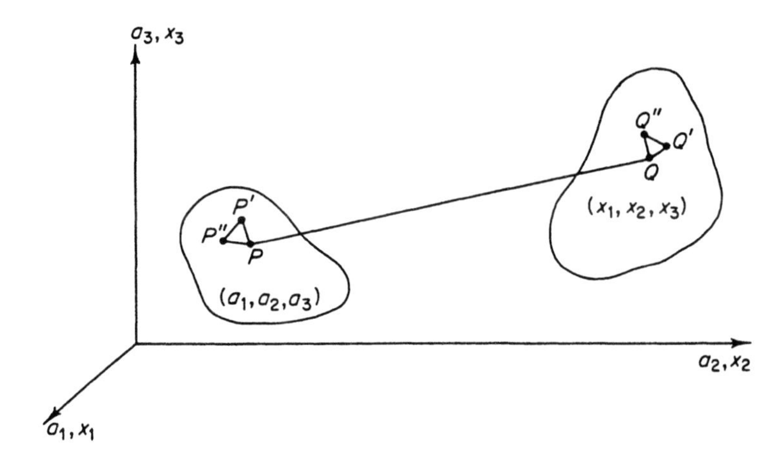

Figure 5.3 Deformation of a body.

and angles of the triangle is completely determined if we know the change in length of the sides. But the "location" of the triangle is undetermined by the change of the sides. Similarly, if the change in length between any two arbitrary points of the body is known, the new configuration of the body will be completely defined, except for the location of the body in space. The description of the change in distance between any two points of the body is the key to the analysis of deformation.

Consider an infinitesimal line element connecting the point $P(a_1, a_2, a_3)$ to a neighboring point $P'(a_1 + da_1, a_2 + da_2, a_3 + da_3)$. The square of the length ds_0 of PP' in the original configuration is given by

$$ds_0^2 = da_1^2 + da_2^2 + da_3^2. \tag{5.2–1}$$

When P and P' are deformed to the points $Q(x_1, x_2, x_3)$ and $Q'(x_1 + dx_1, x_2 + dx_2, x_3 + dx_3)$, respectively, the square of the length ds of the new element QQ' is

$$ds^2 = dx_1^2 + dx_2^2 + dx_3^2. \tag{5.2–2}$$

By Eqs. (5.1–5) and (5.1–6), we have

$$dx_i = \frac{\partial x_i}{\partial a_j} da_j, \qquad da_i = \frac{\partial a_i}{\partial x_j} dx_j. \tag{5.2–3}$$

Hence, on introducing the Kronecker delta, we may write

$$ds_0^2 = \delta_{ij}\, da_i\, da_j = \delta_{ij} \frac{\partial a_i}{\partial x_l} \frac{\partial a_j}{\partial x_m} dx_l\, dx_m, \tag{5.2–4}$$

$$ds^2 = \delta_{ij}\, dx_i\, dx_j = \delta_{ij} \frac{\partial x_i}{\partial a_l} \frac{\partial x_j}{\partial a_m} da_l\, da_m. \tag{5.2–5}$$

The difference between the squares of the length elements may be written, after several changes in the symbols for dummy indices, either as

$$ds^2 - ds_0^2 = \left(\delta_{\alpha\beta} \frac{\partial x_\alpha}{\partial a_i} \frac{\partial x_\beta}{\partial a_j} - \delta_{ij}\right) da_i\, da_j, \tag{5.2–6}$$

or as

$$ds^2 - ds_0^2 = \left(\delta_{ij} - \delta_{\alpha\beta} \frac{\partial a_\alpha}{\partial x_i} \frac{\partial a_\beta}{\partial x_j}\right) dx_i\, dx_j. \tag{5.2–7}$$

We define the *strain tensors*

$$E_{ij} = \frac{1}{2}\left(\delta_{\alpha\beta} \frac{\partial x_\alpha}{\partial a_i} \frac{\partial x_\beta}{\partial a_j} - \delta_{ij}\right), \tag{5.2–8}$$

$$e_{ij} = \frac{1}{2}\left(\delta_{ij} - \delta_{\alpha\beta} \frac{\partial a_\alpha}{\partial x_l} \frac{\partial a_\beta}{\partial x_j}\right), \tag{5.2–9}$$

so that

$$ds^2 - ds_0^2 = 2E_{ij}\, da_i\, da_j, \tag{5.2–10}$$

$$ds^2 - ds_0^2 = 2e_{ij}\, dx_i\, dx_j. \tag{5.2–11}$$

The strain tensor E_{ij} was introduced by Green and St.-Venant and is called Green's strain tensor. The strain tensor e_{ij} was introduced by Cauchy for infinitesimal strains and by Almansi and Hamel for finite strains and is known as Almansi's strain tensor. In analogy with terminology in hydrodynamics, E_{ij} is often referred to as *Lagrangian* and e_{ij} as *Eulerian.*

That E_{ij} and e_{ij} thus defined are tensors in the coordinate systems a_i and x_i, respectively, follows from the quotient rule when it is applied to Eqs. (5.2–10) and (5.2–11). The tensors E_{ij} and e_{ij} are obviously *symmetric*; i.e.,

$$E_{ij} = E_{ji}, \qquad e_{ij} = e_{ji}. \tag{5.2–12}$$

An immediate consequence of Eqs. (5.2–10) and (5.2–11) is that $ds^2 - ds_0^2 = 0$ implies $E_{ij} = e_{ij} = 0$ and vice versa. But a deformation in which the length of every line element remains unchanged is a rigid-body motion. Hence, *the necessary and sufficient condition that a deformation of a body be a rigid-body motion is that all components of the strain tensor E_{ij} or e_{ij} be zero throughout the body.*

5.3 STRAIN COMPONENTS IN TERMS OF DISPLACEMENTS

If we introduce the *displacement vector* $\mathbf{u}$ with components

$$u_\alpha = x_\alpha - a_\alpha, \qquad (\alpha = 1, 2, 3), \tag{5.3–1}$$

then

$$\frac{\partial x_\alpha}{\partial a_i} = \frac{\partial u_\alpha}{\partial a_i} + \delta_{\alpha i}, \qquad \frac{\partial a_\alpha}{\partial x_i} = \delta_{\alpha i} - \frac{\partial u_\alpha}{\partial x_i}, \tag{5.3–2}$$

and the strain tensors reduce to the simple form

$$\begin{aligned} E_{ij} &= \frac{1}{2}\left[\delta_{\alpha\beta}\left(\frac{\partial u_\alpha}{\partial a_i} + \delta_{\alpha i}\right)\left(\frac{\partial u_\beta}{\partial a_j} + \delta_{\beta j}\right) - \delta_{ij}\right] \\ &= \frac{1}{2}\left[\frac{\partial u_j}{\partial a_i} + \frac{\partial u_i}{\partial a_j} + \frac{\partial u_\alpha}{\partial a_i}\frac{\partial u_\alpha}{\partial a_j}\right] \end{aligned} \tag{5.3–3}$$

and

$$\begin{aligned} e_{ij} &= \frac{1}{2}\left[\delta_{ij} - \delta_{\alpha\beta}\left(-\frac{\partial u_\alpha}{\partial x_i} + \delta_{\alpha i}\right)\left(-\frac{\partial u_\beta}{\partial x_j} + \delta_{\beta j}\right)\right] \\ &= \frac{1}{2}\left[\frac{\partial u_j}{\partial x_i} + \frac{\partial u_i}{\partial x_j} - \frac{\partial u_\alpha}{\partial x_i}\frac{\partial u_\alpha}{\partial x_j}\right]. \end{aligned} \tag{5.3–4}$$

In unabridged notations (x, y, z for x_1, x_2, x_3; a, b, c for a_1, a_2, a_3; and u, v, w for u_1, u_2, u_3), we have the typical terms

$$\begin{aligned} E_{aa} &= \frac{\partial u}{\partial a} + \frac{1}{2}\left[\left(\frac{\partial u}{\partial a}\right)^2 + \left(\frac{\partial v}{\partial a}\right)^2 + \left(\frac{\partial w}{\partial a}\right)^2\right], \\ e_{xx} &= \frac{\partial u}{\partial x} - \frac{1}{2}\left[\left(\frac{\partial u}{\partial x}\right)^2 + \left(\frac{\partial v}{\partial x}\right)^2 + \left(\frac{\partial w}{\partial x}\right)^2\right], \\ E_{ab} &= \frac{1}{2}\left[\frac{\partial u}{\partial b} + \frac{\partial v}{\partial a} + \left(\frac{\partial u}{\partial a}\frac{\partial u}{\partial b} + \frac{\partial v}{\partial a}\frac{\partial v}{\partial b} + \frac{\partial w}{\partial a}\frac{\partial w}{\partial b}\right)\right], \\ e_{xy} &= \frac{1}{2}\left[\frac{\partial u}{\partial y} + \frac{\partial v}{\partial x} - \left(\frac{\partial u}{\partial x}\frac{\partial u}{\partial y} + \frac{\partial v}{\partial x}\frac{\partial v}{\partial y} + \frac{\partial w}{\partial x}\frac{\partial w}{\partial y}\right)\right]. \end{aligned} \tag{5.3–5}$$

Note that u, v, w are considered functions of a, b, c, the position of points in the body in unstrained configuration, when the Lagrangian strain tensor is evaluated; whereas they are considered functions of x, y, z, the position of points in the strained configuration, when the Eulerian strain tensor is evaluated.

If the components of displacement u_i are such that their first derivatives are so small that the squares and products of the partial derivatives of u_i are negligible, then e_{ij} reduces to Cauchy's *infinitesimal strain tensor*,

$$e_{ij} = \frac{1}{2}\left[\frac{\partial u_j}{\partial x_i} + \frac{\partial u_i}{\partial x_j}\right]. \tag{5.3–6}$$

In unabridged notation,

$$\begin{aligned} e_{xx} &= \frac{\partial u}{\partial x}, & e_{xy} &= \frac{1}{2}\left(\frac{\partial u}{\partial y} + \frac{\partial v}{\partial x}\right) = e_{yx}, \\ e_{yy} &= \frac{\partial v}{\partial y}, & e_{xz} &= \frac{1}{2}\left(\frac{\partial u}{\partial z} + \frac{\partial w}{\partial x}\right) = e_{zx}, \\ e_{zz} &= \frac{\partial w}{\partial z}, & e_{yz} &= \frac{1}{2}\left(\frac{\partial v}{\partial z} + \frac{\partial w}{\partial y}\right) = e_{zy}. \end{aligned} \tag{5.3–7}$$

In the case of infinitesimal displacement, the distinction between the Lagrangian and Eulerian strain tensor disappears, since then it is immaterial whether the derivatives of the displacements are calculated at the position of a point before or after deformation.

Warning: Notation for Shear Strain

In most books and papers, the strain components are defined as

$$\begin{aligned} e_x &= \frac{\partial u}{\partial x}, & \gamma_{xy} &= 2e_{xy} = \frac{\partial u}{\partial y} + \frac{\partial v}{\partial x}, \\ e_y &= \frac{\partial v}{\partial y}, & \gamma_{yz} &= 2e_{yz} = \frac{\partial v}{\partial z} + \frac{\partial w}{\partial y}, \\ e_z &= \frac{\partial w}{\partial z}, & \gamma_{zx} &= 2e_{zx} = \frac{\partial u}{\partial z} + \frac{\partial w}{\partial x}. \end{aligned}$$

In other words, the shear strains, denoted by γ_{xy}, γ_{yz}, γ_{zx}, are twice as large as the components e_{xy}, e_{yz}, e_{zx}, respectively. We shall not use this notation, because the components e_x, γ_{xy}, etc., together do not form a tensor, and a great deal of mathematical convenience is lost. But beware of this difference when you read other books and papers!

5.4 GEOMETRIC INTERPRETATION OF INFINITESIMAL STRAIN COMPONENTS

Let x, y, z be a set of rectangular Cartesian coordinates. Consider a line element of length dx parallel to the x-axis ($dy = dz = 0$). The change of the square of the length of this element due to deformation is

$$ds^2 - ds_0^2 = 2e_{xx}(dx)^2.$$

Hence,

$$ds - ds_0 = \frac{2e_{xx}(dx)^2}{ds + ds_0}.$$

But $ds = dx$ in this case, and ds_0 differs from ds only by a small quantity of the second order if we assume that the displacements u, v, w, and the strain components e_{ij} are infinitesimal. Hence,

$$\frac{ds - ds_0}{ds} = e_{xx}, \tag{5.4-1}$$

and it is seen that e_{xx} represents the *extension*, or change of length per unit length, of a vector parallel to the x-axis. An application of this discussion to a volume element is illustrated in Fig. 5.4, Case 1.

To see the meaning of the component e_{xy}, let us consider a small rectangle in the body with edges dx, dy. It is evident from Fig. 5.4, Cases 2, 3, and 4, that

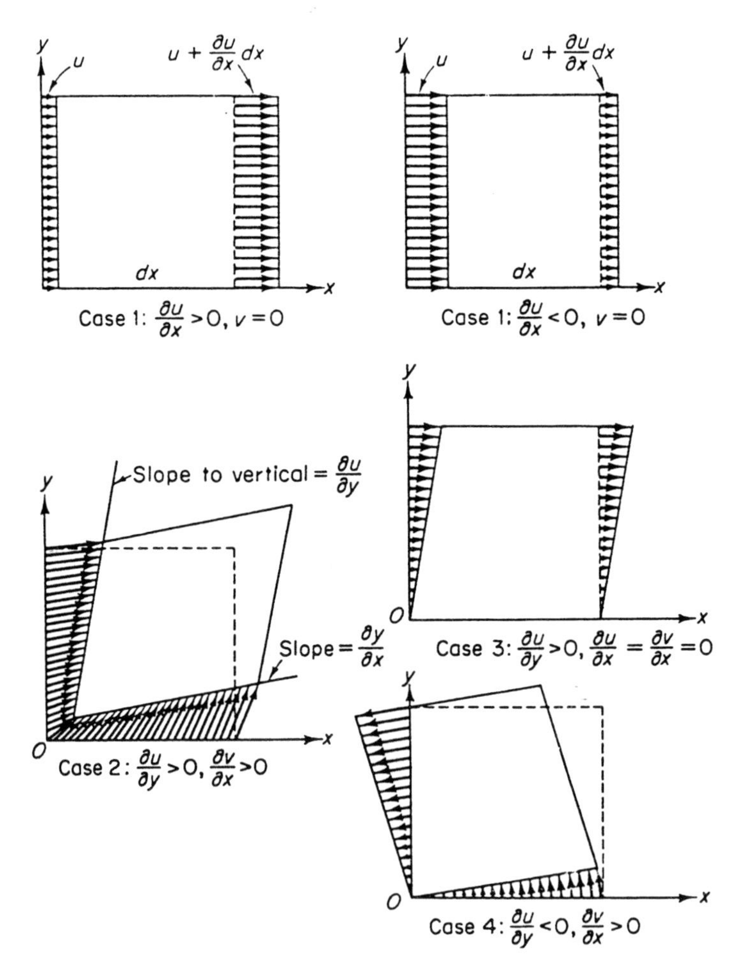

Figure 5.4 Deformation gradients and interpretation of infinitesimal strain components.

the sum $\partial u/\partial y + \partial v/\partial x$ represents the change in the angle xOy, which was originally a right angle. Thus,

$$e_{xy} = \frac{1}{2}\left[\frac{\partial u}{\partial y} + \frac{\partial v}{\partial x}\right] = \frac{1}{2}\tan\text{ (change of angle } xOy). \tag{5.4–2}$$

In engineering usage, the strain components e_{ij} $(i \neq j)$ doubled, i.e., $2e_{ij}$, are called the *shearing strains* or *detrusions*. The name is particularly suggestive in Case 3 of Fig. 5.4, which is called the case of *simple shear*.

5.5 INFINITESIMAL ROTATION

Consider an infinitesimal displacement field $u_i(x_1, x_2, x_3)$. From u_i, form the Cartesian tensor

$$\omega_{ij} = \frac{1}{2}\left(\frac{\partial u_j}{\partial x_i} - \frac{\partial u_i}{\partial x_j}\right), \tag{5.5–1}$$

which is antisymmetric; i.e.,

$$\omega_{ij} = -\omega_{ji}. \tag{5.5–2}$$

Hence, the tensor ω_{ij} has only three independent components—ω_{12}, ω_{23}, and ω_{31}—because ω_{11}, ω_{22}, ω_{33} are zero. From such an antisymmetric tensor, we can always build a *dual vector*

$$\omega_k = \tfrac{1}{2}\epsilon_{kij}\omega_{ij}, \tag{5.5–3}$$

where ϵ_{kij} is the permutation symbol (Sec. 2.3). On the other hand, from Eq. (5.5–3) and the ϵ-δ identity, Eq. (2.3–19), it follows that $\epsilon_{ijk}\omega_k = \frac{1}{2}(\omega_{ij} - \omega_{ji})$, which, by Eq. (5.5–2), is ω_{ij}. Hence,

$$\omega_{ij} = \epsilon_{ijk}\omega_k. \tag{5.5–4}$$

Thus, ω_{ij} may be called the *dual* (antisymmetric) *tensor* of a vector ω_k. We shall call ω_k and ω_{ij}, respectively, the *rotation vector* and *rotation tensor* of the displacement field u_i.

A slight modification of the proof given at the end of Sec. 5.2 will convince us that *the vanishing of the symmetric strain tensor E_{ij} or e_{ij} is a necessary and sufficient condition for a neighborhood of a particle to be moved like a rigid body.* A rigid-body motion consists of a translation and a rotation. The translation is u_i. What is the rotation? We shall show that *in an infinitesimal displacement field for which the strain tensor vanishes at a point P, the rotation of a neighborhood of P is given by the vector ω_i.* To show this, consider a point P' in the neighborhood of P. Let the coordinates of P and P' be x_i and $x_i + dx_i$, respectively. The relative displacement of P' with respect to P is

$$du_i = \frac{\partial u_i}{\partial x_j}dx_j. \tag{5.5–5}$$

This can be written as

$$du_i = \frac{1}{2}\left(\frac{\partial u_i}{\partial x_j} + \frac{\partial u_j}{\partial x_i}\right) dx_j + \frac{1}{2}\left(\frac{\partial u_i}{\partial x_j} - \frac{\partial u_j}{\partial x_i}\right) dx_j.$$

The first quantity in parentheses is the infinitesimal strain tensor, which is zero by assumption. The second quantity in parenthesis may be identified with Eq. (5.5–1). Hence,

$$\begin{aligned} du_i &= -\omega_{ij} dx_j = \omega_{ji}\, dx_j \\ &= -\epsilon_{ijk}\omega_k\, dx_j \qquad \text{[by Eq. (5.5–4)]} \qquad (5.5\text{–}6) \\ &= (\boldsymbol{\omega} \times \mathbf{dx})_i \qquad \text{(by definition).} \end{aligned}$$

Thus, the relative displacement is the vector product of $\boldsymbol{\omega}$ and **dx**. This is exactly what would have been produced by an infinitesimal rotation $|\boldsymbol{\omega}|$ about an axis through P in the direction of $\boldsymbol{\omega}$.

It should be noted that we have restricted ourselves to infinitesimal angular displacements. Angular measures for finite displacements are related to ω_{ij} in a more complicated way.

5.6 FINITE STRAIN COMPONENTS

When the strain components are not small, it is also easy to give simple geometric interpretations for the components of the strain tensors.

Consider a set of rectangular Cartesian coordinates with respect to which the strain components are defined as in Sec. 5.2. Let a line element before deformation be **da**, with components $da_1 = ds_0$, $da_2 = 0$, $da_3 = 0$. Let the extension E_1 of this element be defined by

$$E_1 = \frac{ds - ds_0}{ds_0} \qquad (5.6\text{–}1)$$

or

$$ds = (1 + E_1)\, ds_0. \qquad (5.6\text{–}2)$$

From Eq. (5.2–10), we have

$$ds^2 - ds_0^2 = 2E_{ij}\, da_i\, da_j = 2E_{11}(da_1)^2. \qquad (5.6\text{–}3)$$

Combining Eqs. (5.6–2) and (5.6–3), we obtain

$$(1 + E_1)^2 - 1 = 2E_{11} \qquad (5.6\text{–}4)$$

which gives the meaning of E_{11} in terms of E_1. Conversely,

$$E_1 = \sqrt{1 + 2E_{11}} - 1. \qquad (5.6\text{–}5)$$

This reduces to

$$E_1 \doteq E_{11} \tag{5.6–6}$$

when E_{11} is small compared to 1.

To get the physical significance of the component E_{12}, let us consider two line elements $\mathbf{ds}_0$ and $\mathbf{d\bar{s}}_0$ that are at a right angle in the original state:

$$\begin{aligned} \mathbf{ds}_0&: \quad da_1 = ds_0, \quad da_2 = 0, \quad da_3 = 0; \\ \mathbf{d\bar{s}}_0&: \quad da_1 = 0, \quad da_2 = d\bar{s}_0, \quad da_3 = 0. \end{aligned} \tag{5.6–7}$$

After deformation, these line elements become $\mathbf{ds}$ (with components dx_i) and $\mathbf{d\bar{s}}$ (with components dx_i). Forming the scalar product of the deformed elements, we obtain

$$\begin{aligned} ds\, d\bar{s} \cos\theta &= dx_k\, d\bar{x}_k = \frac{\partial x_k}{\partial a_i} da_i \frac{\partial x_k}{\partial a_j} d\bar{a}_j \\ &= \frac{\partial x_k}{\partial a_1}\frac{\partial x_k}{\partial a_2} ds_0\, d\bar{s}_0. \end{aligned}$$

But according to the definition given in Eq. (5.2–8), we have, since $\delta_{12} = 0$,

$$E_{12} = \frac{1}{2}\frac{\partial x_k}{\partial a_1}\frac{\partial x_k}{\partial a_2}.$$

Hence,

$$ds\, d\bar{s} \cos\theta = 2E_{12}\, ds_0\, d\bar{s}_0. \tag{5.6–8}$$

But, from Eqs. (5.6–1) and (5.6–5), we have

$$ds = \sqrt{1 + 2E_{11}}\, ds_0, \qquad d\bar{s} = \sqrt{1 + 2E_{22}}\, d\bar{s}_0.$$

Hence, Eq. (5.6–8) yields

$$\cos\theta = \frac{2E_{12}}{\sqrt{1 + 2E_{11}}\sqrt{1 + 2E_{22}}}. \tag{5.6–9}$$

The angle θ is the angle between the line elements $\mathbf{ds}$ and $\mathbf{d\bar{s}}$ after deformation. The change of angle between the two line elements, which in the original state are orthogonal, is $\alpha_{12} = \pi/2 - \theta$. From Eq. (5.6–9), we therefore obtain

$$\sin\alpha_{12} = \frac{2E_{12}}{\sqrt{1 + 2E_{11}}\sqrt{1 + 2E_{22}}}. \tag{5.6–10}$$

These equations exhibit the relationship of E_{12} to the angles θ and α_{12}. The interpretation is not as simple as in the infinitesimal case because of the involvement of E_{11} and E_{22} in these equations.

A completely analogous interpretation can be made for the Eulerian strain components. Defining the extension e_1 per unit *deformed* length as

$$e_1 = \frac{ds - ds_0}{ds}, \tag{5.6-11}$$

we find that

$$e_1 = 1 - \sqrt{1 - 2e_{11}}. \tag{5.6-12}$$

Furthermore, if the deviation from a right angle between two elements in the original state which, after deformation, become orthogonal is denoted by β_{12}, we have

$$\sin \beta_{12} = \frac{2e_{12}}{\sqrt{1 - 2e_{11}}\sqrt{1 - 2e_{22}}}. \tag{5.6-13}$$

In case of infinitesimal strains, Eqs. (5.6–10) and (5.6–13) reduce to the familiar results

$$e_1 \doteq e_{11}, \qquad E_1 \doteq E_{11}, \qquad \alpha_{12} \doteq 2E_{12}, \qquad \beta_{12} \doteq 2e_{12}. \tag{5.6-14}$$

5.7 PRINCIPAL STRAINS: MOHR'S CIRCLE

Without much ado, we can extend the results of Secs. 4.1 through 4.8 to the strain, because these properties are derived from the simple fact that the tensor concerned is symmetric. All we have to do is to exchange the word *stress* with *strain*. Thus:

(a) There exist three principal strains e_1, e_2, e_3 that are the roots of the determinantal equation

$$|e_{ij} - e\,\delta_{ij}| = 0. \tag{5.7-1}$$

The roots of this cubic equation are all real numbers.

(b) Associated with each principal strain, say, e_1, there is a principal axis, with direction cosines $\nu_1^{(1)}$, $\nu_2^{(1)}$, $\nu_3^{(1)}$ that are the solutions of the equations

$$(e_{ij} - e_1\,\delta_{ij})\nu_j^{(1)} = 0, \qquad (i = 1, 2, 3). \tag{5.7-2}$$

The three sets of solutions $(\nu_1^{(1)}, \nu_2^{(1)}, \nu_3^{(1)})$, $(\nu_1^{(2)}, \nu_2^{(2)}, \nu_3^{(2)})$, $(\nu_1^{(3)}, \nu_2^{(3)}, \nu_3^{(3)})$ are components of three unit vectors. If the roots e_1, e_2, e_3 of Eq. (5.7–1) are distinct $(e_1 \neq e_2 \neq e_3)$, then the three principal axes are orthogonal to one another. If two of the principal strains are the same, then Eq. (5.7–2) has infinitely many solutions, out of which an infinite number of pairs of orthogonal vectors can be selected and regarded as the principal axes. If all three roots are the same, then any set of three mutually orthogonal unit vectors may be regarded as principal.

(c) A plane perpendicular to a principal axis is called a *principal plane*.

(d) If the coordinate axes x_1, x_2, x_3 coincide with the principal axes, then the strain tensor assumes the canonical form

$$\begin{pmatrix} e_1 & 0 & 0 \\ 0 & e_2 & 0 \\ 0 & 0 & e_3 \end{pmatrix}.$$

(e) We can define a strain deviation tensor $e'_{ij} = e_{ij} - \frac{1}{3} e_{\alpha\alpha} \delta_{ij}$. Tensors e_{ij} and e'_{ij} have the following independent strain invariants:

$$\begin{aligned} I_1 &= e_{ij}\delta_{ij}. & J_1 &= e'_{ij}\delta_{ij} = 0, \\ I_2 &= \tfrac{1}{2} e_{ik} e_{ik}, & J_2 &= \tfrac{1}{2} e'_{ik} e'_{ik}, \\ I_3 &= \tfrac{1}{3} e_{ik} e_{km} e_{mi}. & J_3 &= \tfrac{1}{3} e'_{ik} e'_{km} e'_{mi}. \end{aligned} \qquad (5.7\text{–}3)$$

(f) Mohr's circle may be used for the graphical analysis of strain. Lamé's ellipsoid is also applicable to strain.

5.8 INFINITESIMAL STRAIN COMPONENTS IN POLAR COORDINATES

As we indicated in Sec. 3.6, it is often desirable to introduce curvilinear coordinates for reference. The strain components can be referred to a local rectangular frame of reference oriented in the direction of the curvilinear coordinates. For example, in polar coordinates r, θ, z, the strain components may be designated e_{rr}, $e_{\theta\theta}$, e_{zz}, $e_{r\theta}$, e_{rz}, $e_{z\theta}$, and they are related to e_{xx}, e_{yy}, e_{zz}, e_{xy}, e_{yz}, e_{zx} by the tensor transformation law, as in the cases of stresses. (See Sec. 3.6.)

However, if displacement vectors are resolved into components in the directions of the curvilinear coordinates, the strain-displacement relationship involves derivatives of the displacement components and therefore is influenced by the curvature of the coordinate system. The strain-displacement relations may appear quite different from the corresponding formulas in rectangular coordinates.

A truly general method for handling curvilinear coordinates is that of general tensor analysis. The reader is referred to more advanced treatises. An introduction is given in the author's *Foundations of Solid Mechanics*, (Y. C. Fung, 1965, Prentice Hall, Englewood Cliffs, N.J.). Limiting ourselves in the present book to Cartesian tensors, we must treat each set of curvilinear coordinates in an ad hoc manner.

We shall illustrate two ad hoc approaches in the case of cylindrical polar coordinates: by transformation of coordinates and by detailed enumeration. The former will be discussed in this section, the latter in Sec. 5.9.

In the first approach, we start from the relations between the polar coordinates r, θ, z and the rectangular coordinates x, y, z:

$$\begin{cases} x = r\cos\theta, & \theta = \tan^{-1}\dfrac{y}{x}, \\ y = r\sin\theta, & r^2 = x^2 + y^2, \end{cases} \qquad z = z. \tag{5.8–1}$$

$$\frac{\partial r}{\partial x} = \frac{x}{r} = \cos\theta, \qquad \frac{\partial r}{\partial y} = \frac{y}{r} = \sin\theta, \tag{5.8–2}$$

$$\frac{\partial\theta}{\partial x} = -\frac{y}{r^2} = -\frac{\sin\theta}{r}, \qquad \frac{\partial\theta}{\partial y} = \frac{x}{r^2} = \frac{\cos\theta}{r}. \tag{5.8–3}$$

It follows that any derivative with respect to x and y in the Cartesian equations may be transformed into derivatives with respect to r and θ by

$$\begin{aligned} \frac{\partial}{\partial x} &= \frac{\partial r}{\partial x}\frac{\partial}{\partial r} + \frac{\partial\theta}{\partial x}\frac{\partial}{\partial\theta} = \cos\theta\frac{\partial}{\partial r} - \frac{\sin\theta}{r}\frac{\partial}{\partial\theta}, \\ \frac{\partial}{\partial y} &= \frac{\partial r}{\partial y}\frac{\partial}{\partial r} + \frac{\partial\theta}{\partial y}\frac{\partial}{\partial\theta} = \sin\theta\frac{\partial}{\partial r} + \frac{\cos\theta}{r}\frac{\partial}{\partial\theta}. \end{aligned} \tag{5.8–4}$$

Now, in polar coordinates, we denote the components of the displacement vector $\mathbf{u}$ by u_r, u_θ, u_z, as shown in Fig. 5.5. The components of the same vector resolved

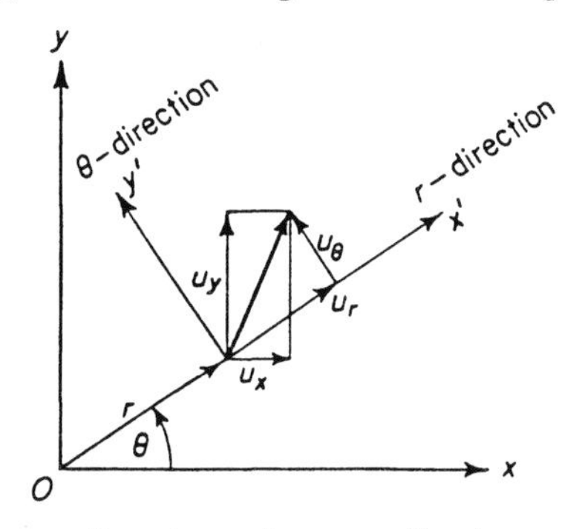

Figure 5.5 Displacement vector in polar coordinates.

in the directions of rectangular coordinates are u_x, u_y, u_z. From the figure, it is seen that these displacements are related by the equations

$$\begin{aligned} u_x &= u_r\cos\theta - u_\theta\sin\theta, \\ u_y &= u_r\sin\theta + u_\theta\cos\theta, \\ u_z &= u_z. \end{aligned} \tag{5.8–5}$$

The strain components in polar coordinates are designated as

$$\begin{pmatrix} e_{rr} & e_{r\theta} & e_{rz} \\ e_{\theta r} & e_{\theta\theta} & e_{\theta z} \\ e_{zr} & e_{z\theta} & e_{zz} \end{pmatrix} \tag{5.8–6}$$

These are really the strain components referred to a local frame of rectangular coordinates $x'y'z'$, with x' coinciding with the r-direction, y' coinciding with the θ-direction, and z' with z. The direction cosines between the two sets of coordinates are:

	x	y	z
r or x'	$\cos\theta$	$\sin\theta$	0
θ or y'	$-\sin\theta$	$\cos\theta$	0
z or z'	0	0	1

(5.8–7)

The tensor transformation law holds, and we have

$$\begin{aligned} e_{rr} &= e_{xx}\cos^2\theta + e_{yy}\sin^2\theta + e_{xy}\sin 2\theta, \\ e_{\theta\theta} &= e_{xx}\sin^2\theta + e_{yy}\cos^2\theta - e_{xy}\sin 2\theta, \\ e_{r\theta} &= (e_{yy} - e_{xx})\cos\theta\sin\theta + e_{xy}(\cos^2\theta - \sin^2\theta), \\ e_{zr} &= e_{zx}\cos\theta + e_{zy}\sin\theta, \\ e_{z\theta} &= -e_{zx}\sin\theta + e_{zy}\cos\theta, \\ e_{zz} &= e_{zz}. \end{aligned} \tag{5.8–8}$$

Finally, we have

$$\begin{aligned} e_{xx} &= \frac{\partial u_x}{\partial x}, \qquad e_{yy} = \frac{\partial u_y}{\partial y}, \qquad e_{zz} = \frac{\partial u_z}{\partial z}, \\ e_{xy} &= \frac{1}{2}\left(\frac{\partial u_x}{\partial y} + \frac{\partial u_y}{\partial x}\right), \qquad e_{yz} = \frac{1}{2}\left(\frac{\partial u_y}{\partial z} + \frac{\partial u_z}{\partial y}\right), \\ e_{zx} &= \frac{1}{2}\left(\frac{\partial u_z}{\partial x} + \frac{\partial u_x}{\partial z}\right). \end{aligned} \tag{5.8–9}$$

Now, a substitution of Eqs. (5.8–4) and (5.8–3) into Eq. (5.8–9) yields

$$\begin{aligned} e_{xx} &= \left(\cos\theta\frac{\partial}{\partial r} - \frac{\sin\theta}{r}\frac{\partial}{\partial\theta}\right)(u_r\cos\theta - u_\theta\sin\theta) \\ &= \cos^2\theta\frac{\partial u_r}{\partial r} + \sin^2\theta\left(\frac{u_r}{r} + \frac{1}{r}\frac{\partial u_\theta}{\partial\theta}\right) - \cos\theta\sin\theta\left(\frac{\partial u_\theta}{\partial r} + \frac{\partial u_r}{r\partial\theta} - \frac{u_\theta}{r}\right), \\ e_{yy} &= \sin^2\theta\frac{\partial u_r}{\partial r} + \cos^2\theta\left(\frac{u_r}{r} + \frac{\partial u_\theta}{r\partial\theta}\right) + \cos\theta\sin\theta\left(\frac{\partial u_\theta}{\partial r} + \frac{\partial u_r}{r\partial\theta} - \frac{u_\theta}{r}\right), \\ e_{xy} &= \frac{\sin^2\theta}{2}\left(\frac{\partial u_r}{\partial r} - \frac{\partial u_\theta}{r\partial\theta} - \frac{u_r}{r}\right) + \frac{\cos^2\theta}{2}\left(\frac{\partial u_\theta}{\partial r} + \frac{\partial u_r}{r\partial\theta} - \frac{u_\theta}{r}\right). \end{aligned} \tag{5.8–10}$$

Substituting these and similar results into Eq. (5.8–8) and reducing, we obtain

$$
\begin{aligned}
e_{rr} &= \frac{\partial u_r}{\partial r}, \\
e_{\theta\theta} &= \frac{u_r}{r} + \frac{1}{r}\frac{\partial u_\theta}{\partial \theta}, \\
e_{r\theta} &= \frac{1}{2}\left(\frac{1}{r}\frac{\partial u_r}{\partial \theta} + \frac{\partial u_\theta}{\partial r} - \frac{u_\theta}{r}\right), \\
e_{zr} &= \frac{1}{2}\left(\frac{\partial u_r}{\partial z} + \frac{\partial u_z}{\partial r}\right), \\
e_{z\theta} &= \frac{1}{2}\left(\frac{1}{r}\frac{\partial u_z}{\partial \theta} + \frac{\partial u_\theta}{\partial z}\right), \\
e_{zz} &= \frac{\partial u_z}{\partial z}.
\end{aligned}
\tag{5.8–11}
$$

Thus, we see that the method of transformation of coordinates is tedious but straightforward. Note that the structures of Eqs. (5.8–11) and (5.8–9) are different. In the language of tensor analysis, the difference is caused by the differences in the fundamental metric tensors of the two coordinate systems.

The reader should be warned again that we have adopted the tensor notations for the strain, so that the shear strain components $e_{r\theta}$, e_{rz}, $e_{z\theta}$ are one-half of those ordinarily given as $\gamma_{r\theta}$, γ_{rz}, $\gamma_{z\theta}$ in most books.

5.9 DIRECT DERIVATION OF THE STRAIN-DISPLACEMENT RELATIONS IN POLAR COORDINATES

The results of the preceding section can be derived directly from the geometric definition of the infinitesimal strain components. Recall that the normal strain components mean the ratio of change of length per unit length, whereas the shearing strain components mean one-half of the change of a right angle. For infinitesimal displacements, these changes can be seen directly from drawings such as those shown in Fig. 5.6.

Consider first the displacement in the r-direction, u_r. We see from Fig. 5.6(a) that

$$
e_{rr} = \frac{u_r + (\partial u_r/\partial r)dr - u_r}{dr} = \frac{\partial u_r}{\partial r}. \tag{5.9–1}
$$

From the same figure, we see also that a radial displacement of a circumferential element causes an elongation of that element and, hence, a strain in the θ-direction. The element ab, which was originally of length $r\,d\theta$, is displaced to $a'b'$

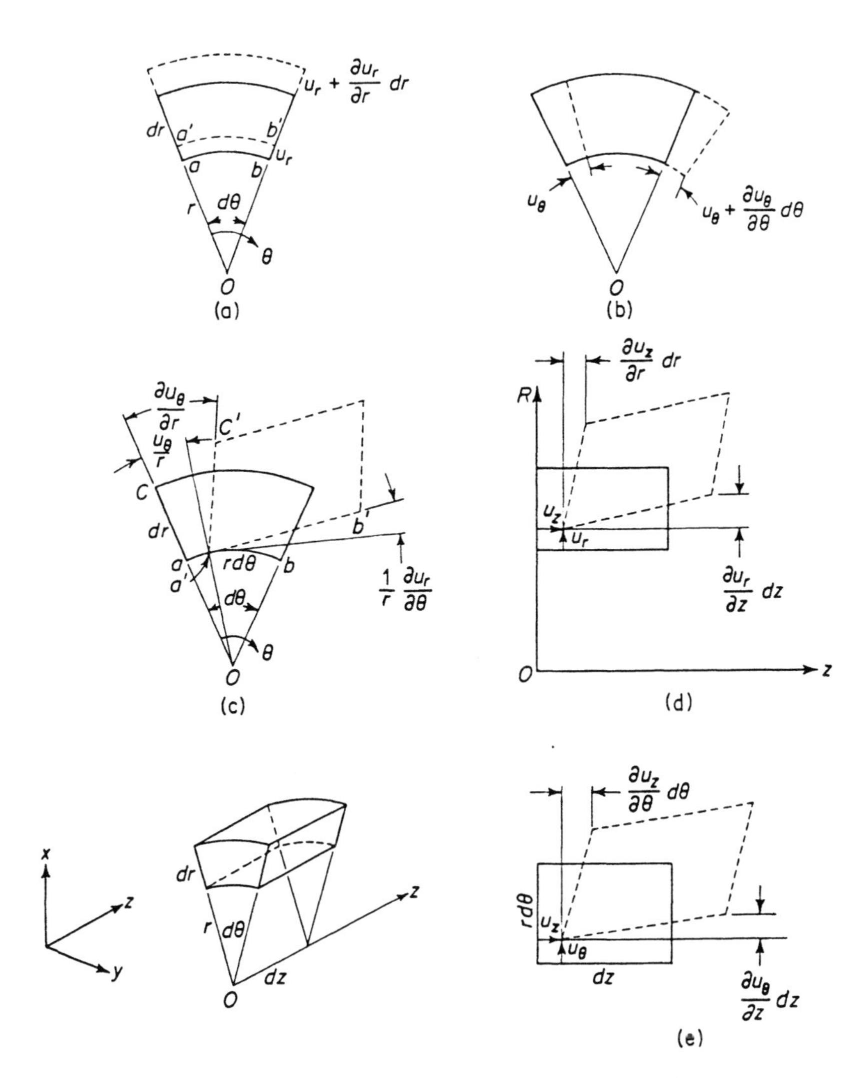

Figure 5.6 Displacement in cylindrical polar coordinates. (From E. E. Sechler, *Elasticity in Engineering*, Courtesy Mrs. Magaret Sechler.) A free-body diagram of an infinitesimal element of material and two systems of coordinates are shown at the lower left corner. (a) Radial strain due to variation of the radial displacement field in the radial direction. (b) Circumferential strain due to variation of circumferential displacement in the circumferential direction. (c) $\partial u_\theta / \partial r$ and $(1/r)\partial u_r / \partial \theta$ cause shear strain $e_{r\theta}$. (d) $\partial u_z / \partial r$ and $\partial u_r / \partial z$ cause shear strain e_{rz}. (e) $(1/r)\partial u_z / \partial \theta$ and $\partial u_\theta / \partial z$ cause shear strain $e_{z\theta}$.

and becomes of length $(r + u_r)\, d\theta$. The tangential strain due to this radial displacement is, therefore,

$$e_{\theta\theta}^{(1)} = \frac{(r + u_r)\, d\theta - r\, d\theta}{r\, d\theta} = \frac{u_r}{r}. \tag{5.9–2}$$

On the other hand, as shown in Fig. 5.6(b), the tangential displacement u_θ gives rise to a tangential strain equal to

$$e_{\theta\theta}^{(2)} = \frac{u_\theta + (\partial u_\theta/\partial\theta)\, d\theta - u_\theta}{r\, d\theta} = \frac{1}{r}\frac{\partial u_\theta}{\partial\theta}. \tag{5.9–3}$$

The total tangential strain is

$$e_{\theta\theta} = \frac{u_r}{r} + \frac{1}{r}\frac{\partial u_\theta}{\partial\theta}. \tag{5.9–4}$$

The normal strain in the axial direction is

$$e_{zz} = \frac{\partial u_z}{\partial z}, \tag{5.9–5}$$

as in the case of rectangular coordinates.

The shearing strain $e_{r\theta}$ is equal to one-half of the change of angle $\angle C'a'b' - \angle Cab$, as illustrated in Fig. 5.6(c). A direct examination of the figure shows that

$$e_{r\theta} = \frac{1}{2}\left(\frac{1}{r}\frac{\partial u_r}{\partial\theta} + \frac{\partial u_\theta}{\partial r} - \frac{u_\theta}{r}\right). \tag{5.9–6}$$

The first term comes from the change in the radial displacement in the θ-direction, the second term comes from the change in the tangential displacement in the radial direction, and the last term appears since part of the change in slope of the line $a'C'$ comes from the rotation of the element as a solid body about the axis through O.

The remaining strain components, $e_{z\theta}$ and e_{zr} can be derived with reference to Figs. 5.6(d) and (e). We have

$$e_{z\theta} = \frac{1}{2}\left[\frac{(\partial u_z/\partial\theta)d\theta}{r\, d\theta} + \frac{(\partial u_\theta/\partial z)dz}{dz}\right] = \frac{1}{2}\left[\frac{1}{r}\frac{\partial u_z}{\partial\theta} + \frac{\partial u_\theta}{\partial z}\right] \tag{5.9–7}$$

and

$$e_{zr} = \frac{1}{2}\left[\frac{(\partial u_r/\partial z)dz}{dz} + \frac{(\partial u_z/\partial r)dr}{dr}\right] = \frac{1}{2}\left[\frac{\partial u_r}{\partial z} + \frac{\partial u_z}{\partial r}\right]. \tag{5.9–8}$$

These equations are, of course, the same as Eq. (5.8–11). Indeed, the direct geometric method of derivation provides a much clearer mental picture than the algebraic method of the preceding section.

5.10 OTHER STRAIN MEASURES

We must not think that the strain tensors we have defined are the only ones suitable for the description of deformation. They are the most natural ones when we base our analysis of deformation on the change of the square of the distances between any two particles (Sec. 5.2). The *square* of distances is a convenient starting point because we have Pythagoras's theorem, which states that the square of the hypothenuse of a right triangle is equal to the sum of the squares of the legs. Using this theorem, we state that the square of the distance between two points x_i and $x_i + dx_i$, with coordinates referred to a rectangular Cartesian frame of reference, is

$$ds^2 = dx_1^2 + dx_2^2 + dx_3^2,$$

In Sec. 5.2, we based our analysis on this equation; the result was a natural definition of strain tensors.

Deformation does not, however, have to be described this way. For example, we may insist on using the change of distance ds (instead of ds^2) as our starting point, or on using the set of nine first derivatives of the displacement field:

$$\begin{pmatrix} \dfrac{\partial u}{\partial x} & \dfrac{\partial u}{\partial y} & \dfrac{\partial u}{\partial z} \\ \dfrac{\partial v}{\partial x} & \dfrac{\partial v}{\partial y} & \dfrac{\partial v}{\partial x} \\ \dfrac{\partial w}{\partial x} & \dfrac{\partial w}{\partial y} & \dfrac{\partial w}{\partial z} \end{pmatrix}. \tag{5.10–1}$$

Indeed, these derivatives, called "deformation gradients," are quite convenient. We may separate the matrix $(\partial u_i/\partial x_j)$ into a sum of a symmetric part and an antisymmetric part:

$$\begin{pmatrix} \dfrac{\partial u}{\partial x} & \dfrac{1}{2}\left(\dfrac{\partial u}{\partial y} + \dfrac{\partial v}{\partial x}\right) & \dfrac{1}{2}\left(\dfrac{\partial u}{\partial z} + \dfrac{\partial w}{\partial x}\right) \\ \dfrac{1}{2}\left(\dfrac{\partial v}{\partial x} + \dfrac{\partial u}{\partial y}\right) & \dfrac{\partial v}{\partial y} & \dfrac{1}{2}\left(\dfrac{\partial v}{\partial z} + \dfrac{\partial w}{\partial y}\right) \\ \dfrac{1}{2}\left(\dfrac{\partial w}{\partial x} + \dfrac{\partial u}{\partial z}\right) & \dfrac{1}{2}\left(\dfrac{\partial w}{\partial y} + \dfrac{\partial v}{\partial z}\right) & \dfrac{\partial w}{\partial z} \end{pmatrix}$$

$$+ \begin{pmatrix} 0 & \dfrac{1}{2}\left(\dfrac{\partial u}{\partial y} - \dfrac{\partial v}{\partial x}\right) & \dfrac{1}{2}\left(\dfrac{\partial u}{\partial z} - \dfrac{\partial w}{\partial x}\right) \\ -\dfrac{1}{2}\left(\dfrac{\partial u}{\partial y} - \dfrac{\partial v}{\partial x}\right) & 0 & \dfrac{1}{2}\left(\dfrac{\partial v}{\partial z} - \dfrac{\partial w}{\partial y}\right) \\ -\dfrac{1}{2}\left(\dfrac{\partial u}{\partial z} + \dfrac{\partial w}{\partial x}\right) & -\dfrac{1}{2}\left(\dfrac{\partial v}{\partial z} - \dfrac{\partial w}{\partial y}\right) & 0 \end{pmatrix}. \tag{5.10–2}$$

Then it is evident that the symmetric part of the deformation gradient matrix is the matrix of the infinitesimal strain, as defined in Sec. 5.3.

Other well-known strain measures are Cauchy's strain tensors and Finger's strain tensors. When the mapping is given by Eqs. (5.1–5) and (5.1–6), Cauchy's strain tensors are

$$C_{ij} = \frac{\partial a_k}{\partial x_i}\frac{\partial a_k}{\partial x_j}, \qquad \overline{C}_{ij} = \frac{\partial x_k}{\partial a_i}\frac{\partial x_k}{\partial a_j}, \tag{5.10–3}$$

whereas Finger's strain tensors are

$$B_{ij} = \frac{\partial x_i}{\partial a_k}\frac{\partial x_j}{\partial a_k}, \qquad \overline{B}_{ij} = \frac{\partial a_i}{\partial x_k}\frac{\partial a_j}{\partial x_k}. \tag{5.10–4}$$

For these tensors, the absence of strain is indicated, not by the vanishing of C_{ij} or B_{ij}, but by $C_{ij} = \delta_{ij}$, $B_{ij} = \delta_{ij}$.

We shall not discuss these strain measures any further, except to note that they may be convenient for some special purposes in advanced theories of continua.

PROBLEMS

5.1 A blood vessel is incompressible, i.e., its volume does not change. Under normal conditions, a blood vessel can be considered as a circular cylinder. Suppose a person has his or her blood pressure increased for some reason, and the inner radius of the blood vessel increases from a to $a + \Delta a$, while the axial length is unchanged. Compute the changes in the circumferential and radial strains throughout the blood vessel due to the increase in blood pressure.

5.2 (a) A state of deformation in which the displacement field u_i is a linear function of the coordinates x_i is called a *homogeneous deformation*. What is the equation of a surface that will become a sphere $x^2 + y^2 + z^2 = r^2$ *after* a homogeneous deformation? [Use an equation of the type $f(x, y, z) = 0$, in which x, y, z are rectangular Cartesian coordinates.]

(b) As a special case of homogeneous deformation, consider the following linear transformations of coordinates from (x, y, z) to (x', y', z'), both of which refer to the same Cartesian frame of reference. (See Fig. P5.2.)

(1) *Pure shear:* $x' = kx$, $y' = k^{-1}y$, $z' = z$.
(2) *Simple shear:* $x' = x + 2sy$, $y' = y$, $z' = z$.

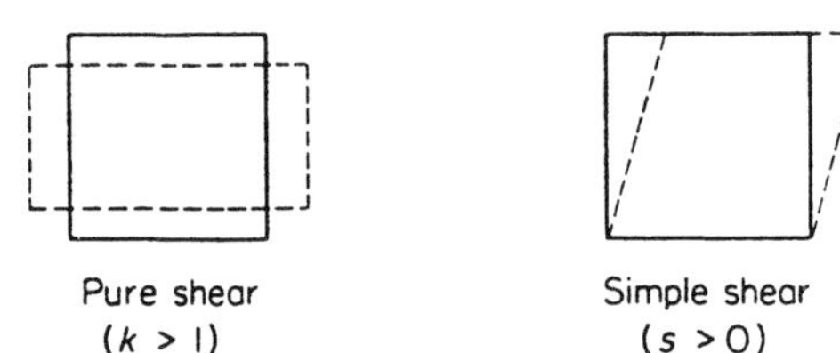

Figure P5.2 Pure shear and simple shear.

We may regard (x, y, z) as the coordinates of a material particle before a deformation is imposed on it and (x', y', z') as the coordinates after deformation. Show that a pure shear may be regarded as a simple shear referred to axes inclined at $\tan^{-1}(k^{-1})$

with Ox, Oy if $s = \frac{1}{2}(k - k^{-1})$. Equivalently, a simple shear may be regarded as a pure shear with $k = \sqrt{(s^2 + 1)} + s$ and the major axis of the strain ellipsoid inclined at $\frac{1}{4}\pi - \frac{1}{2}\tan^{-1} s = \tan^{-1}(k^{-1})$ with Ox.

(Drawings of the strain ellipses for these two cases can be found in J. C. Jaeger, *Elasticity, Fracture and Flow*. London: Methuen & Co., 1956, p. 32.)

Solution: We define a *homogeneous deformation* as a deformation in which the displacement field u_i is a linear function of the coordinates, so that a point x_i is moved to x_i' under the transformation

$$x_i' = x_i + u_i = x_i + u_i^{(0)} + a_{ik}x_k, \tag{1}$$

where $u_i^{(0)}$ and a_{ik} are constants. Under this transformation, a sphere $x'^2 + y'^2 + z'^2 = r^2$ corresponds to an ellipsoid

$$[u_i^{(0)} + x_i + a_{ik}x_k][u_i^{(0)} + x_i + a_{ik}x_k] = r^2. \tag{2}$$

Now, pure shear and simple shear are defined by the following equations and can be represented graphically for a square, as shown in Fig. P5.2.

Pure shear:

$$x' = kx, \qquad y' = y/k, \qquad z' = z. \tag{3}$$

Simple shear:

$$x' = x + 2sy, \qquad y' = y, \qquad z' = z. \tag{4}$$

The two transformations appear quite different in the figures. But in fact, they are similar. The similarity is best shown by considering the strain ellipsoids.

Since $z' = z$, it is sufficient to consider transformations of curves in the xy-plane. By Eq. (3), a circle $x'^2 + y'^2 = 1$ is transformed into an ellipse

$$k^2x^2 + \frac{y^2}{k^2} = 1, \tag{5}$$

whereas by Eq. (4), the same circle is transformed into another ellipse

$$x^2 + 4sxy + (1 + 4s^2)y^2 = 1. \tag{6}$$

Let us simplify Eq. (6) by a rotation of coordinates. By Eq. (2.4–2), if x, y is rotated to ξ, η through an angle θ, we have

$$x = \xi\cos\theta - \eta\sin\theta, \qquad y = \xi\sin\theta + \eta\cos\theta. \tag{7}$$

On substituting into Eq. (6) and simplifying, we obtain

$$\begin{aligned} &\xi^2[\cos^2\theta + 4s\cos\theta\sin\theta + (1 + 4s^2)\sin^2\theta] + \eta^2[\sin^2\theta \\ &-4s\sin\theta\cos\theta + (1 + 4s^2)\cos^2\theta] + \xi\eta[-2\cos\theta\sin\theta \\ &+ 4s(\cos^2\theta - \sin^2\theta) + 2\cos\theta\sin\theta(1 + 4s^2)] = 1. \end{aligned} \tag{8}$$

The coefficient of $\xi\eta$ vanishes if $s = -\cot 2\theta$, or $\theta = -\frac{1}{2}\tan^{-1}(1/s)$. With this value of θ, the coefficient of ξ^2 in Eq. (8) becomes

$$\begin{aligned} &\cos^2\theta - 2\cot 2\theta\sin 2\theta + (1 + 4\cot^2 2\theta)\sin^2\theta \\ &= 1 - 2\cos 2\theta + \cos^2 2\theta/\cos^2\theta = \tan^2\theta. \end{aligned}$$

Similarly, the coefficient of η^2 in Eq. (8) can be reduced to $\cot^2\theta$. Therefore, Eq. (8) becomes

$$\tan^2\theta\,\xi^2 + \cot^2\theta\,\eta^2 = 1. \tag{9}$$

If we write $k = \tan\theta$, then Eq. (9) is reduced exactly to Eq. (5). Therefore, these two strain ellipsoids are equal; one is rotated from the other by an angle θ. This verifies the equivalence of pure shear and simple shear.

To find the relation between k and s, we note that

$$\cot 2\theta = \frac{\cos 2\theta}{\sin 2\theta} = \frac{\cos^2\theta - \sin^2\theta}{2\sin\theta\cos\theta} = \frac{1}{2}[\cot\theta - \tan\theta].$$

Therefore, since $s = -\cot 2\theta$ and $k = \tan\theta$, we have

$$-s = \frac{1}{2}\left[\frac{1}{k} - k\right], \qquad \text{and} \qquad k = s + \sqrt{(s^2 + 1)}. \tag{10}$$

5.3 A steel pipe of length 60 cm, diameter 6 cm, and wall thickness 0.12 cm is stretched 0.010 cm axially, expanded 0.001 cm in diameter, and twisted through 1°. Determine the strain components in the pipe.

5.4 For the truss shown in Fig. P5.4, determine

(a) The loads in the rods.

(b) The stresses in the rods.

(c) Assume a one-dimensional stress-strain relationship $e = \sigma/E$ for the rods, and assume that Young's modulus for steel is $E = 207$ GPa (3×10^7 psi). Determine the longitudinal strain e in the rods.

(d) Determine the displacement vector at the point of loading, B.

Answer: (b) $\sigma_{AB} = 503$ MPa (72,000 lb/in^2), $\sigma_{BC} = -88.2$ MPa ($-12{,}800$ lb/in^2)
(c) $e_{AB} = 2.4 \times 10^{-3}$, $\epsilon_{BC} = -4.25 \times 10^{-4}$
(d) 0.640 cm (0.252 in)

Solution: The loads in the rods are determined by static equilibrium, as in Chapter 1. We obtain a tension of 6428 kg ($\sqrt{2} \times 10^4$ lbs) in AB and a compression of 4545 kg ($-10{,}000$ lb) in BC. The stresses are obtained by dividing the loads by the cross-sectional area of the members. A further division by Young's modulus gives the strains $e_{AB} = 2.405 \times 10^{-3}$ and $e_{BC} = -0.425 \times 10^{-3}$.

To determine the displacement at B, we note that the steel rods are pin ended. As a consequence of shortening the rod BC, the point B moves to the left, but the rod B can swing around C, so that the locus of the possible location of B lies on an arc of a circle with C as center and $\overline{BC}$ as radius. For very small $\Delta\,\overline{BC}$ (as compared with $\overline{BC}$), this locus is a line segment perpendicular to $\overline{BC}$. Similarly, the bar AB extends $\Delta\,\overline{AB}$ in length, and the locus of B on AB lies on an arc perpendicular to AB. The intersection of these arcs, B^*, is the final location of the displaced joint B.

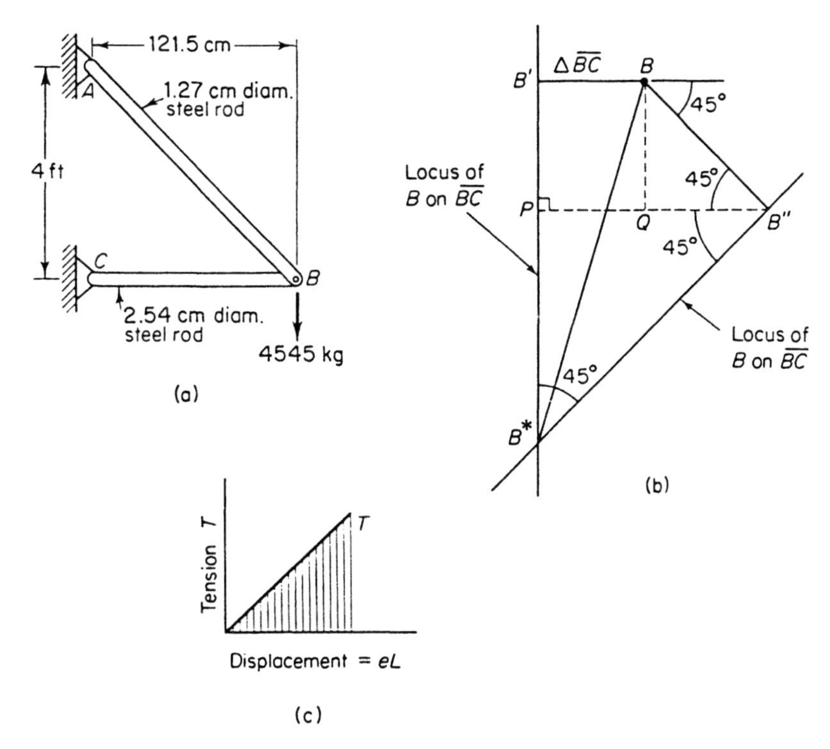

Figure P5.4 A simple truss and a method of determining the displacement at the joint B. (a) A simple truss. (b) The locus of the movement of the point B on the arm BC, namely, $BB'B^*$; and that on the arm AB, namely, $BB''B^*$. (c) The area of the shaded region is proportional to the strain energy stored in a bar when it is stretched.

To compute the displacement $\overline{BB^*}$, we see from Fig. P5.4(b) that

$$\overline{BB^*} = \sqrt{\overline{BB'}^2 + \overline{B'B^*}^2} = \sqrt{\overline{BB'}^2 + (\overline{B'P} + \overline{PB^*})^2}$$

$$= \sqrt{\overline{BB'}^2 + (\overline{B'P} + \overline{PB''})^2} = \sqrt{\overline{BB'}^2 + (\overline{B'P} + \overline{PQ} + \overline{QB''})^2}$$

$$= \sqrt{\overline{BB'}^2 + (\overline{BQ} + \overline{BB'} + \overline{QB''})^2}.$$

Now,

$$\overline{BB'} = |e_{BC}| \cdot 121 \text{ cm} = 5.26 \times 10^{-2} \text{ cm}$$

$$\overline{BQ} = \overline{BB''} \cos 45° = e_{AB} \overline{AB} \cos 45° = 0.293 \text{ cm}$$

Similarly, $\overline{QB''} = \overline{BQ} = 0.293$ cm. Hence, we obtain $\overline{BB^*} = 0.640$ cm by substitution.

Note: Alternative Method of Finding Displacement at B. The work done by the load is equal to the strain energy stored in the rods. When a rod is subject to a gradually increasing tension from zero to T, its length changes by amount $eL = TL/EA$, where L is the length of the rod and A is its cross-sectional area. The strain energy stored in the rod is equal to $\frac{1}{2}(T^2L/AE)$. [See Fig. P5.4(c).] Now, when a load W is gradually applied onto the bracket, the work done by it is equal to $\frac{1}{2}W\delta$ where δ is the displacement in the direction of the load, i.e., the vertical component of the displacement. The factor $\frac{1}{2}$ is necessary because, the structure being linearly elastic, the force-deflection relationship is linear, so that the area under the curve, which represents the work done, is $\frac{1}{2}$ of load $\times$ deflection. Hence, on equating the work done with the strain energy stored, we obtain

$$\frac{1}{2} \cdot W\,\delta = \frac{1}{2}\frac{T_{AB}^2 L_{AB}}{EA_{AB}} + \frac{1}{2}\frac{T_{BC}^2 L_{BC}}{EA_{BC}}.$$

On substituting numerical values into this equation, we obtain $\delta = 0.635$ cm. The total displacement of the joint B is $(\delta^2 + \Delta\,\overline{BC}^2)^{1/2} = 0.640$ cm.

5.5 A rocket-launching tower is affected by thermal deflection caused by nonuniform heating of the rocket under the sun (Fig. P5.5). Assume that the body of the rocket is a circular cylinder, and estimate the horizontal displacement of the tip A if the following assumptions hold:

(a) The linear thermal coefficient of expansion is $\alpha = 10^{-5}/°F = 0.555 \times 10^{-5}/°C$.

(b) The maximum temperature on the body of the rocket on the side facing the sun is 20°F hotter than the minimum temperature on the shady side.

(c) The temperature distribution is uniform along the length (longitudinal axis) of the rocket, but varies linearly along the x-axis.

(d) As a consequence of (c), a plane section of the rocket remains plane in thermal expansion.

(e) The rocket is unloaded and is free to deform.

Hint: Compute the thermal strain and then integrate to obtain the deflection.

Answer: Thermal strain difference from two sides $= \alpha T = 20 \times 10^{-5}$. Tip deflection $= 26.3$ cm.

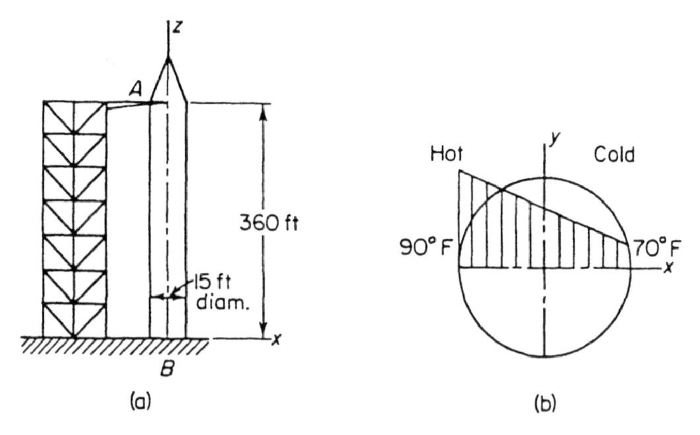

Figure P5.5 Thermal deflection of a launching tower. (a) The dimensions of the rocket. (b) The temperature distribution.

5.6 Derive an expression for the change in volume of an element of unit volume subjected to small strains e_{ij}. Show that the invariant $I_1 = e_{11} + e_{22} + e_{33}$ represents the change of volume per unit volume when the strains are small.

Solution: According to Sec. 5.7, we can find a set of rectangular Cartesian coordinates with respect to which the strain tensor assumes the form $e_k \delta_{ik}$ (k not summed), where e_1, e_2, e_3 are the principal strains. Let us consider a body undergoing strain and choose a unit cube whose edges are oriented along the principal axes of strain. Each edge, originally of length 1, becomes $1 + e_i$ after deformation. The new volume is, therefore,

$$(1 + e_1)(1 + e_2)(1 + e_3) = 1 + e_1 + e_2 + e_3 + \text{higher order terms.}$$

Hence, on ignoring the higher order terms, we see that the change of volume per unit volume is $e_1 + e_2 + e_3$.

We know from Eq. (5.7–3) that $I_1 = e_{ij}\delta_{ij}$ is invariant. It is equal to $e_1 + e_2 + e_3$ with reference to the principal axes. Hence, $I_1 = e_1 + e_2 + e_3$ with reference to any Cartesian frame of reference. Thus, $I_1 = e_{ij}\delta_{ij}$ means the change of volume per unit volume when the strains are small.

5.7 Given a stress field σ_{ij}, with components referred to a system of coordinates x_1, x_2, x_3,
(a) What is the definition of principal stresses?
(b) What is the definition of principal axes?
(c) Describe briefly how the principal directions (i.e., the directions of the principal axes) can be determined in principle.
(d) Consider a strain tensor e_{ij} referred to the same coordinate axes. How do you determine the principal strains and the corresponding principal directions?
(e) If the stress and strain tensors are related by the relation

$$\sigma_{ij} = \lambda e_{kk}\delta_{ij} + 2\mu e_{ij},$$

where λ and μ are constants, prove that the principal axes of stress coincide with the principal axes of strain.

5.8 In a study of earthquakes, Lord Rayleigh investigated a solution of the linearized equations of elasticity in the form

$$u = Ae^{-by}\exp[ik(x - ct)],$$
$$v = Be^{-by}\exp[ik(x - ct)],$$
$$w = 0.$$

If the plane xz represents the ground, while y represents the depth into the earth, and u, v, w are the displacements of the particles of the earth, then Rayleigh's solution represents a wave propagating in the x-direction with a speed c and an amplitude that decreases exponentially from the ground surface. The wave is assumed to be generated inside the earth. The ground surface is free; i.e., the stress vector acting on the ground surface is zero. After checking the equations of motion and the boundary conditions, Rayleigh found the constants A, B, b, and c and obtained the solution

$$u = A(e^{-0.8475ky} - 0.5773e^{-0.3933ky})\cos k(x - c_R t),$$
$$v = A(-0.8475e^{-0.8475ky} + 1.4679e^{-0.3933ky})\sin k(x - c_R t),$$
$$w = 0.$$

The constant c_R is the so-called Rayleigh wave speed, which is equal to 0.9194 times the shear wave speed if Poisson's ratio is $\frac{1}{4}$. This solution satisfies the conditions of a wave propagating in a semiinfinite elastic solid with a free surface $y = 0$. The particles move in the xy-plane, with amplitude decreasing as the distance from the free surface increases (see Fig. P5.8). The Rayleigh wave represents one of the most prominent waves that can be seen on a seismograph when there is an earthquake.
(a) Sketch the waveform.
(b) Sketch the path of motion of particles on the free surface $y = 0$ at several values of x. Do the same for several particles at different values of $y > 0$.
(c) Show that the motion of the particles is retrograde.
(d) Determine the places where the maximum principal strain occurs at any given instant and the value of this strain.

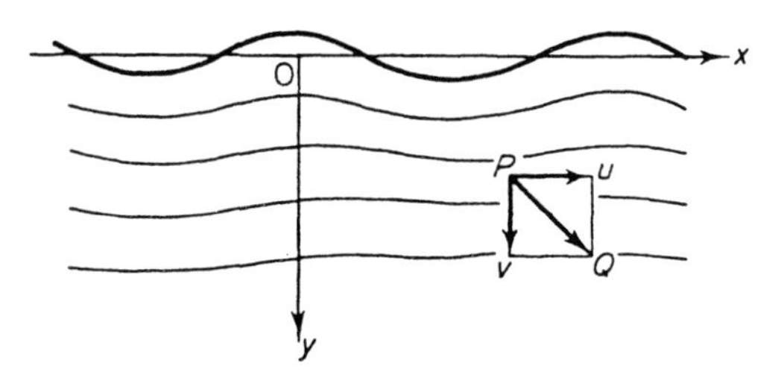

Figure P5.8 Rayleigh surface wave.

Partial Solution:
(d) Since $w = 0$, only the strain components e_{xx}, e_{yy}, e_{xy} are not identically zero. The exponential function e^{-by}, with $b > 0$, shows that the largest values of u, v, w and their derivatives will occur at $y = 0$. On this plane and at $t = 0$, we have

$$e_{xx} = \frac{\partial u}{\partial x} = -Ak(1 - 0.5773) \sin kx,$$

$$e_{yy} = \frac{\partial v}{\partial y} = Ak[(0.8475)^2 - 1.4679 \times 0.3933] \sin kx,$$

$$e_{xy} = \frac{Ak}{2} [(-0.8475 + 0.5773 \times 0.3933) + (-0.8475 + 1.4679)] \cos kx = 0.$$

Hence, the maximum principal strains are

$$e_{xx} = \pm 0.4227\, Ak, \qquad e_{yy} = \pm 0.14094\, Ak.$$

5.9 Consider a square plate of unit size deformed as shown in Fig. P5.9. Find the strain components.

Solution: The deformation can be described by the following equations:

$$x_1 = a_1 + \frac{1}{\sqrt{3}} a_2, \qquad x_2 = a_2, \qquad x_3 = a_3,$$

or

$$a_1 = x_1 - \frac{1}{\sqrt{3}} x_2, \qquad a_2 = x_2, \qquad a_3 = x_3.$$

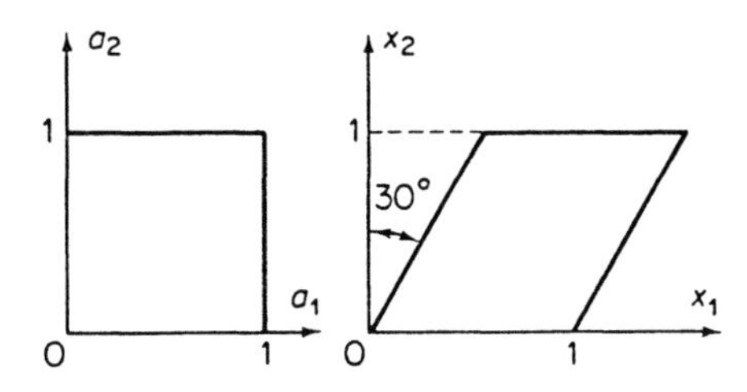

Figure P5.9 Deformation of a square plate.

Hence,

$$ds^2 - ds_0^2$$

$$= \left\{\left[\left(\frac{\partial x_1}{\partial a_1}\right)^2 - 1\right]da_1^2 + 2\frac{\partial x_1}{\partial a_1}\frac{\partial x_1}{\partial a_2}da_1\,da_2 + \left[\left(\frac{\partial x_1}{\partial a_2}\right)^2 + \left(\frac{\partial x_2}{\partial a_2}\right)^2 - 1\right]da_2^2\right\}$$

$$= \left\{\left[1 - \left(\frac{\partial a_1}{\partial x_1}\right)^2\right]dx_1^2 - 2\frac{\partial a_1}{\partial x_1}\frac{\partial a_1}{\partial x_2}dx_1\,dx_2 + \left[1 - 1 - \left(\frac{\partial a_1}{\partial x_2}\right)^2\right]dx_2^2\right\}$$

$$= \frac{2}{\sqrt{3}}da_1\,da_2 + \left(\frac{1}{3}\right)da_2^2 = \frac{2}{\sqrt{3}}dx_1\,dx_2 - \frac{1}{3}dx_2^2.$$

But by Eq. (5.2–10) this is $2\,E_{11}da_1^2 + 2\,(E_{12} + E_{21})\,da_1\,da_2 + 2\,E_{22}da_2^2$. Hence,

$$E_{12} = \frac{1}{2\sqrt{3}}, \qquad E_{22} = \frac{1}{6}; \qquad e_{12} = \frac{1}{2\sqrt{3}}, \qquad e_{22} = -\frac{1}{6},$$

whereas all other components of strain are zero.

5.10 Consider the square plate again, but this time shear to the right only a very small amount, so that

$$x_1 = a_1 + 0.01a_2, \qquad a_1 = x_1 - 0.01x_2, \qquad x_2 = a_2, \qquad x_3 = a_3.$$

Then

$$ds^2 - ds_0^2 = 0.01\,da_1\,da_2 + (0.01)^2\,da_2^2 = 0.01\,dx_1\,dx_2 - (0.01)^2\,dx_2^2.$$

Hence,

$$E_{12} = 0.0025, \qquad E_{22} = 5 \times 10^{-5}, \qquad e_{12} = 0.0025, \qquad e_{22} = -5 \times 10^{-5}.$$

In this case, the E_{ij} and e_{ij} measures are approximately the same.

5.11 A square plate is deformed uniformly from configuration (a) to configuration (b), as shown in the three cases in Fig. P5.11. Determine the strain components E_{11}, E_{22}, E_{12} and e_{11}, e_{22}, e_{12}.

Answer: The transformation that leads from configuration (a) to configuration (b) in Case 1 is $x_1 = 1.4a_1$, $x_2 = 1.2a_2$, $x_3 = a_3$. That in Case 2 is $x_1 = 1.2a_1 + 0.5a_2$, $x_2 = 1.2a_2$, $x_3 = a_3$. In Case 3, we have $x_1 = 1.01a_1 + 0.02a_2$, $x_2 = 1.01a_2$, $x_3 = a_3$. From these, the strain components are obtained from Eq. (5.3–5). Case 3 qualifies for "infinitesimal" strains, as given by Eq. (5.3–7).

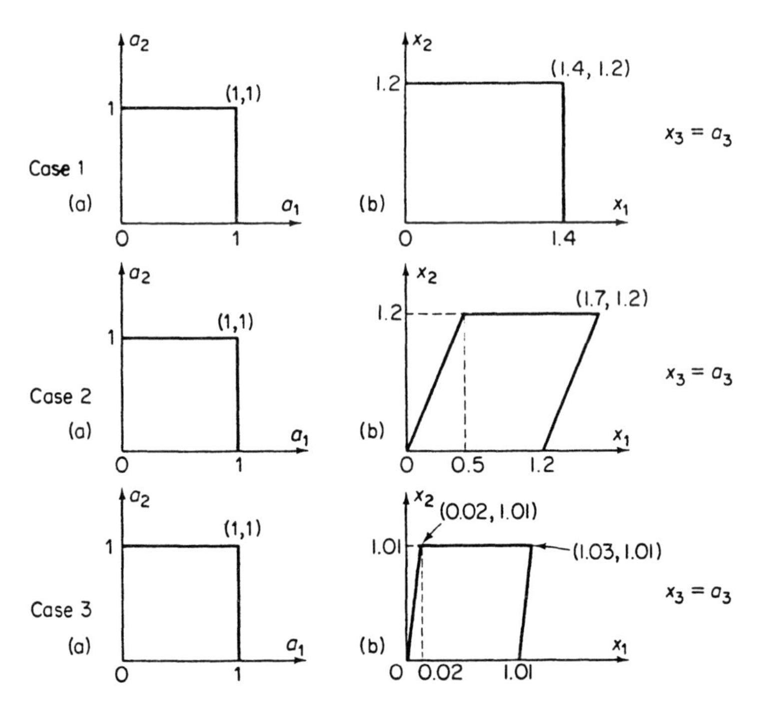

Figure P5.11 Three patterns of deformation of a square plate.

5.12 A unit square $OABC$ is distorted to $OA'B'C'$ in three ways, as shown in Fig. P5.12. In each of the cases, write down the displacement field u_1, u_2 of every point in the square as a function of the location (a_1, a_2) of the point in the original position. Then determine the strains E_{ij}, e_{ij}. Assume that $u_3 = 0$ and that u_1, u_2 are independent of x_3 and a_3. In Cases (b) and (c), assume that the lengths of OA, OA', OC, and OC' are all 1. Also, obtain the simplified expressions of the strains e_{ij} if ϵ_1, ϵ_2, θ, ψ are infinitesimal.

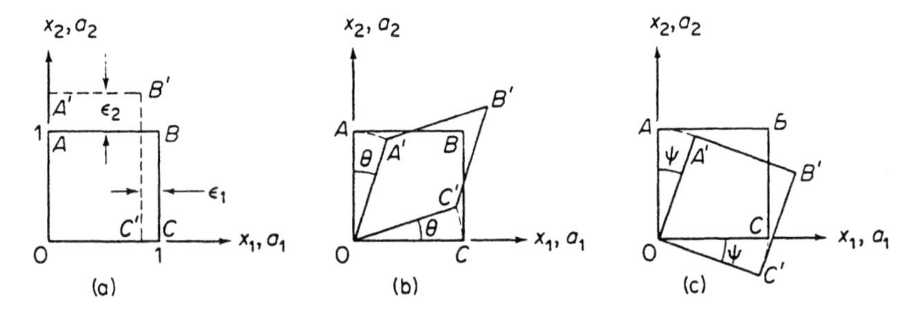

Figure P5.12 Deformation of $OABC$ to $OA'B'C'$.

5.13 A unit square $OABC$ is first subjected to a stretching, as shown in Fig. P5.12(a), then to a distorsion, as shown in Fig. P5.12(b), and finally to a rotation, as shown in Fig. P5.12(c). After the three steps in succession, what are the values of the strains E_{ij}, e_{ij}?

Answer this problem first for finite values of ϵ_1, ϵ_2, θ, ψ and then for infinitesimal values of ϵ_1, ϵ_2, θ, ψ.

5.14 Find the strain components E_{ij} and e_{ij} when one of the wedges in Fig. P5.14 is transformed into the other. The first wedge has an apex angle of 30°; the other is 90°. The radii are the same.

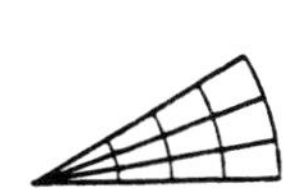

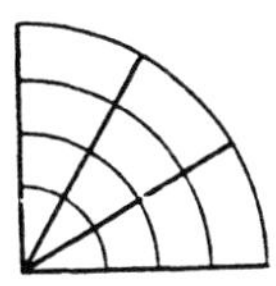

Figure P5.14 Wedge changing angle.

5.15 Let $ABCD$ be a unit square in the xy-plane (Fig. P5.15). $ABCD$ is a part of a large deformable body subjected to a small strain that is uniform in the entire body and is given by

$$\begin{pmatrix} 1 & 2 & 3 \\ 2 & 1 & 0 \\ 3 & 0 & 2 \end{pmatrix} \times 10^{-3}.$$

What is the change in length of the lines AC and AE?

Answer: AC changes by 0.00423; AE changes by 0.00290.

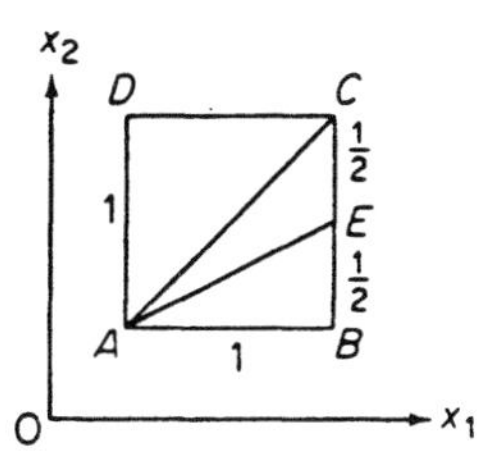

Figure P5.15 Change of length of line segments in a plate of known strain.

5.16 A square membrane, $-1 \le x \le 1$, $-1 \le y \le 1$, is stretched in such a manner that the displacement is described by

$$u = a(x^2 + y^2),$$

$$v = bxy,$$

$$w = 0.$$

What are the strain components at (x, y)? What is the principal strain at the origin (0, 0). Assume the constants a, b to be infinitesimal.

5.17 A pin-jointed truss is shown in Fig. P5.17, where L is the length of the vertical and horizontal members. The cross-sectional area of all the members is the same, namely A. The material of all the members is the same, with Young's modulus E. The truss is loaded at the center by a load P. What would be the vertical deflection of the point under the load?

Answer: Solve the problem by the strain-energy method illustrated in Prob. 5.4. The deflection is $5.828PL/AE$.

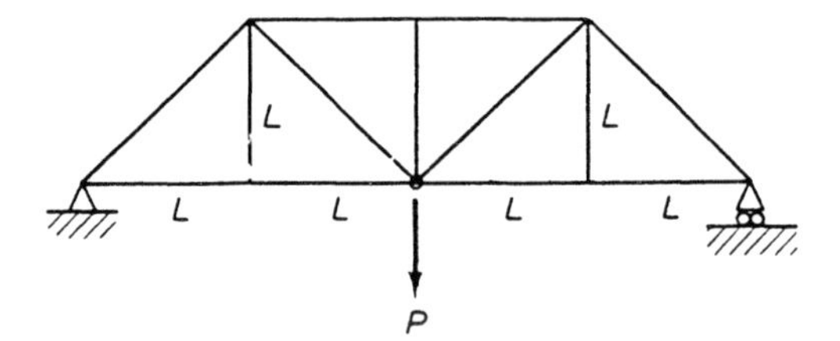

Figure P5.17 Calculation of the vertical deflection of a joint of a truss.

5.18 The following may happen in a number of situations, such as the flow of water, the forming of metals, and in cell membranes. The material is incompressible. The displacement component w in the z-direction vanishes. The displacements u, v are infinitesimal and are functions of x, y. If, in a certain domain, we know that

$$u = (1 - y^2)(a + bx + cx^2),$$

where a, b, c are constants, compute the displacement v in the y-direction.
Hint: Use the facts shown in Prob. 5.6.

5.19 In the problem of the torsion of a cylindrical bar of steel with an elliptic cross section (Fig. P5.19), it was found that the displacement can be described by the equations

$$u = \alpha zy, \; v = \alpha zx, \; w = -\frac{a^2 - b^2}{a^2 + b^2}\alpha xy$$

where α is the angle of twist in radians per unit length of the bar. Let $a = 2$ cm and $b = 1$ cm. Compute the strain that acts at the point A ($x = 0$, $y = b$). What is the maximum shear strain at A? On what plane do the maximum tension and maximum shear act?

Partial Answer:

$$e_{xz} = -\frac{8}{5}\alpha, \; e_{yz} = 0, \text{ max shear strain } = -\frac{8}{5}\alpha, \text{ max normal strain } = \pm\frac{8}{5}x.$$

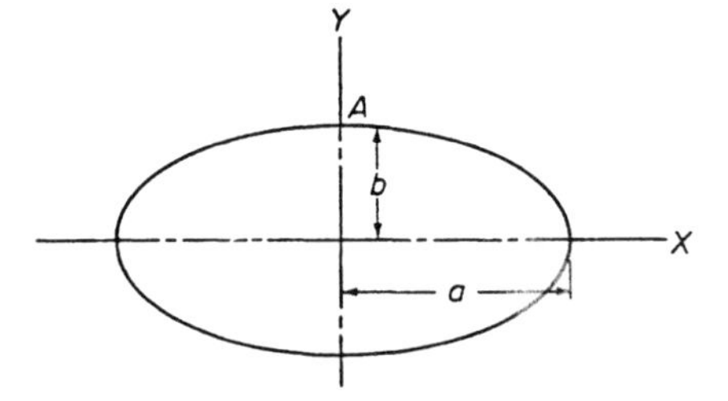

Figure P5.19 An elliptic cross section.

5.20 By differentiating an arbitrary analytic function $\phi(x_1, x_2, x_3)$, one can obtain a displacement field $\mathbf{u}(u_1, u_2, u_3)$ defined by the equation

$$u_i = \frac{\partial \phi}{\partial x_i}.$$

What are the strain components and strain invariants of this displacement field? Consider a special case

$$\phi = \frac{C}{R^2} + DR^2$$

where

$$R^2 = x_1^2 + x_2^2 + x_3^2$$

and C, D are constants. Apply it to a hollow sphere whose inner radius is a and outer radius is b. What are the values of the strains in this sphere?

5.21 Figure P5.21 is another classical drawing from Borelli's 300 year old book (see Prob. 1.23). Here is shown Borelli's observations on the arrangement of muscle fibers on

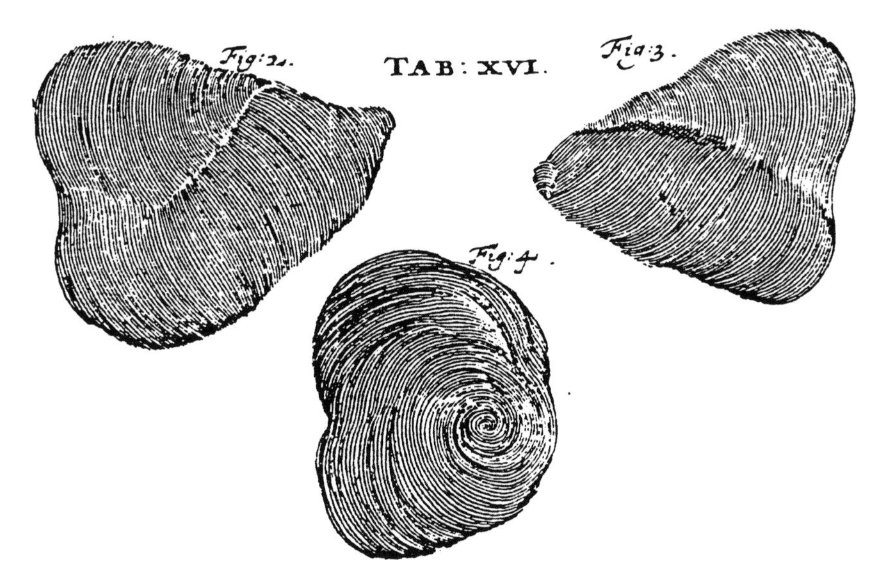

Figure P5.21 A classic drawing by Borelli.

the surface of the heart. Muscle fibers contract or lengthen, change the shape and strain of the organ. Geometric arrangement of fibers has a great influence on the function of the muscle. If a muscle is shaped like a parallelogram shown in the "simple shear" case of Fig. P5.2 (Problem 5.2), with fibers parallel to the inclined borders, then a contraction of the muscle which turns the muscle block into a rectangle would leave the volume and width unchanged. On the other hand, if the muscle fibers were parallel to the y-axis as shown in Fig. P5.2's "pure shear" case, then on contraction the width of the bundle will bulge out. When the heart works to pump blood, the chambers of the left ventricle and right ventricle should expand and contract, but the myocardium does not change its volume, nor bulges out locally (unlike the biceps flexor muscle of the front of the upper arm). Borelli thought he knew how it could be done.

Take up this suggestion and discuss it. Add theoretical or experimental details as far as you can. Modern data on heart muscle fiber structure can be found in Streeter, D. Jr., "Gross morphology and fiber geometry of the heart," In *Handbook of Physiology*, Sec. 2 *Cardiovascular System*, Vol. 1 *The Heart*. (Berne, R. M. and Sperelakis, N., eds.), American Physiological Society, Bethesda, MD, pp. 61–112. Expand the observation to explain how a crab or lobster can move its powerful claws by muscles encased in a rigid shell. What kind of strains these muscles must have?

6 VELOCITY FIELDS AND COMPATIBILITY CONDITIONS

We shall consider the velocity field and define the strain-rate tensor. Then we shall study the question of compatibility of the strain components or the strain-rate components.

6.1 VELOCITY FIELDS

For the study of fluid flow, we are generally concerned with the velocity field, i.e., with the velocity of every particle in the body of the fluid. We refer the location of each fluid particle to a frame of reference O-xyz; then the field of flow is described by the velocity vector field $\mathbf{v}(x, y, z)$, which defines the velocity at every point (x, y, z). In terms of components, the velocity field is expressed by the functions

$$u(x, y, z), \qquad v(x, y, z), \qquad w(x, y, z),$$

or, if index notations are used, by $v_i(x_1, x_2, x_3)$.

For a continuous flow, we consider the continuous and differentiable functions $v_i(x_1, x_2, x_3)$. There are occasions, however, in which we must study the relationship between velocities at neighboring points. Let the particles P and P' be located instantaneously at x_i and $x_i + dx_i$, respectively. The difference in velocities at these two points is

$$dv_i = \frac{\partial v_i}{\partial x_j} dx_j, \tag{6.1–1}$$

where the partial derivatives $\partial v_i / \partial x_j$ are evaluated at the particle P. Now,

$$\frac{\partial v_i}{\partial x_j} = \frac{1}{2}\left(\frac{\partial v_i}{\partial x_j} + \frac{\partial v_j}{\partial x_i}\right) - \frac{1}{2}\left(\frac{\partial v_j}{\partial x_i} - \frac{\partial v_i}{\partial x_j}\right). \tag{6.1–2}$$

Let us define the *rate-of-deformation* tensor V_{ij} and the *spin tensor* Ω_{ij} as

$$V_{ij} \equiv \frac{1}{2}\left(\frac{\partial v_i}{\partial x_j} + \frac{\partial v_j}{\partial x_i}\right), \tag{6.1–3}$$

$$\Omega_{ij} \equiv \frac{1}{2}\left(\frac{\partial v_j}{\partial x_i} - \frac{\partial v_i}{\partial x_j}\right). \tag{6.1–4}$$

Then

$$\frac{\partial v_i}{\partial x_j} = V_{ij} - \Omega_{ij}. \tag{6.1–5}$$

It is evident that V_{ij} is symmetric and Ω_{ij} is antisymmetric; i.e.,

$$V_{ij} = V_{ji}, \qquad \Omega_{ij} = -\Omega_{ji}, \tag{6.1–6}$$

Hence, the Ω_{ij} tensor has only three independent elements, and there exists a vector $\mathbf{\Omega}$ *dual* to Ω_{ij}; that is,

$$\Omega_k \equiv \epsilon_{kij}\Omega_{ij}; \qquad \text{i.e.,} \qquad \mathbf{\Omega} = \text{curl } \mathbf{v}, \tag{6.1–7}$$

where ϵ_{kij} is the permutation tensor defined in Sec. 2.3, Eq. (2.3–16). The vector $\mathbf{\Omega}$ is called the *vorticity* vector.

Equations (6.1–7) and (6.1–1) are similar to Eqs. (5.5–3) and (5.5–5). Their geometric interpretations are also similar. Therefore, the analysis of the velocity field is very much like the analysis of an infinitesimal deformation field. Indeed, if we multiply v_i by an infinitesimal interval of time dt, the result is an infinitesimal displacement $u_i = v_i\, dt$. Hence, whatever we learned about the infinitesimal strain field can be immediately extended correspondingly to the *rate of change* of strain, with the word *velocity* replacing the word *displacement*.

6.2 THE COMPATIBILITY CONDITION

Suppose we were given a set of two partial differential equations for one unknown function $u(x, y)$, such as

$$\frac{\partial u}{\partial x} = x + 3y, \qquad \frac{\partial u}{\partial y} = x^2. \tag{6.2–1}$$

We know that these equations cannot be solved: We have too many equations that are mutually inconsistent. The inconsistency can be clarified if we compute the second derivative $\partial^2 u/\partial x\, \partial y$ from the two equations: The first yields 3, the second $2x$. They are unequal.

Therefore, when partial differential equations are given, the question of integrability arises. The differential equations

$$\frac{\partial u}{\partial x} = f(x, y), \qquad \frac{\partial u}{\partial y} = g(x, y) \tag{6.2–2}$$

cannot be integrated unless the condition

$$\frac{\partial f}{\partial y} = \frac{\partial g}{\partial x} \tag{6.2–3}$$

is satisfied. This condition, a *condition of integrability*, is also called an *equation of compatibility*.

Now consider a plane state of strain, such as may exist in the solid propellant grain of a rocket. Suppose that an engineer made a laboratory model and obtained, by various instruments such as strain gauges, photoelastic equipment, laser holography combined with Moire pattern analysis, etc., a set of strain data that may be presented as

$$e_{xx} = f(x, y), \qquad e_{yy} = g(x, y), \qquad e_{xy} = h(x, y), \qquad e_{zz} = e_{zx} = e_{zy} = 0. \tag{6.2–4}$$

The question arises whether the data are self-consistent. Could the consistency be checked? And if they are consistent, can we compute the displacements $u(x, y)$ and $v(x, y)$ from these data?

If the strain is small, the last question can be formulated as a mathematical question of integrating the differential equations

$$\begin{aligned} \frac{\partial u}{\partial x} &= f(x, y), \qquad (= e_{xx}) \\ \frac{\partial v}{\partial y} &= g(x, y), \qquad (= e_{yy}) \\ \frac{\partial u}{\partial y} + \frac{\partial v}{\partial x} &= 2h(x, y). \qquad (= 2e_{xy}) \end{aligned} \tag{6.2–5}$$

Now, if we differentiate the first equation with respect to y twice, the second with respect to x twice, and the third with respect to x and y once each, we obtain

$$\frac{\partial^3 u}{\partial x\, \partial y^2} = \frac{\partial^2 f}{\partial y^2}, \qquad \frac{\partial^3 v}{\partial x^2\, \partial y} = \frac{\partial^2 g}{\partial x^2}, \tag{6.2–6}$$

$$\frac{\partial^3 u}{\partial x\, \partial y^2} + \frac{\partial^3 v}{\partial x^2\, \partial y} = 2\frac{\partial^2 h}{\partial x\, \partial y}. \tag{6.2–7}$$

Substituting Eqs. (6.2–6) into Eq. (6.2–7), we have

$$\frac{\partial^2 f}{\partial y^2} + \frac{\partial^2 g}{\partial x^2} = 2\frac{\partial^2 h}{\partial x\, \partial y}. \tag{6.2–8}$$

The experimental data must satisfy this equation. If not, the data are not consistent and there must have been errors.

Expressing the preceding results in terms of strain components, we have

$$\frac{\partial^2 e_{xx}}{\partial y^2} + \frac{\partial^2 e_{yy}}{\partial x^2} = 2\frac{\partial^2 e_{xy}}{\partial x\, \partial y}, \tag{6.2–9}$$

which is the equation of compatibility for a plane state of strain.

A similar discussion applies to a two-dimensional velocity field of a fluid. The components of the rate-of-strain tensor may be measured, for example, by the method of optical birefringence, if the fluid is birefringent. Or a set of strain rates may have been obtained theoretically. To check the consistency, we must have

$$\frac{\partial^2 V_{xx}}{\partial y^2} + \frac{\partial^2 V_{yy}}{\partial x^2} = 2\frac{\partial^2 V_{xy}}{\partial x\, \partial y}, \tag{6.2–10}$$

where V_{ij} are the components of the rate-of-strain tensor. (See Sec. 6.1.) In fluid mechanics, however, this equation is referred to as the *condition of integrability*. Thus, compatibility and integrability mean the same thing.

6.3 COMPATIBILITY OF STRAIN COMPONENTS IN THREE DIMENSIONS

Extending the question discussed in the previous section to three dimensions, how do we integrate the differential equations

$$e_{ij} = \frac{1}{2}\left[\frac{\partial u_j}{\partial x_i} + \frac{\partial u_i}{\partial x_j}\right] \tag{6.3–1}$$

to determine u_i?

Inasmuch as there are six equations for three unknown functions u_i, the system of Eq. (6.3–1) will have a single-valued solution only if the functions e_{ij} satisfy the conditions of compatibility.

By differentiation of Eq. (6.3–1), we have

$$e_{ij,kl} = \tfrac{1}{2}(u_{i,jkl} + u_{j,ikl}), \tag{6.3–2}$$

where the indices k and l following a comma indicate successive partial differentiations with respect to x_k and x_l. Interchanging subscripts, we get

$$e_{kl,ij} = \tfrac{1}{2}(u_{k,lij} + u_{l,kij}),$$

$$e_{jl,ik} = \tfrac{1}{2}(u_{j,lik} + u_{l,jik}),$$

$$e_{ik,jl} = \tfrac{1}{2}(u_{i,kjl} + u_{k,ijl}).$$

From these, we verify at once that

$$e_{ij,kl} + e_{kl,ij} - e_{ik,jl} - e_{jl,ik} = 0. \qquad \blacktriangle \tag{6.3–3}$$

This is the *equation of compatibility* of St. Venant for infinitesimal strains.

Of the 81 equations represented by Eq. (6.3–3), only 6 are essential. The rest are either identities or repetitions, on account of the symmetry of e_{ij} with respect to i, j and of $e_{ij,kl}$ with respect to k, l. The 6 equations, written in unabridged notation, are

$$
\begin{aligned}
\frac{\partial^2 e_{xx}}{\partial y\,\partial z} &= \frac{\partial}{\partial x}\left(-\frac{\partial e_{yz}}{\partial x} + \frac{\partial e_{zx}}{\partial y} + \frac{\partial e_{xy}}{\partial z}\right),\\
\frac{\partial^2 e_{yy}}{\partial z\,\partial x} &= \frac{\partial}{\partial y}\left(-\frac{\partial e_{zx}}{\partial y} + \frac{\partial e_{xy}}{\partial z} + \frac{\partial e_{yz}}{\partial x}\right),\\
\frac{\partial^2 e_{zz}}{\partial x\,\partial y} &= \frac{\partial}{\partial z}\left(-\frac{\partial e_{xy}}{\partial z} + \frac{\partial e_{yz}}{\partial x} + \frac{\partial e_{zx}}{\partial y}\right),\\
2\,\frac{\partial^2 e_{xy}}{\partial x\,\partial y} &= \frac{\partial^2 e_{xx}}{\partial y^2} + \frac{\partial^2 e_{yy}}{\partial x^2},\\
2\,\frac{\partial^2 e_{yz}}{\partial y\,\partial z} &= \frac{\partial^2 e_{yy}}{\partial z^2} + \frac{\partial^2 e_{zz}}{\partial y^2},\\
2\,\frac{\partial^2 e_{zx}}{\partial z\,\partial x} &= \frac{\partial^2 e_{zz}}{\partial x^2} + \frac{\partial^2 e_{xx}}{\partial z^2}.
\end{aligned}
\qquad \blacktriangle \quad (6.3\text{–}4)
$$

For finite strains, a compatibility condition can be derived from Riemann's theorem by the fact that the deformed body remains in a Euclidean space. Riemann has given the necessary and sufficient conditions for the metric tensor (related to strains) to represent a Euclidean space. See Ref. at end of chapter.

Equations (6.3–3) or (6.3–4) are necessary conditions. Are they sufficient? That is, would the six compatibility conditions, together with the six differential equations given by Eq. (6.3–1), guarantee the existence of a set of functions $u_1(x, y, z)$, $u_2(x, y, z)$, $u_3(x, y, z)$ that are single valued and continuous in a continuum? To answer this question, we note first that since strain components only determine the relative positions of points in a body, and since any rigid-body motion corresponds to zero strain, we expect that the solution u_i can be determined only up to an arbitrary rigid-body motion. Next, if e_{ij} were specified arbitrarily, we could expect cases similar to those shown in Fig. 6.1 to exist. Here, a rectangular portion of material is given, of which the legs AB, BC, AD, and DE (C and E are the same point) are composed of successive small rectangular elements (each element similar to those illustrated in Fig. 5.4). Each element is deformed according to the specified strains. By glueing the deformed elements together, first along AB and BC and then along AD and DE, we might end at the points C and E separated, either with a gap between them or with an overlapping of material somewhere. For a single-valued continuous solution to exist (up to a rigid-body motion), the ends C and E must meet perfectly in the strained configuration. This cannot be guaranteed unless the specified strain field obeys certain conditions.

Following this kind of reasoning, one may construct a line integral starting from an arbitrary point A in the body to find the displacement (u_1, u_2, u_3) at point C along two arbitrarily different paths and demand that the results are the same.

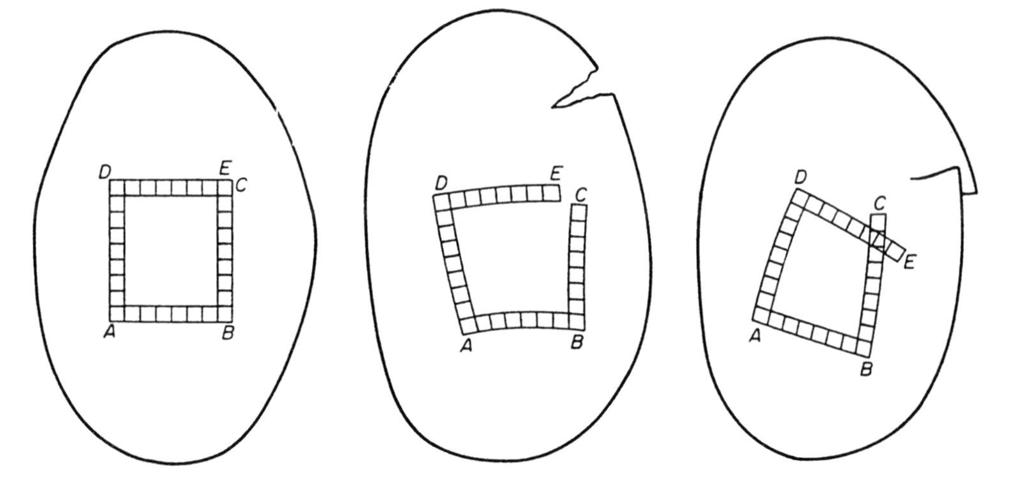

Figure 6.1 Illustration for the requirement of compatibility. The figure on the left consists of a sequence of rectangular elements making up a continuous region in the unstrained state. If strains were specified for each of the little rectangles, and they deformed according to the specifications, then on putting the deformed little rectangles together, the conditions illustrated in the middle or right-hand figure might occur. A sufficient condition, discussed in the text, is necessary to prevent these situations from happening.

Cesaro (1906) has shown that the sufficient conditions for the uniqueness of a solution are exactly Eqs. (6.3–4) if the region enclosed by the arbitrary paths is *simply connected*. However, if the region is multiply connected, additional conditions of sufficiency are required. (See Fung, *Foundations of Solid Mechanics*, Englewood Cliffs, N.J., Prentice Hall, 1965, pp. 101–108, for details.)

PROBLEMS

6.1 Consider the motion of a body fluid with velocity components u and v derived from a potential Φ

$$u = \frac{\partial \Phi}{\partial x}, \qquad v = \frac{\partial \Phi}{\partial y},$$

while the component w is identically zero. Sketch the velocity field for the following potentials:

(a) $\Phi = \dfrac{1}{4\pi} \log (x^2 + y^2) = \dfrac{1}{2\pi} \log r, \qquad (r^2 = x^2 + y^2)$

(b) $\Phi = x$

(c) $\Phi = Ar^n \cos n\theta, \qquad \left(\theta = \tan^{-1} \dfrac{y}{x}\right)$

(d) $\Phi = \dfrac{\cos \theta}{r}$

Note: A flow field whose velocity components are derived from a potential function $\Phi(x, y, z)$ is called a *potential flow*. In the examples given in this problem, we have several cases in which Φ is expressed in terms of the polar coordinates r, θ. If we notice that the velocity vector (u, v) is exactly the gradient of the scalar function $\Phi(x, y)$ (see Chapter 2), we see from vector analysis that the velocity components in the polar coordinates are

$$u_r = \frac{\partial \Phi(r, \theta)}{\partial r}, \qquad u_\theta = \frac{1}{r}\frac{\partial \Phi(r, \theta)}{\partial \theta} \tag{1}$$

in the radial and tangential directions, respectively. (See Fig. P6.1.)

These relations can be derived formally as follows. Since

$$r^2 = x^2 + y^2, \qquad \theta = \tan^{-1}\frac{y}{x}, \tag{2}$$

$$x = r\cos\theta, \qquad y = r\sin\theta. \tag{3}$$

$$\frac{\partial x}{\partial r} = \cos\theta, \qquad \frac{\partial y}{\partial r} = \sin\theta,$$

$$\frac{\partial x}{\partial \theta} = -r\sin\theta, \qquad \frac{\partial y}{\partial \theta} = r\cos\theta,$$

we have

$$\frac{\partial \Phi}{\partial r} = \frac{\partial \Phi}{\partial x}\frac{\partial x}{\partial r} + \frac{\partial \Phi}{\partial y}\frac{\partial y}{\partial r} = u\cos\theta + v\sin\theta. \tag{4}$$

$$\frac{1}{r}\frac{\partial \Phi}{\partial \theta} = -\frac{\partial \Phi}{\partial x}\sin\theta + \frac{\partial \Phi}{\partial y}\cos\theta = -u\sin\theta + v\cos\theta. \tag{5}$$

But it is seen from Fig. P6.1 that

$$u_r = u\cos\theta + v\sin\theta, \qquad u_\theta = -u\sin\theta + v\cos\theta. \tag{6}$$

Hence, Eq. (1) follows from Eqs. (4) and (5).

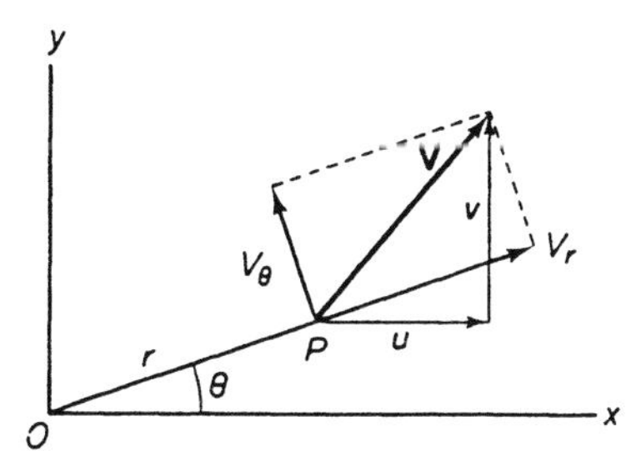

Figure P6.1 Velocity components in polar coordinates.

6.2 The motion of an incompressible fluid in two dimensions may be derived from a stream function ψ as follows:

$$u = -\frac{\partial \psi}{\partial y}, \qquad v = \frac{\partial \psi}{\partial x}, \qquad w = 0.$$

Sketch the lines ψ = const. for the following functions, and compare them with the results of the preceding problem:

(a) $\psi = c\theta$
(b) $\psi = y$
(c) $\psi = Ar^n \sin n\theta$
(d) $\psi = -\dfrac{\sin\theta}{r}$

6.3 For the flows described by the potentials listed in Prob. 6.1 and stream functions listed in Prob. 6.2,

(a) Show that the vorticity vanishes in every case.
(b) Derive expressions for the rate-of-strain tensor.

Solution: The vorticity Ω, given by Eq. (6.1–7), has the components given in Eq. (6.1–4). In a two-dimensional flow, there is only one component of vorticity $\Omega_{12} = (\partial v/\partial x - \partial u/\partial y)/2$, that is not identically zero. If $u = \partial\phi/\partial x$ and $v = \partial\phi/\partial y$, then $\Omega_{12} \equiv 0$. Hence, all potential flow is irrotational. If

$$u = -\frac{\partial\psi}{\partial y}, \qquad v = \frac{\partial\psi}{\partial x}, \tag{1}$$

then Ω_{12} is

$$\text{vorticity} = \frac{1}{2}\left(\frac{\partial^2\psi}{\partial x^2} + \frac{\partial^2\psi}{\partial y^2}\right). \tag{2}$$

In polar coordinates, we have

$$\text{vorticity} = \frac{1}{2}\left(\frac{\partial^2\psi}{\partial r^2} + \frac{1}{r}\frac{\partial\psi}{\partial r} + \frac{1}{r^2}\frac{\partial^2\psi}{\partial\theta^2}\right). \tag{3}$$

That all the cases in Prob. 6.2 are irrotational can be verified by direct substitution.

For part (b), the components of the rate-of-strain tensor in polar coordinates can be derived by the transformation of coordinates shown in Sec. 5.8, p. 126, or by direct derivation, as shown in Sec. 5.9, p. 128. By a slight change of notations, we obtain, according to Eq. (6.1–3).

$$V_{rr} = \frac{\partial u_r}{\partial r}, \qquad V_{\theta\theta} = \frac{u_r}{r} + \frac{1}{r}\frac{\partial u_\theta}{\partial \theta},$$

$$V_{r\theta} = \frac{1}{2}\left(\frac{1}{r}\frac{\partial u_r}{\partial\theta} + \frac{\partial u_\theta}{\partial r} - \frac{u_\theta}{r}\right).$$

With these equations, the problem is easily solved.

6.4 Suppose we were given the following displacement field defined in a unit circle,

$$u = ax^2 + bxy + c,$$
$$v = by^2 + cx + mz,$$
$$w = mz^3.$$

Is there any question of compatibility?

6.5 Suppose the displacement field in a unit circle is the following. Is it compatible?

$$u = ar\log\theta,$$
$$v = ar^2 + c\sin\theta,$$
$$w = 0.$$

6.6 In a two-dimensional, plane-strain field, the displacements are described by $u(x, y)$, $v(x, y)$, whereas that along the z-axis, w, is identically zero, x, y, z being a set of rectangular Cartesian coordinates.
(a) Express the strain components e_{xx}, e_{xy}, e_{yy} in terms of u, v.
(b) Derive the equation of compatibility for the strain system e_{xx}, e_{xy}, e_{yy}.
(c) Is the following strain system compatible?

$$e_{xx} = k(x^2 - y^2), \qquad e_{yy} = kxy, \qquad e_{xy} = k'xy,$$

where k, k' are constants. All other strain components are zero.

Answer to (c): It is, if $k' = -k$.

6.7 A rectangular plate of width a and height b rests on a *rigid* base, Fig. P6.7. The plate material is isotropic and obeys Hooke's law. Its density is ρ. The plate is subjected to a uniformly distributed pressure on top and gravitational load in the vertical direction. (a) State a set of possible boundary conditions. (b) Derive a possible stress distribution that satisfies the equation of equilibrium and those boundary conditions in which the stress is specified. (c) Compute the strains and check whether the conditions of compatibility are satisfied. (d) Determine the displacements in the plate. Are the displacements continuous and single valued? (e) Are all the boundary conditions specified in (a) satisfied? If they are, can you claim that the stress distribution you have just obtained an exact solution of the problem? If all the boundary conditions are not satisfied, then obviously you did not obtain a solution. Two things can be done. First, can you modify the boundary conditions in such a way that you can now claim to have found a solution of a different problem? Next, returning to the original problem, and step (b), can you find a different stress distribution that has a chance to be an exact solution? Is there a general method to obtain an exact solution? Are there restrictions to the statement of boundary conditions in step (a)? Can we say that some boundary-value problems are well posed, whereas others are not well posed? What should be the criterion for well-posedness? For (c), use Hooke's law, Eq. (7.4–7), p. 158.

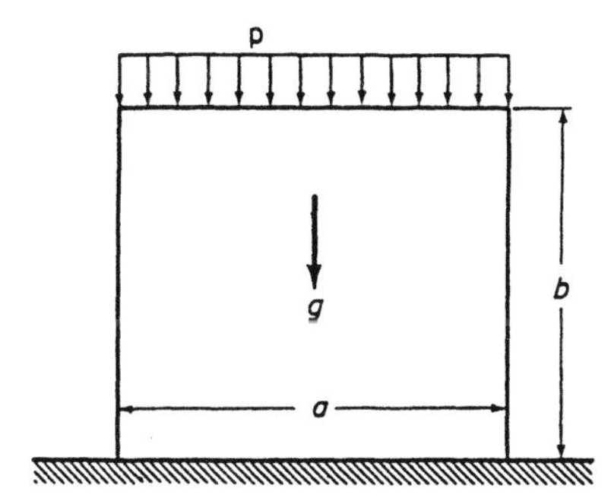

Figure P6.7 A plate loaded by weight and pressure.

FURTHER READING

ERINGEN, A. C., *Nonlinear Theory of Continuous Media*. New York, McGraw-Hill, 1962, pp. 44–46.

TRUESDELL, C., AND TOUPIN, R., *The Classical Field Theories*. In *Handbuch der Physik*, Vol III/1. Springer-Verlag, Berlin, 1960, Art. 34, footnotes.

7 CONSTITUTIVE EQUATIONS

The three most commonly used constitutive equations are presented. They are mathematical abstractions and are given here in the barest outline to exhibit their similarities and differences. They can be simplified greatly if the material is isotropic. Since the concept of isotropy is very important and is usually passed over too lightly by beginners, we shall devote Chapter 8 to it. The properties of some real materials are discussed in Chapter 9.

7.1 SPECIFICATION OF THE PROPERTIES OF MATERIALS

The properties of materials are specified by constitutive equations. A wide variety of materials exists. Thus, we are not surprised that there are a great many constitutive equations describing an almost infinite variety of materials. What is surprising, is the fact that three simple, idealized stress-strain relationships give a good description of the mechanical properties of many materials around us, namely, the nonviscous fluid, the Newtonian viscous fluid, and the perfectly elastic solid. We shall describe these idealized relations in this chapter, but we hasten to add that the properties of real materials differ more or less from the properties described by these idealized laws. When the differences are great, we speak of real gases, non-Newtonian viscous fluids, viscoelastic solids, plasticity, etc., which will be considered in Chapter 9.

An equation that describes a property of a material is called a *constitutive equation* of that material. A stress-strain relationship describes the mechanical property of a material and is therefore a constitutive equation. Our main objective in this chapter is to discuss the stress-strain relationship. There are other constitutive equations, such as those describing heat transfer characteristics, electric resistance, mass transport, etc., but they are not our immediate concern.

7.2 THE NONVISCOUS FLUID

A nonviscous fluid is a fluid for which the stress tensor is isotropic, i.e., of the form

$$\sigma_{ij} = -p\delta_{ij}, \qquad \blacktriangle \quad (7.2\text{–}1)$$

where δ_{ij} is the Kronecker delta and p is a scalar called *pressure*. In matrix form, the components of stress in a nonviscous fluid may be displayed as

$$(\sigma_{ij}) = \begin{pmatrix} -p & 0 & 0 \\ 0 & -p & 0 \\ 0 & 0 & -p \end{pmatrix}. \qquad (7.2\text{–}2)$$

The pressure p in an *ideal gas* is related to the density ρ and temperature T by the equation of state

$$\frac{p}{\rho} = RT, \qquad (7.2\text{–}3)$$

where R is the gas constant. For a real gas or a liquid, it is often possible to obtain an equation of state

$$f(p, \rho, T) = 0. \qquad (7.2\text{–}4)$$

An anomaly exists in the case of an *incompressible fluid*, for which the equation of state is merely

$$\rho = \text{const.} \qquad (7.2\text{–}5)$$

Thus, the pressure p is left as an arbitrary variable for an incompressible fluid. It is determined solely by the equations of motion and the boundary conditions. For example, an incompressible fluid in the cylinder of a hydraulic press can assume any pressure, depending on the force applied to the piston.

Since hydrodynamics is concerned mostly with incompressible fluids, we shall see that pressure is controlled by boundary conditions, whereas variations in pressure (the pressure gradient) are calculated from the equations of motion.

Air and water can be treated as nonviscous in many problems. For example, in problems concerning tides around the earth, waves in the ocean, the flight of an airplane, flow in a jet, and combustion in an automobile engine, excellent results can be obtained by ignoring the viscosity of the medium and treating it as a nonviscous fluid. On the other hand, there are important problems in which the viscosity of the medium, though small, must not be neglected. Among such problems are those of determining the drag force acting on an airplane, whether a flow is turbulent or laminar, the heating of a reentry spacecraft, and the cooling of an automobile engine.

7.3 NEWTONIAN FLUID

A Newtonian fluid is a viscous fluid for which the shear stress is linearly proportional to the rate of deformation. For a Newtonian fluid, the stress-strain relationship is specified by the equation

$$\sigma_{ij} = -p\delta_{ij} + \mathcal{D}_{ijkl}V_{kl}. \qquad \blacktriangle \ (7.3\text{–}1)$$

where σ_{ij} is the stress tensor, V_{kl} is the rate-of-deformation tensor, $\mathcal{D}_{ijkl}$ is a tensor of viscosity coefficients of the fluid, and p is the *static pressure*. The term $-p\delta_{ij}$ represents the state of stress possible in a fluid at rest (when $V_{kl} = 0$). The static pressure p is assumed to depend on the density and temperature of the fluid, according to an equation of state. For Newtonian fluids, we assume that the elements of the tensor $\mathcal{D}_{ijkl}$ may depend on the temperature, but not on the stress or the rate of deformation. The tensor $\mathcal{D}_{ijkl}$, of rank 4, has $3^4 = 81$ elements. Not all of these constants are independent. A study of the theoretically possible number of independent elements can be made by examining the symmetry properties of the tensors σ_{ij}, V_{kl} and the symmetry that may exist in the atomic constitution of the fluid. We shall not pursue it here, because we know of no fluid that has been examined in such detail as to have all the constants in the tensor $\mathcal{D}_{ijkl}$ determined. Most fluids appear to be isotropic, for which the structure of $\mathcal{D}_{ijkl}$ is greatly simplified, as will be seen shortly. Those readers who are interested in the general structure of $\mathcal{D}_{ijkl}$ should read Sec. 7.4 and the references referred to therein, because the tensor of elastic constants C_{ijkl} has a similar structure.

If the fluid is *isotropic*, i.e., if the tensor $\mathcal{D}_{ijkl}$ has the same array of components in any system of rectangular Cartesian coordinates, then $\mathcal{D}_{ijkl}$ can be expressed in terms of two independent constants λ and μ (see Sec. 8.4) as

$$\mathcal{D}_{ijkl} = \lambda\delta_{ij}\delta_{kl} + \mu(\delta_{ik}\delta_{jl} + \delta_{il}\delta_{jk}), \qquad (7.3\text{–}2)$$

and we obtain

$$\sigma_{ij} = -p\delta_{ij} + \lambda V_{kk}\delta_{ij} + 2\mu V_{ij}. \qquad \blacktriangle \ (7.3\text{–}3)$$

A contraction of Eq. (7.3–3) gives

$$\sigma_{kk} = -3p + (3\lambda + 2\mu)V_{kk}. \qquad (7.3\text{–}4)$$

If it is assumed that the mean normal stress $\frac{1}{3}\sigma_{kk}$ is independent of the rate of dilation V_{kk}, then we must set

$$3\lambda + 2\mu = 0; \qquad (7.3\text{–}5)$$

thus, the constitutive equation becomes

$$\sigma_{ij} = -p\delta_{ij} + 2\mu V_{ij} - \tfrac{2}{3}\mu V_{kk}\delta_{ij}. \qquad \blacktriangle \ (7.3\text{–}6)$$

This formulation is due to George G. Stokes, and a fluid that obeys Eq. (7.3–6) is called a *Stokes fluid*, for which one material constant μ, the coefficient of viscosity, suffices to define its property.

If a fluid is *incompressible*, then $V_{kk} = 0$, and we have the constitutive equation for an *incompressible* viscous fluid:

$$\sigma_{ij} = -p\delta_{ij} + 2\mu V_{ij}. \qquad \blacktriangle \quad (7.3\text{–}7)$$

If $\mu = 0$, we obtain the constitutive equation of the *nonviscous fluid*:

$$\sigma_{ij} = -p\delta_{ij}. \qquad (7.3\text{–}8)$$

The presence of the static pressure term p marks a fundamental difference between fluid mechanics and elasticity. To accommodate this new variable, it is often assumed that an *equation of state* exists which relates the pressure p, the density ρ, and the absolute temperature T, i.e.,

$$f(p, \rho, T) = 0. \qquad (7.3\text{–}9)$$

For example, for an *ideal gas*, Eq. (7.2–3) applies; for a real gas, Eq. (9.1–3) may be used. For fresh water and sea water, Tait (1888) and Li (1967) have obtained their equations of state (see refs. at the end of Chap. 9.) An *incompressible* fluid specified by Eq. (7.2–5) is again a special case, for which the pressure p is a variable to be determined by the equations of motion and boundary conditions.

Fluids obeying Eq. (7.3–1) or Eq. (7.3–3), whose viscosity effects are represented by terms that are linear in the components of the rate of deformation, are called *Newtonian fluids*. Fluids that behave otherwise are said to be *non-Newtonian*. For example, a fluid whose coefficient of viscosity depends on the basic invariants of V_{ij} is non-Newtonian. (See Sec. 9.8 for further discussion.)

7.4 HOOKEAN ELASTIC SOLID

A Hookean elastic solid is a solid that obeys Hooke's law, which states that the stress tensor is linearly proportional to the strain tensor; i.e.,

$$\sigma_{ij} = C_{ijkl}e_{kl}, \qquad \blacktriangle \quad (7.4\text{–}1)$$

where σ_{ij} is the stress tensor, e_{kl} is the strain tensor, and C_{ijkl} is a tensor of *elastic constants*, or *moduli*, which are independent of stress or strain. The tensorial quality of the constants C_{ijkl} follows the quotient rule (Sec. 2.9).

As a tensor of rank 4, C_{ijkl} has $3^4 = 81$ elements; but inasmuch as $\sigma_{ij} = \sigma_{ji}$, we must have

$$C_{ijkl} = C_{jikl}. \qquad (7.4\text{–}2)$$

Furthermore, since $e_{kl} = e_{lk}$, and in Eq. (7.4–1) the indices k and l are dummies for contraction, we can always symmetrize C_{ijkl} with respect to k and l without altering the sum. Thus, we can always write Eq. (7.4–1) as

$$\sigma_{ij} = \tfrac{1}{2}(C_{ijkl} + C_{ijlk})e_{kl} = C'_{ijkl}e_{lk}, \qquad (7.4\text{–}3)$$

with the property

$$C'_{ijkl} = C'_{ijlk}. \tag{7.4–4}$$

If such a symmetrization has been done, then under the conditions (7.4–2) and (7.4–4), C_{ijkl} has a maximum of 36 independent constants.

That the total number of elastic constants cannot be more than 36 can be seen if we recall that because $\sigma_{ij} = \sigma_{ji}$ and $e_{ij} = e_{ji}$, there are only six independent elements in the stress tensor σ_{ij} and six in the strain e_{ij}. Hence, if each element of σ_{ij} is linearly related to all elements of e_{ij}, or vice versa, there will be six equations with 6 constants each and, hence, 36 constants in total.

For most elastic solids, the number of independent elastic constants is far smaller than 36. The reduction is caused by the existence of material symmetry. (See the excellent discussions on this subject in the classical books on the theory of elasticity by Love, and by Green and Adkins, listed at the end of this chapter.)

The greatest reduction in the number of elastic constants is obtained when the material is *isotropic*, i.e., when the elastic properties are identical in all directions. More precisely, isotropy for a material is defined by the requirement that the array of numbers C_{ijkl} has exactly the same numerical values, no matter how the coordinate system is oriented. Because of the importance of the concept of isotropy, we shall discuss it in greater detail in Chapter 8. It will be shown that for any isotropic material, exactly *two* independent elastic constants characterize the material. Hooke's law for an isotropic elastic solid reads

$$\sigma_{ij} = \lambda e_{\alpha\alpha}\delta_{ij} + 2\mu e_{ij}. \qquad \blacktriangle \tag{7.4–5}$$

The constants λ and μ are called the *Lamé constants*. In engineering literature, the second Lamé constant μ is practically always written as G and identified as the *shear modulus*.

It will be useful to write out Eq. (7.4–5) *in extenso*. With x, y, z as rectangular Cartesian coordinates, we have Hooke's law for an isotropic elastic solid:

$$\begin{aligned}
\sigma_{xx} &= \lambda(e_{xx} + e_{yy} + e_{zz}) + 2Ge_{xx} \\
\sigma_{yy} &= \lambda(e_{xx} + e_{yy} + e_{zz}) + 2Ge_{yy} \\
\sigma_{zz} &= \lambda(e_{xx} + e_{yy} + e_{zz}) + 2Ge_{zz} \\
\sigma_{xy} &= 2Ge_{xy}, \qquad \sigma_{yz} = 2Ge_{yz}, \qquad \sigma_{zx} = 2Ge_{zx}.
\end{aligned} \qquad \blacktriangle \tag{7.4–6}$$

These equations can be solved for e_{ij}. But customarily, the inverted form is written as

$$\begin{aligned}
e_{xx} &= \frac{1}{E}[\sigma_{xx} - \nu(\sigma_{yy} + \sigma_{zz})], \qquad & e_{xy} &= \frac{1+\nu}{E}\sigma_{xy} = \frac{1}{2G}\sigma_{xy}, \\
e_{yy} &= \frac{1}{E}[\sigma_{yy} - \nu(\sigma_{zz} + \sigma_{xx})], \qquad & e_{yz} &= \frac{1+\nu}{E}\sigma_{yz} = \frac{1}{2G}\sigma_{yz}, \\
e_{zz} &= \frac{1}{E}[\sigma_{zz} - \nu(\sigma_{xx} + \sigma_{yy})], \qquad & e_{zx} &= \frac{1+\nu}{E}\sigma_{zx} = \frac{1}{2G}\sigma_{zx},
\end{aligned} \qquad \blacktriangle \tag{7.4–7}$$

Or, in index notation,

$$e_{ij} = \frac{1 + \nu}{E} \sigma_{ij} - \frac{\nu}{E} \sigma_{\alpha\alpha} \delta_{ij}. \qquad \blacktriangle \quad (7.4\text{–}8)$$

The constants E, ν, and G are related to the Lamé constants λ and G (or μ). (See Eq. (9.6–9) on p. 000.) E is called *Young's modulus*, ν is called *Poisson's ratio*, and G is called the *modulus of elasticity in shear*, or *shear modulus*. In the one-dimensional case, in which σ_{xx} is the only nonvanishing component of stress, we have used the simplified version of these equations in Chapter 5, viz., Eq. (5.1–3) and (5.1–4).

It is very easy to remember Eq. (7.4–7). Recall the one-dimensional case, Eq. (5.1–3). Apply it to the simple block as illustrated in Fig. 1.9, p. 00. When the block is compressed in the z-direction, it shortens by a strain

$$e_{zz} = \frac{1}{E} \sigma_{zz}. \qquad (7.4\text{–}9)$$

In the meantime, the lateral sides of the block will bulge out somewhat. For a linear material, the bulging strain is proportional to σ_{zz} and is in a sense opposite to the stress: A compression induces lateral bulging; a tension induces lateral shrinking. Hence, we write

$$e_{xx} = -\frac{\nu}{E} \sigma_{zz}, \qquad e_{yy} = -\frac{\nu}{E} \sigma_{zz}. \qquad (7.4\text{–}10)$$

This is the case in which σ_{zz} is the only nonvanishing stress. If the block is subjected also to σ_{xx}, σ_{yy}, as is illustrated in Fig. 3.1, p. 00, and if the material is isotropic and linear (so that causes and effects are linearly superposable), then the influence of σ_{xx} on e_{yy}, e_{zz} and σ_{yy} on e_{xx}, e_{zz} must be the same as the influence of σ_{zz} on e_{xx}, e_{yy}. Hence, Eq. (7.4–9) becomes

$$e_{zz} = \frac{1}{E} \sigma_{zz} - \frac{\nu}{E} \sigma_{xx} - \frac{\nu}{E} \sigma_{yy},$$

which is one of the equations of Eq. (7.4–7), and similarly for other equations in Eq. (7.4–7). For the shear stress and shear strain, each component produces its own effect.

Other Forms of Hooke's Law

For an *isotropic elastic material*, Hooke's law may be stated in the form

$$\sigma_{\alpha\alpha} = 3K e_{\alpha\alpha}, \qquad (7.4\text{–}11)$$

$$\sigma'_{ij} = 2G e'_{ij}, \qquad (7.4\text{–}12)$$

where K and G are constants and σ'_{ij} and e'_{ij} are the stress deviation and strain deviation, respectively; i.e.,

$$\sigma'_{ij} = \sigma_{ij} - \tfrac{1}{3}\sigma_{\alpha\alpha}\delta_{ij}, \tag{7.4–13}$$

$$e'_{ij} = e_{ij} - \tfrac{1}{3}e_{\alpha\alpha}\delta_{ij}. \tag{7.4–14}$$

We have seen before that $\frac{1}{3}\sigma_{\alpha\alpha}$ is the mean stress at a point and that, if the strain were infinitesimal, $e_{\alpha\alpha}$ is the change in volume per unit volume: Both are invariants. Thus, Eq. (7.4–11) states that the change in volume of the material is proportional to the mean stress. In the special case of hydrostatic compression, i.e., when

$$\sigma_{xx} = \sigma_{yy} = \sigma_{zz} = -p, \qquad \sigma_{xy} = \sigma_{yz} = \sigma_{zx} = 0,$$

we have $\sigma_{\alpha\alpha} = -3p$, and Eq. (7.4–11) may be written, in the case of infinitesimal strain, with V and ΔV denoting volume and change in volume, respectively, as

$$\frac{\Delta V}{V} = -\frac{p}{K}. \qquad \blacktriangle \tag{7.4–15}$$

Thus, the coefficient K is appropriately called the *bulk modulus* of the material.

The strain deviation e'_{ij} describes a deformation without any change in volume. The stress deviation is simply proportional to the strain deviation. The relationships between the elastic constants are*

$$\lambda = \frac{2G\nu}{1-2\nu} = \frac{G(E-2G)}{3G-E} = K - \frac{2}{3}G = \frac{E\nu}{(1+\nu)(1-2\nu)}$$
$$= \frac{3K\nu}{1+\nu} = \frac{3K(3K-E)}{9K-E},$$
$$G = \frac{\lambda(1-2\nu)}{2\nu} = \frac{3}{2}(K-\lambda) = \frac{E}{2(1+\nu)} = \frac{3K(1-2\nu)}{2(1+\nu)} = \frac{3KE}{9K-E},$$
$$\nu = \frac{\lambda}{2(\lambda+G)} = \frac{\lambda}{(3K-\lambda)} = \frac{E}{2G} - 1 = \frac{3K-2G}{2(3K+G)} = \frac{3K-E}{6K},$$
$$E = \frac{G(3\lambda+2G)}{\lambda+G} = \frac{\lambda(1+\nu)(1-2\nu)}{\nu} = \frac{9K(K-\lambda)}{3K-\lambda)} \tag{7.4–16}$$
$$= 2G(1+\nu) = \frac{9KG}{3K+G} = 3K(1-2\nu),$$
$$K = \lambda + \frac{2}{3}G = \frac{\lambda(1+\nu)}{3\nu} = \frac{2G(1+\nu)}{3(1-2\nu)} = \frac{GE}{3(3G-E)} = \frac{E}{3(1-2\nu)}.$$
$$\frac{G}{\lambda+G} = 1-2\nu, \qquad \frac{\lambda}{\lambda+2G} = \frac{\nu}{1-\nu}.$$

*Data can be found in the *American Institute of Physics Handbook*, New York: McGraw-Hill Book Company (1957), pp. 2-56–2-60.

When the Poisson's ratio ν is 1/4, $\lambda = G$. When $\nu = 1/2$, then $G = E/3$, $1/K = 0$, and $e_{\alpha\alpha} = 0$.

7.5 EFFECT OF TEMPERATURE

In the preceding sections, the stress-strain or strain-rate relations are determined at a given temperature. The viscosity of a fluid, however, varies with temperature as does the elastic modulus of a solid. In other words, the coefficients $\mathcal{D}_{ijkl}$ in Eq. (7.3–1) and C_{ijkl} in Eq. (7.4–1) are functions of temperature and are determined under an isothermal experiment.

Heat induces thermal expansion and affects the zero-stress state of a solid or liquid. If a body has no stress at a temperature T_0, and the stress remains at zero when the temperature is changed to T, then the linear law

$$e_{ij} = \alpha_{ij}(T - T_0) \tag{7.5–1}$$

states that the body will have a strain e_{ij} relative to the state at T_0. Conversely, if the configuration of the body is so restrained that $e_{ij} = 0$ when the temperature changes from T_0 to T, then a stress

$$\sigma_{ij} = -\beta_{ij}(T - T_0) \tag{7.5–2}$$

is induced in the body. α_{ij} and β_{ij} are symmetric tensors of material constants measured at zero stress and zero strain, respectively, at temperature T_0.

When Eq. (7.5–2) is combined with Hooke's law, we obtain the *Duhamel-Neumann law* for thermoelasticity:

$$\sigma_{ij} = C_{ijkl}e_{kl} - \beta_{ij}(T - T_0), \qquad \blacktriangle \tag{7.5–3}$$

For an isotropic material, the second-order tensor β_{ij} must also be isotropic. It follows that β_{ij} must be of the form $\beta\delta_{ij}$ (see Sec. 8.2). Hence, for an isotropic Hookean solid,

$$\sigma_{ij} = \lambda e_{kk}\delta_{ij} + 2Ge_{ij} - \beta(T - T_0)\delta_{ij}. \tag{7.5–4}$$

Here, λ and G are Lamé constants measured at constant temperature. (Further details can be found in Fung, *Foundations of Solid Mechanics*, Chapter 12, esp. p. 355.)

7.6 MATERIALS WITH MORE COMPLEX MECHANICAL BEHAVIOR

As we have said before, the nonviscous fluids, the Newtonian fluids, and the Hookean elastic solids are abstractions. No real material is known to behave exactly as any one of them, although in limited ranges of temperature, stress, and strain, some materials may follow one of these laws quite well.

Real materials may have more complex behavior. For fluids, household paints and varnish are non-Newtonian, as are wet clay and mud. Most colloidal solutions are non-Newtonian also. For solids, most structural materials are, fortunately, Hookean in the useful range of stresses and strains; but beyond certain limits, Hooke's law no longer applies. For example, virtually every known solid material can be broken (fractured) in one way or another, under sufficiently large stresses or strains; but to break is to disobey Hooke's law.

Nevertheless, the vast literature on continuum mechanics is centered around these idealized materials, and the results have been remarkably useful. We shall discuss more complex behavior of liquids and solids in Chapter 9, but we shall leave the mathematical treatment of the non-Newtonian, nonlinearly elastic or inelastic solids to specialized treatises.

PROBLEMS

7.1 You were given a fluid and asked to determine whether it is Newtonian or ideal. What experiment would you do to provide an accurate answer?

7.2 The viscosity of a fluid can be measured in a number of ways. Propose two ways to do it, and in each case present a sketch, an explanation of the design, the method of calculating the coefficient of viscosity, the applicability of the instruments, and pros and cons of the two instruments in comparison.

7.3 To a civil engineer building a large dam, the viscoelastic behavior of the concrete is a very serious matter. Make a speculative assessment of the possible consequences of the "flow" of the concrete in a large dam. Design an experiment to verify the constitutive equation of the concrete. Propose an instrumental system to do this.

7.4 Much can be learned of the constitutive equations of materials by looking into things in our kitchen. Take a stalk of fresh celery, or a fresh carrot. Bend them and they snap crisply. Let the celery and carrot dry out a few days. Then bend them and they would not break. Why? How would the constitutive equations of the carrot and celery reflect these observations?

7.5 Sour dough and spaghetti are two excellent materials for rheological investigation. Feel them with your hands and fingers and propose a mathematical description of their constitutive equations.

7.6 Take a string and a pair of scissors. If the string is slack, and the scissors are dull, you may find that the string is not easily cut. Now, stretch the string taut, and then apply the scissors. The cutting is then very easy. Why?

7.7 A group of physiologists climbed Mt. Everest, the highest peak of the world. They wanted to collect samples of Himalayan air at high altitude to bring back to their laboratories for detailed analysis. How can you do that? Please invent a way. One suggestion was to bring a number of glass pipettes, and use an electric current to seal the ends. Could that be a practical way?

7.8 Getting the constitutive equations and equations of state of fresh and sea water is very important for the understanding of oceanography. How can you obtain samples of sea water at great depth and over a wide area of the sea for testing in the laboratory? One of the earliest to do it was P. G. Tait. See references to Tait (1888) and Li (1967) at the end of Chapter 9. Design a modern way to do it yourself.

Discuss the importance of the data with regard to wave motion, marine life, underwater acoustics, antisubmarine warfare.

7.9 The explosions of great volcanos are nature's display of continuum mechanics. What are the materials involved? What are their constitutive equations? How to study these constitutive equations?

7.10 People living in the Gobi desert in Sinkiang and Mongolia describe the occasional terrible sandstorms. How can sand flow like a fluid? Speculate on a constitutive equation, and design an experiment for testing. Design an instrument that can be taken there to do some measurements.

7.11 I trust that you have visited a sand dune somewhere. Now, with the knowledge of continuum mechanics, formulate a mathematical problem that will enable you to predict the shape of the sand dune.

7.12 In geological terms, glaciers flow, rock beds bend, mountains move, continents collide. All depend on forces, structures, and constitutive equations. Design some experiments to investigate the constitutive equations of the ice, rock, mountains, ocean floors, and the continents.

7.13 We would certainly like to know the constitutive equations of the earth's mantle and core, and of the planets, the sun, and stars. How can we get a handle on this? What kind of observations would help us deducing information on the constitutive equations of the materials of these objects?

7.14 An astronaut went to the moon and brought back some rock samples. We are interested in the mechanical properties of these rocks. Design a program of experiments to get as much information out of the small amount of rocks as possible.

7.15 A way to explore the moon is to use unmanned rockets with remote-controlled instruments. Suppose that a landing is planned to study the mechanical properties of the materials of the lunar surface. Design an instrument package that will yield the desired information.

7.16 Arthritis of the knee, hip, elbow, and finger joints afflicts many people. In terms of the constitutive equation of the articular cartilage, what happens?

7.17 Suppose that you are planning a biomechanics laboratory to investigate the mechanical properties of the muscles from the point of view of determining that muscles' constitutive equations. Make a list of the desired properties to be investigated. Make a list of the experiments that should be done. I am pretty sure that all the instruments needed for all the desirable experiments do not exist. Here is your opportunity to invent. Select some instruments of key significance and go ahead to invent. Make sketches, designs, calculations. Consider the feasibility, the cost, and the pay off.

7.18 There are three very different kinds of muscles: the skeletal, the heart, and the smooth muscles of the blood vessels, ureter, bladder, uterus, and other internal organs. There are differences in the availability of test specimens from animals and from humans. There are differences between the testing of isolated specimens and in vivo specimens. So the answer to the preceding problem has to be narrowed down to more specific categories. As an investigator, this narrowing down is indeed a crucial step which requires wisdom, experience, and ambition. The selection of an appropriate target of research will reflect one's training and personality. Give some thought to this, make your selection, explain the reasoning to yourself, write it down, and review it one year from today.

7.19 A similar problem can be formulated for other tissues for investigation in a biomechanics laboratory. No living tissue is unimportant. People take the health of their bodies for granted until disease strikes. Realizing this, plan an investigation of a tissue other than the muscles.

7.20 Every constitutive equation must be tensorially correct. Based on this requirement, consider the question of generalizing Starling's hypothesis, which is well known in physiology, but is formulated without regard to stresses other than pressure. Starling's hypothesis states that the rate of transfer of water across a membrane is governed by the following formula:

$$\dot{m} = k(p_1 - p_2 - \pi_1 + \pi_2)$$

where $\dot{m}$ is the rate of movement of water (g/sec/m^2), p_1 and π_1 are, respectively, the hydrostatic and osmotic pressures on one side of the membrane, p_2 and π_2 are those on the other side, and k is the permeability constant, with units of seconds per meter. In considering water movement across the endothelium of the blood vessel, we realize that the flowing blood will impose a shear stress on the membrane, and the endothelial cells will respond to the shear stress with internal stresses. Many recent papers have reported the important effects of the shear stress on blood vessel remodeling, and on the transport of ions and enzymes across the endothelium. It may play a role on the transport of water also. Hence, propose a generalization of Starling's law to include the stresses in the media on both sides of the membrane, and describe a plan to verify the proposal experimentally.

Discussion. Let τ_{ij} be the stress-deviation tensor. Since $\dot{m}$ is a scalar, any involvement of τ_{ij} must be in the form of scalar invariants, such as $\tau'_{ij}\tau'_{ij}$, $\tau'_{ij}V'_{ij}$, $\tau'_{ij}e'_{ij}$, or $c_{ij}\tau_{ij}$, where V'_{ij} is the rate of the deformation deviation tensor, e'_{ij} is the strain deviation tensor, and c_{ij} is a tensorial set of constants. Hence, we might have the hypothetical relationships

$$\dot{m} = k(\Delta p - \Delta\pi) + c\,\Delta\tau'_{ij}\tau'_{ij}.$$

$$\dot{m} = k(\Delta p - \Delta\pi) + c\,\Delta\tau'_{ji}V'_{ij}.$$

or

$$\dot{m} = k(\Delta p - \Delta\pi) + c\,\Delta\tau'_{ij}e'_{ij}.$$

$$\dot{m} = k(\Delta p - \Delta\pi) + \Delta c_{ij}\tau'_{ij}.$$

Here, Δ means the difference of the quantities on the two sides of the membrane.

FURTHER READING

ERINGEN, A. C., *Mechanics of Continua*. New York: Wiley (1967), p. 145.

GREEN, A. E., AND J. E. ADKINS, *Large Elastic Deformations*. Oxford: University Press (1960), Chap. 1, esp. pp. 11–35.

LOVE, A. E. H., *A Treatise on the Mathematical Theory of Elasticity*. Cambridge: University Press, 1st ed. (1892), 4th ed. (1927); New York: Dover Publications (1963), Chapter 6, esp. pp. 151–165.

8 ISOTROPY

The concept of isotropy is used frequently as a simplifying assumption in continuum mechanics. First, we shall define material isotropy and isotropic tensors. Then we shall determine isotropic tensors of ranks 2, 3, and 4 and apply them to the constitutive equations of isotropic materials.

8.1 THE CONCEPT OF MATERIAL ISOTROPY

Materials whose mechanical properties do not depend on directions are said to be *isotropic*. For example, if we make a tension test on a metal and find that the result does not depend on the direction the tension specimen was cut from the ingot and that the lateral contraction is the same in every direction perpendicular to the direction of pulling, we may suspect that the metal is isotropic.

To give a precise definition, we make use of the constitutive equation: *A material is isotropic if its constitutive equation (the stress-strain-history law) is unaltered under orthogonal transformations of coordinates* (Sec. 2.4). For example, if the constitutive equation is $\sigma_{ij} = C_{ijkl}e_{kl}$, we demand that, after an orthogonal transformation, the law read $\bar{\sigma}_{ij} = C_{ijkl}\bar{e}_{kl}$, where the barred quantities refer to the new coordinates.

Since orthogonal transformations consist of translations, rotations, and reflections of coordinate axes, the definition requires that the mathematical form of the constitutive equation remain unchanged, no matter how the axes are translated, rotated, or reflected. In particular, the array of material constants must have the same values in any right-handed or left-handed system of rectangular Cartesian coordinates.

8.2 ISOTROPIC TENSOR

Definitions

An *isotropic tensor* in Euclidean space is a tensor whose components in any rectangular Cartesian system are unaltered by orthogonal transformations of coordinates.

By definition (Sec. 2.4), an orthogonal transformation from x_1, x_2, x_3 to $\bar{x}_1, \bar{x}_2, \bar{x}_3$ is

$$\bar{x}_i = \beta_{ij} x_j + \alpha_i, \qquad (i = 1, 2, 3) \tag{8.2–1}$$

where β_{ij} and α_i are constants, under the restriction that

$$\beta_{ik}\beta_{jk} = \delta_{ij}. \tag{8.2–2}$$

An orthogonal transformation is said to be *proper* if a right-handed system of coordinate axes is transformed into a right-handed one. For a transformation to be proper, the Jacobian must be positive (see Sec. 2.5). For the orthogonal transformation (8.2–1), the Jacobian is the determinant $|\beta_{ij}|$, which, according to Eq. (8.2–2), must have the value ± 1. Hence, for an orthogonal transformation to be proper, we must have

$$\det |\beta_{ij}| = 1. \tag{8.2–3}$$

For example, all rotations of coordinate axes are proper, but a reflection in the $x_2 x_3$-plane

$$\begin{cases} \bar{x}_1 = -x_1, \\ \bar{x}_2 = x_2, \\ \bar{x}_3 = x_3, \end{cases} \qquad (\beta_{ij}) = \begin{pmatrix} -1 & 0 & 0 \\ 0 & 1 & 0 \\ 0 & 0 & 1 \end{pmatrix}, \qquad |\beta_{ij}| = -1 \tag{8.2–4}$$

is orthogonal, but improper, because it turns a right-handed system into a left-handed one.

Connection between Isotropic Tensors and Isotropic Material

We shall prove that if the relation

$$\sigma_{ij} = C_{ijkl} e_{kl} \tag{8.2–5}$$

is isotropic, then C_{ijkl} is an isotropic tensor.

Proof: By the quotient rule (Sec. 2.9), C_{ijkl} is a tensor of rank 4. Hence, C_{ijkl} transforms according to the tensor transformation rule. Now, transforming Eq. (8.2–5) into new coordinates $\bar{x}_i$, we have

$$\bar{\sigma}_{ij} = \bar{C}_{ijkl} \bar{e}_{kl}. \tag{8.2–6}$$

But the definition of material isotropy requires that

$$\bar{\sigma}_{ij} = C_{ijkl} \bar{e}_{kl}. \tag{8.2–7}$$

Hence, by comparing Eq. (8.2–6) with Eq. (8.2–7), we obtain

$$\bar{C}_{ijkl} = C_{ijkl}. \tag{8.2–8}$$

Thus, C_{ijkl} is an isotropic tensor.

Isotropic Tensors of Ranks 0, 1, and 2

All scalars are, of course, isotropic. But there is no isotropic tensor of rank 1. For, if the vector A_i were isotropic, then it would have to satisfy the equation

$$\overline{A}_i = A_i = \beta_{ij} A_j \tag{8.2–9}$$

for all possible orthogonal transformations. In particular, for a 180° rotation about the x_1-axis, we would have

$$\begin{cases} \overline{x}_1 = x_1, \\ \overline{x}_2 = -x_2, \\ \overline{x}_3 = -x_3, \end{cases} \qquad (\beta_{ij}) = \begin{pmatrix} 1 & 0 & 0 \\ 0 & -1 & 0 \\ 0 & 0 & -1 \end{pmatrix}, \tag{8.2–10}$$

Eq. (8.2–9) then becomes

$$A_1 = A_1, \qquad A_2 = -A_2, \qquad A_3 = -A_3.$$

Hence, $A_2 = A_3 = 0$. Similarly, by the same process but with the role of x_1, x_2, x_3 permuted, we obtain $A_1 = 0$. Thus, the nonexistence of any isotropic tensor of rank 1 is proved.

For tensors of rank 2, the Kronecker delta δ_{ij} is an isotropic tensor, because

$$\begin{aligned} \overline{\delta}_{ij} &= \beta_{im}\beta_{jn}\delta_{mn} && \text{(by the definition of a tensor)} \\ &= \beta_{im}\beta_{jm} && (\text{since } \delta_{mn} = 0 \text{ if } m \neq n) \\ &= \delta_{ij}. && \text{[by Eq. (8.2–2)].} \end{aligned}$$

We propose to show that every isotropic tensor of rank 2 may be reduced to the form $p\delta_{ij}$, where p is a scalar.

For the proof, we note first that if a tensor B_{ij} is isotropic, it must be diagonal. For, imposing the 180° rotation about the x_1-axis, as specified by Eq. (8.2–10), we obtain

$$\overline{B}_{12} = \beta_{1m}\beta_{2n}B_{mn} = -B_{12}.$$

But isotropy requires that $\overline{B}_{12} = B_{12}$. Hence, $B_{12} = 0$. Similarly, $B_{ij} = 0$ if $i \neq j$. Hence, B_{ij} is symmetric and diagonal.

Next, let ϵ_{ijk} be the permutation tensor, and consider the transformation

$$\overline{x}_j = (\delta_{ij} + d\theta\epsilon_{3ij})x_i,$$

$$(\beta_{ij}) = (\delta_{ij} + d\theta\epsilon_{3ij}) = \begin{pmatrix} 1 & d\theta & 0 \\ -d\theta & 1 & 0 \\ 0 & 0 & 1 \end{pmatrix}, \tag{8.2–11}$$

which represents a rotation about the x_3-axis with an infinitesimal angle of rotation $d\theta$.* The definition of tensors furnishes the relation

$$
\begin{aligned}
\overline{B}_{ij} &= (\delta_{im} + d\theta\epsilon_{3im})(\delta_{jn} + d\theta\epsilon_{3jn})B_{mn} \\
&= \delta_{im}\delta_{jn}B_{mn} + d\theta(\epsilon_{3im}\delta_{jn}B_{mn} + \epsilon_{3jn}\delta_{im}B_{mn}) + d\theta^2\epsilon_{3im}\epsilon_{3jn}B_{mn} \qquad (8.2\text{–}12) \\
&= B_{ij} + d\theta(\epsilon_{3im}B_{mj} + \epsilon_{3jn}B_{in}) + O(d\theta^2).
\end{aligned}
$$

But if B_{ij} is isotropic, we must have $\overline{B}_{ij} = B_{ij}$. Hence, for small but arbitrary $d\theta$, we must have

$$\epsilon_{3im}B_{mj} + \epsilon_{3jn}B_{in} = 0. \qquad (8.2\text{–}13)$$

Take $i = 1, j = 1$; then we have

$$\epsilon_{312}B_{21} + \epsilon_{312}B_{12} = B_{21} + B_{12} = 0.$$

But B_{ij} is symmetric, as we have shown. Hence, $B_{12} = B_{21} = 0$. This agrees with what we have just learned, but no new knowledge is gained.

Now take $i = 1, j = 2$; then we have

$$\epsilon_{312}B_{22} + \epsilon_{321}B_{11} = B_{22} - B_{11} = 0.$$

Hence, $B_{11} = B_{22}$. It is evident that an entirely similar rotation about the x_1-axis would yield $B_{23} = 0$, $B_{22} = B_{33}$, and a rotation about the x_2-axis would yield $B_{31} = 0$, $B_{33} = B_{11}$. Hence, the isotropic tensor B_{ij} is reduced to the form $B_{11}\delta_{ij}$. Writing p for B_{11}, we obtain $B_{ij} = p\delta_{ij}$.

Now any rotation from one rectangular Cartesian coordinate system to another can be performed by repeated infinitesimal rotations about coordinate axes. Hence, the conditions just examined are the only conditions imposed by isotropy with respect to proper orthogonal transformations. Thus, $B_{ij} = p\delta_{ij}$ for all proper orthogonal transformations.

For the second-rank isotropic tensor $p\delta_{ij}$, a reflection in the x_2x_3-plane, Eq. (8.2–4), does not change the value of the tensor. By the argument of arbitrary rotation, we conclude that a reflection in any plane would not affect its value. Hence, the form we have found is isotropic with respect to all orthogonal transformations. Q.E.D.

This proof is due to Jeffreys, see Ref. on p. 180. Note that for an isotropic tensor, the coordinate axes may be labeled in an arbitrary order. Thus, a cyclic permutation of the indices 1, 2, 3 cannot affect the values of the components of a tensor that is isotropic with respect to rotation of the coordinate axes. Hence, $B_{12} = 0$ implies $B_{31} = 0$. If the tensor is isotropic also with respect to reflection, then an arbitrary permutation of the indices 1, 2, 3 will not affect the values of the components. Use of these arguments may shorten the proof.

*See the rotation matrix of Eq. (2.4–5), and note that when θ is very small, $\cos\theta \doteq 1$, $\sin\theta \doteq 0$. Identifying the angle θ of Sec. 2.4 with $d\theta$ here furnishes the geometric interpretation of the transformation of Eq. (8.2–11).

8.3 ISOTROPIC TENSORS OF RANK 3

For tensors of rank 3, we can verify that the permutation tensor ϵ_{ijk} is isotropic with respect to rotation of coordinate axes (proper orthogonal transformations). It is not isotropic with respect to reflection in a coordinate plane, because a reflection such as Eq. (8.2–4) turns $\epsilon_{123} = 1$ into $\bar{\epsilon}_{123} = -1$.

We can show that with respect to all rotations of coordinates, the only isotropic tensors of rank 3 are scalar multiples of ϵ_{ijk}. The proof can be constructed similarly to that for the second-rank tensor. Let u_{ijk} be an isotropic tensor of rank 3. Consider an infinitesimal rotation of an angle $d\theta$ about an arbitrary axis ξ (a vector with components ξ_k) passing through the origin:

$$\bar{x}_j = (\delta_{ij} + d\theta\, \xi_k \epsilon_{kij})x_i. \tag{8.3–1}$$

Then, according to the tensor transformation law,

$$\begin{aligned}\bar{u}_{ijk} &= (\delta_{im} + d\theta \xi_s \epsilon_{sim})(\delta_{jn} + d\theta \xi_s \epsilon_{sjn})(\delta_{kp} + d\theta \xi_s \epsilon_{skp})u_{mnp} \\ &= u_{ijk} + d\theta\{\xi_s \epsilon_{sim} u_{mjk} + \xi_s \epsilon_{sjn} u_{ink} + \xi_s \epsilon_{skp} u_{ijp}\} + O(d\theta^2).\end{aligned}$$

By isotropy, $\bar{u}_{ijk} = u_{ijk}$; hence, for small $d\theta$, the quantity in the braces must vanish. (We can ignore quantities of the higher order.) Thus, for all i, j, k,

$$\xi_s \epsilon_{sim} u_{mjk} + \xi_s \epsilon_{sjn} u_{ink} + \xi_s \epsilon_{skp} u_{ijp} = 0. \tag{8.3–2}$$

Take $i = j = 1$. Then

$$\begin{aligned}&-\xi_2 u_{31k} + \xi_3 u_{21k} - \xi_2 u_{13k} + \xi_3 u_{12k} + \xi_s \epsilon_{sk1} u_{111} \\ &\qquad + \xi_s \epsilon_{sk2} u_{112} + \xi_s \epsilon_{sk3} u_{113} = 0.\end{aligned} \tag{8.3–3}$$

Now put $k = 2$. Then, since ξ_1, ξ_2, ξ_3 are arbitrary, their coefficients must vanish, and we obtain

$$\begin{aligned}u_{212} + u_{122} &= u_{111}, \\ u_{312} + u_{132} &= 0, \\ u_{113} &= 0.\end{aligned} \tag{8.3–4}$$

From the last equation, and by symmetry, $u_{ijk} = 0$ if two of i, j, k are equal and the third is unequal. Then, by the first equation of Eq. (8.3–4), u_{ijk} is also zero if all of i, j, k are equal. The second equation shows that

$$u_{ijk} = -u_{jik}.$$

If, in Eq. (8.3–3), we put $k = 1$, then every term vanishes, yielding no new information.

Finally, consider the case in which i, j, k are all different in Eq. (8.3–2). We note that u_{mjk} is zero when $m = j$. Then it is clear that Eq. (8.3–2) holds because all the coefficients vanish. It follows that the only isotropic tensors of rank 3 (isotropic with respect to rotations, not reflections) are scalar multiples of ϵ_{ijk}.

Q.E.D.

8.4 ISOTROPIC TENSORS OF RANK 4

Isotropic tensors of rank 4 are of particular interest to the constitutive equations of materials. It is readily seen that since the unit tensor δ_{ij} is isotropic, the tensors

$$\delta_{ij}\delta_{kl}, \qquad \delta_{ik}\delta_{jl} + \delta_{il}\delta_{jk}, \qquad \delta_{ik}\delta_{jl} - \delta_{il}\delta_{jk} = \epsilon_{sij}\epsilon_{skl} \tag{8.4–1}$$

are isotropic. We propose to show that if u_{ijkl} is an isotropic tensor of rank 4, then it is of the form

$$\lambda\delta_{ij}\delta_{kl} + \mu(\delta_{ik}\delta_{jl} + \delta_{il}\delta_{jk}) + \nu(\delta_{ik}\delta_{jl} - \delta_{il}\delta_{jk}), \tag{8.4–2}$$

where λ, μ, and ν are scalars. Furthermore, if u_{ijkl} has the symmetry properties

$$u_{ijkl} = u_{jikl}, \qquad u_{ijkl} = u_{ijlk}, \tag{8.4–3}$$

then

$$u_{ijkl} = \lambda\delta_{ij}\delta_{kl} + \mu(\delta_{ik}\delta_{jl} + \delta_{il}\delta_{jk}). \tag{8.4–4}$$

Proof: We shall establish the results for isotropy with respect to both rotation of coordinate axes and reflections in coordinate planes.

First, we note that the coordinate axes may be labeled in an arbitrary order. Thus, a permutation in the indices 1, 2, 3 cannot affect the values of the components of an isotropic tensor. Hence,

$$\begin{aligned}
&u_{1111} = u_{2222} = u_{3333}, \\
&u_{1122} = u_{2233} = u_{3311} = u_{1133} = u_{2211} = u_{3322}, \\
&u_{1212} = u_{2323} = u_{3131} = u_{1313} = u_{2121} = u_{3232}, \\
&u_{1221} = u_{2332} = u_{3113} = u_{2112} = u_{3223} = u_{1331}.
\end{aligned} \tag{8.4–5}$$

Next, we note that a rotation of 180° about the x_1-axis, corresponding to the transformation given by Eq. (8.2–10), changes the sign of any term with an odd number of the index 1. But these terms must not change sign on account of isotropy. Hence, they are zero. For example,

$$u_{1222} = u_{1223} = u_{2212} = 0. \tag{8.4–6}$$

By symmetry, this is true for any index i.

These conditions reduce the maximum number of numerically distinct components of the tensor u_{ijkl} to four, namely, u_{1111}, u_{1122}, u_{1212}, n_{1221}.

Now, let us impose the transformation given by Eq. (8.2–11) corresponding to an infinitesimal rotation about the x_3-axis. The tensor transformation law requires that

$$\bar{u}_{pqrs} = u_{pqrs} + d\theta\{\epsilon_{3ip}u_{iqrs} + \epsilon_{3iq}u_{pirs} + \epsilon_{3ir}u_{pqis} + \epsilon_{3is}u_{pqri}\} + O(d\theta^2). \tag{8.4–7}$$

Since, for an isotropic tensor $\bar{u}_{pqrs} = u_{pqrs}$, the terms in the braces must vanish,

$$\epsilon_{3ip}u_{iqrs} + \epsilon_{3iq}u_{pirs} + \epsilon_{3ir}u_{pqis} + \epsilon_{3is}u_{pqri} = 0. \tag{8.4–8}$$

Because there are only three possible values (1, 2, 3) for each of the four indices $pqrs$, at least two of them must be equal. Hence, we may consider separately the cases where (a) all four are equal, (b) three are equal, (c) two are equal and the other two are unequal, and (d) two are equal and the other two are equal.

In Case a, take $p = q = r = s = 1$. Then we see that all terms in Eq. (8.4–8) vanish on account of Eq. (8.4–6). Similarly, $p = q = r = s = 2$ or 3 yields no information.

In Case b, take $p = q = r = 1, s = 2$. We get

$$-u_{2112} - u_{1212} - u_{1122} + u_{1111} = 0. \tag{8.4–9}$$

No new information is obtained by setting $p = q = r = 2, s = 1$, because this merely amounts to an interchange of indices 1 and 2, which has been considered in Eq. (8.4–5). The case $p = q = r = 3$ is trivial because the ϵ_{3ip} terms vanish.

Cases c and d yield conditions contained in Eqs. (8.4–5) and (8.4–6).

Since a rotation from one rectangular coordinate system to another with the same origin can be obtained by repeated infinitesimal rotations about coordinate axes, no additional conditions are imposed on u_{pqrs} by isotropy.

Now let

$$\begin{aligned} u_{1122} &= \lambda, \\ u_{1212} &= \mu + \nu, \\ u_{2112} &= \mu - \nu. \end{aligned} \tag{8.4–10}$$

Then Eq. (8.4–9) says

$$u_{1111} = \lambda + 2\mu. \tag{8.4–11}$$

There appear, therefore, to be three independent isotropic tensors of order 4, obtainable by taking each of λ, μ, ν in turn equal to 1 and the others to 0.

The tensor obtained by taking $\lambda = 1$, $\mu = \nu = 0$ has components $u_{ijkl} = 1$ if $i = j$, $k = l$ and vanishes in all other cases. Therefore, it is equivalent to

$$u_{ijkl} = \delta_{ij}\delta_{kl}. \tag{8.4–12}$$

In the tensor obtained by taking $\mu = 1$, $\lambda = \nu = 0$, the component $u_{ijkl} = 1$ if $i = k$, $j = l$, $i \neq j$, and if $i = l$, $j = k$, $i \neq j$; whereas $u_{ijkl} = 2$ if $i = j = k = l$. Other components are zero. This is exactly

$$u_{ijkl} = \delta_{ik}\delta_{jl} + \delta_{il}\delta_{jk}. \tag{8.4–13}$$

The tensor obtained by taking $\lambda = \mu = 0$, $\nu = 1$ has elements $u_{ijkl} = 1$ when $i = k, j = l, i \neq j$; and $u_{ijkl} = -1$ when $i = l, j = k, i \neq j$. All other components are zero. Hence,

$$u_{ijkl} = \delta_{ik}\delta_{jl} - \delta_{il}\delta_{jk}. \tag{8.4–14}$$

The general isotropic tensor of rank 4 is therefore given by Eq. (8.4–2). From this equation, Eq. (8.4–4) follows under the symmetry condition given in Eq. (8.4–3).

Q.E.D.

8.5 ISOTROPIC MATERIALS

If an elastic solid is isotropic, the tensor C_{ijkl} in Eq. (8.1–1),

$$\sigma_{ij} = C_{ijkl}e_{kl}, \tag{8.5–1}$$

must be isotropic (Sec. 8.2). Furthermore, it has been shown generally that $C_{ijkl} = C_{jikl}$ because the stress tensor is symmetric and that $C_{ijkl} = C_{ijlk}$ because the strain tensor is symmetric and the sum $C_{ijkl}e_{kl}$ is symmetrizable without loss of generality. Hence, according to Eq. (8.4–4),

$$C_{ijkl} = \lambda\delta_{ij}\delta_{kl} + \mu(\delta_{ik}\delta_{jl} + \delta_{il}\delta_{jk}) \tag{8.5–2}$$

and Eq. (8.5–1) becomes

$$\sigma_{ij} = \lambda e_{kk}\delta_{ij} + 2\mu e_{ij}. \tag{8.5–3}$$

This is the most general form of the stress-strain relationship for an isotropic elastic solid for which the stresses are linear functions of the strains. Therefore, an isotropic elastic solid is characterized by two material constants: λ and μ.

Similarly, an isotropic viscous fluid (Sec. 7.3) is governed by the relationship

$$\sigma_{ij} = -p\delta_{ij} + \lambda V_{kk}\delta_{ij} + 2\mu V_{ij}. \tag{8.5–4}$$

8.6 COINCIDENCE OF PRINCIPAL AXES OF STRESS AND OF STRAIN

An important attribute of the isotropy of an elastic body (or a viscous fluid) is that the principal axes of stress and the principal axes of strain (or strain rate) coincide. This follows from Eq. (8.5–3) or Eq. (8.5–4), because the direction cosines of the principal axes of stress and strain are, respectively, the solutions of the equations (Secs. 4.5 and 5.7)

$$(\sigma_{ji} - \sigma\delta_{ji})\nu_j = 0, \qquad |\sigma_{ji} - \sigma\delta_{ji}| = 0, \tag{8.6–1}$$

$$(e_{ji} - e\delta_{ji})\nu_j = 0, \qquad |e_{ji} - e\delta_{ji}| = 0. \tag{8.6–2}$$

By Eq. (8.5–3), Eq. (8.6–1) becomes

$$(\lambda e_{kk}\delta_{ij} + 2\mu e_{ij} - \sigma\delta_{ji})\nu_j = 0, \tag{8.6–3}$$

or

$$2\mu(e_{ji} - \sigma'\delta_{ji})\nu_j = 0 \tag{8.6–4}$$

if we introduce a new variable

$$\sigma' = \frac{\sigma - \lambda e_{kk}}{2\mu}. \tag{8.6–5}$$

But Eq. (8.6–4) is of precisely the same form as Eq. (8.6–2). Thus, although the eigenvalues (principal stresses and strains) are different, the principal directions, given by the solutions ν_j, are the same.

There are other ways of recognizing the coincidence of the principal directions of stress and strain. For example, we recognize in Mohr's circle construction (Secs. 4.3, 5.7) that the angle between the principal axes and the x-axis does not depend on the location of the center of the circle. The principal angle can be determined if the center is translated to the origin. Such a translation is accomplished by setting $\sigma_{kk} = 0$, $e_{kk} = 0$, under which condition the stress-strain relationship becomes simply

$$\sigma'_{ij} = 2\mu e'_{ij}.$$

The coincidence of principal directions is then evident because 2μ is just a numerical factor.

8.7 OTHER METHODS OF CHARACTERIZING ISOTROPY

There are other ways to characterize isotropy. For example, one may define the property of an elastic body through the strain-energy function $W(e_{11}, e_{12}, \ldots, e_{33})$, which is a function of the strain components and which defines the stress components by the relation

$$\sigma_{ij} = \frac{\partial W}{\partial e_{ij}}. \tag{8.7–1}$$

Then isotropy may be stated as the fact that the strain-energy function depends only on the *invariants* of the strain. For example, using the strain invariants

$$I_1 = e_{ii},$$

$$I_2 = \tfrac{1}{2} e_{ij} e_{ji}, \qquad \text{or } J_2 = \tfrac{1}{2} e'_{ij} e'_{ji}$$

$$I_3 = \tfrac{1}{2} e_{ij} e_{jk} e_{ki},$$

we may specify $W(e_{11}, e_{12}, \ldots, e_{33})$ to be a function

$$W(I_1, I_2, I_3). \tag{8.7–2}$$

Since the invariants retain their form (and value) under all rotations of coordinates, the same attribute applies to Eq. (8.7–1).

8.8 CAN WE RECOGNIZE A MATERIAL'S ISOTROPY FROM ITS MICROSTRUCTURE?

A material is said to be isotropic if its stress-strain relationship does not change when the frame of reference is rotated. If you cut a test specimen from an isotropic material and perform a test (e.g., a strip for a tensile test, a block for a compression test, a beam for a bending test, a shaft for a torsion test, a plate for a biaxial loading test, a plate with holes or notches for a stress concentration or fatigue strength test, or a cube or a cylinder for a triaxial loading test), then the results relating the measured stresses and strains should be the same no matter in which orientation you cut the specimen, provided that the size of the specimen is large enough so that the stresses and strains are well defined according to the limit concept discussed in Secs. 1.5 and 1.6. If the material is spatially nonhomogeneous, so that its mechanical properties vary from one place to another, then it is usually advisable to cut the test specimens sufficiently small in size so that the properties may be considered uniform in each specimen. This wish may not be realizable in some cases; for example, in biology, the skin, the blood vessel wall, and the cell membrane are layered materials with different mechanical properties in different layers, but in general, we cannot peel these layers off by surgery without damaging the tissue.

Now, one can examine the structure of the material with a light or electron microscope or an X-ray diffraction machine, a nuclear magnetic resonance equipment, or a positron device. With increasing power of magnification, one can cross the allowable lower limit of size for the definition of stress and strain discussed in Sec. 1.6. The details of the ultrastructure at scales smaller than the lower limit that defines the stress and strain are, however, irrelevant to the mechanics of the material. Nevertheless, we are often interested in learning about the ultrastructure of a material to gain a greater understanding of the material's mechanical properties. Sometimes it is even possible to derive the constitutive equation of a material at a given range of sizes from the ultrastructure at a lower level of sizes. To illustrate this kind of approach, let us consider a few examples.

Example 1. Crystalline Solid with a Cubic Lattice

Consider a crystal with atoms arranged in a cubic lattice, as illustrated in Fig. 8.1 Let the length of each side of the cube be one unit of measure. Let an orthogonal frame of reference (x_1, x_2, x_3) be chosen, and let the material be subjected to a stress described by a tensor

$$\begin{pmatrix} \sigma_{11} & \sigma_{12} & \sigma_{13} \\ \sigma_{21} & \sigma_{22} & \sigma_{23} \\ \sigma_{31} & \sigma_{32} & \sigma_{33} \end{pmatrix}. \tag{8.8–1}$$

In response to the stress, the crystal deforms. In particular, the crystal lengthens in the x_1-direction due to σ_{11} and shortens in the x_1-direction due to σ_{22} and σ_{33}. Suppose that it was found theoretically or experimentally that

$$e_{11} = \frac{\sigma_{11}}{E} - \frac{\nu}{E}(\sigma_{22} + \sigma_{33}), \tag{8.8–2}$$

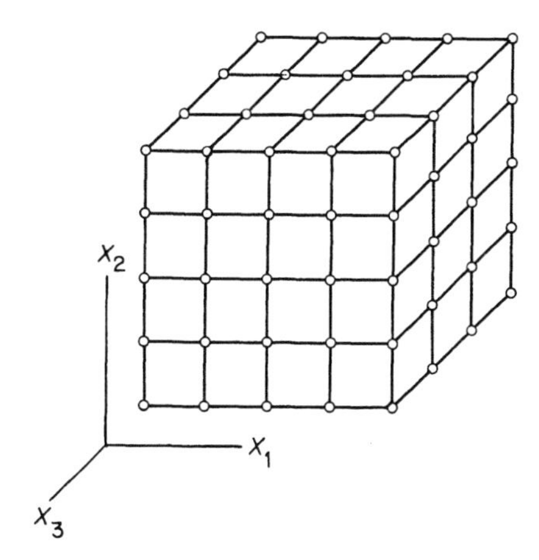

Figure 8.1 Mechanical model of a crystal with atoms arranged in a cubic lattice.

and that, similarly, the deformations in the x_2- and x_3-directions yield the strains

$$e_{22} = \frac{\sigma_{22}}{E} - \frac{\nu}{E}(\sigma_{33} + \sigma_{11}) \tag{8.8–3}$$

and

$$e_{33} = \frac{\sigma_{33}}{E} - \frac{\nu}{E}(\sigma_{11} + \sigma_{22}). \tag{8.8–4}$$

The shear stress σ_{ij} produces the shear strain e_{ij}. Suppose that it was found that

$$\sigma_{12} = 2Ge_{12}, \qquad \sigma_{23} = 2Ge_{23}, \qquad \sigma_{31} = 2Ge_{31} \tag{8.8–5}$$

where G is a constant. Now, can we assert that the mechanical property is isotropic? The answer is, in general, no. We have three material constants: E, ν, and G. According to Sec. 8.5, and isotropic constitutive equation can have only two independent constants. Indeed, if the stress-strain relationship were isotropic, then the constants G, E, and ν would have to be related by the equation

$$G = \frac{E}{2(1 + \nu)}. \tag{8.8–6}$$

(See Eq. 7.4–16.) If experimental results show that Eq. (8.8–6) does hold, then Eqs. (8.8–1) to (8.8–6) coincide with an isotropic constitutive equation in one particular coordinate system. We can, however, claim more in this case, because any Cartesian frame of reference can be transformed into the crystal frame of Fig. 8.1 by translation and rotation, and a stress tensor in any arbitrary frame of reference can be transformed into the form of Eq. (8.8–1) in the crystal frame. Thus, we can claim that if Eq. (8.8–6) holds, the material composed of a cubic lattice of atoms is isotropic.

Example 2. Lung Tissue

The lung tissue proper is a composite structure similar to an open-pored foam rubber. (See Figs. 1.3–1.6 in Chap. 1.) A carefully validated model of the microstructure of the lung proposed by Fung (1988) is shown in Fig. 8.2. Each terminal unit of the airway is called an *alveolus*. An assembly of alveoli of the same shape and size fills the entire space of the lung. The walls of the alveoli are called the *interalveolar septa*, which are thin membranes enclosing (as in a sandwich) capillary blood vessels. The capillary blood vessels of the lung fill 80% of the sandwich space

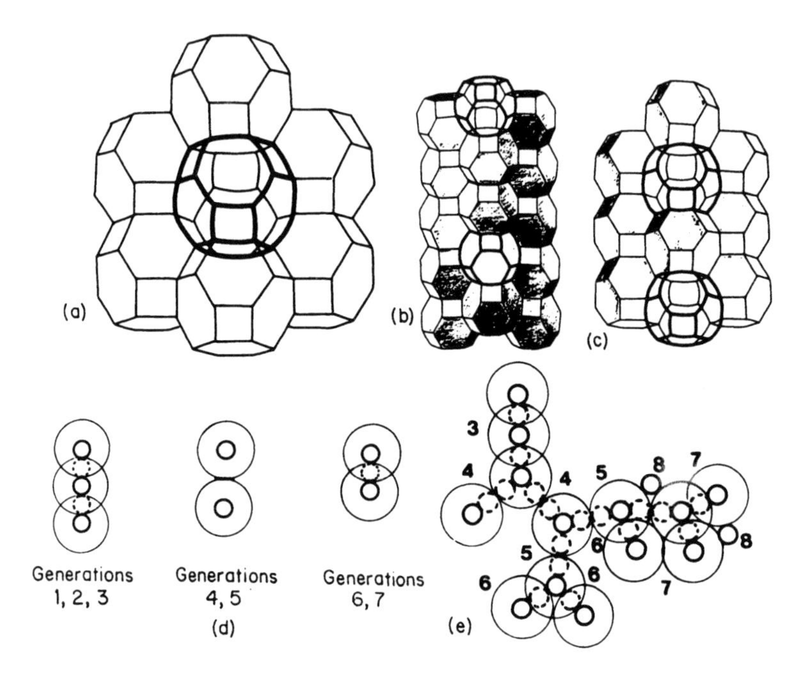

Figure 8.2 A mathematical model of the pulmonary alveolar duct according to Y. C. Fung, "A Model of the Lung Structure and Its Validation," *J. Appl. Physiology* 64(5):2132–41, 1988. (a) Basic unit of alveolar duct, consisting of fourteen 14-sided polyhedra surrounding a central 14-sided polyhedron without walls. Each wall is a membrane or interalveolar septum. (b) Two units stacked together, with one membrane in contact removed, to form a longer duct. (c) Two basic units forming a shorter duct by removing a few more membranes in common. (d) Pulmonary alveolar ducts of generations 1, 2, 3 are made of three basic units, with each neighboring pair structured as in part (c). Ducts of generations 4 and 5 are units shown in part (b). Ducts of generations 6 and 7 are units shown in part (c). (e) A ductal tree formed by ducts of various generations. A number of single 14-sided units, given a generation number of 8, are needed to fill the entire space. The lung tissue is filled by these trees, which converge on bronchioles, then bronchi, and, finally, the trachea.

of the interalveolar septa. A good model of the alveolus is a 14-sided tetrakaidecahedron, or 14-hedron. Fourteen 14-hedra enclosing a central 14-hedron whose walls are all perforated form a second-order 14-hedron, which is the basic unit of the alveolar duct. All alveoli of the basic second-order 14-hedron unit ventilate to the central space. Two or three basic units joined together, with a proper number of interalveolar septa removed for ventilation, make up the branches of the alveolar ducts of the lung. Ducts are ventilated to bronchioles, which in turn are ventilated to bronchi, to trachea, and, finally, to the nose and mouth. The lung tissue is an assemblage of first- and second-order 14-hedra.

For such a structure, stress can be defined with respect to areas with diameters much larger than the diameter of the individual alveolus. (In humans, the alveolar diameter is about 100–300 μm; hence, a plane area of 1 cm^2 will intersect 1,000–10,000 alveoli, and stress can be defined quite well in such an area.) Similarly, strain can be defined in bodies with volumes on the order of 1 cm^3.

In the lung, each interalveolar septum is made of capillary blood vessels and connective tissues whose main structural components are collagen and elastin fibers. To maintain the structural integrity, the edges of the interalveolar septa that were ventilated were seen to be reinforced with additional collagen and elastin. The quantity, size, and curvature of these collagen and elastin fibers in human lung have been measured. (See Fung, *Biomechanics* [1990], pp. 405–416.) Under stress, these fibers and connective tissues will deform, leading to the overall stress-and-strain relationship of the lung tissue. Such a relationship is useful for understanding the stress and strain distribution in the lung under a gravitational load in normal life, under zero gravity in space flight, under acceleration in sports, and under conditions of disease, as well as analyzing the distribution of ventilation and blood flow in the lung.

In spite of the rather complex geometric structure of the lung tissue, the basic cubic symmetry is clear, because each 14-hedron is obtained by cutting off the eight corners of a cube, and the assembly retains the intrinsic cubic character. It follows, from Example 1, that in the small-strain, linear range, the stress-strain relationship of the lung could be, but is not necessarily, isotropic, depending on whether the shear modulus, Young's modulus, and Poisson's ratio obey Eq. (8.8–6) or not.

The lung tissue, however, is capable of large deformation, and normally works in a finite strain range relative to the zero-stress condition. At a finite strain, the constitutive equation of the lung tissue is nonlinear. Knowledge of its initial isotropy in the neighborhood of the zero-stress state, however, goes a long way, because if the lung tissue is initially isotropic in the linear, small-strain range, then it is initially isotropic also in the nonlinear finite-strain range.

PROBLEMS

8.1 Distinguish the words *homogeneous* and *isotropic*. Consider the atmosphere of the earth:

(a) If you are concerned with a high-altitude sounding rocket, would you call the atmosphere homogeneous or isotropic?

(**b**) If the problem is concerned with the flow around the immediate neighborhood of the rocket, which is flying at such a speed that no shock wave is generated, could the air be treated as homogeneous or isotropic?
(**c**) If shock waves are generated in part (b), what then?

8.2 Show that the theorem proved in Sec. 8.4 may be restated as "The most general isotropic tensor of rank 4 has the form

$$u_{ijkm} = \alpha\delta_{ij}\delta_{km} + \beta\delta_{ik}\delta_{jm} + \gamma\delta_{im}\delta_{jk},$$

where α, β, γ are constants."

8.3 Show that the tensor $\epsilon_{ijk}\delta_{lm}$ is isotropic. Are there other isotropic tensors of rank 5?

8.4 Form some isotropic tensors of rank 6. Generalize to isotropic tensors of even order $2n$.

8.5 Name three liquids that are not isotropic.

8.6 Name five solids that are not isotropic.

8.7 Isotropy is a special feature of the constitutive equation. Hence, an experimental testing of isotropy would require the same equipment and instruments used in the determination of mechanical properties. Suppose that you were asked to determine whether a certain metal is isotropic. Design a testing program that will enable you to provide a definitive answer.

8.8 If you are concerned with a biological material, such as the human skin, the desired testing program would most likely be different from the one designed for metals. The skin is obviously not isotropic in three dimensions. But it may very well be "transversely" isotropic: isotropic in its plane. Design a new testing program for the skin.

8.9 For industrial materials, such as concrete and plastics, it is very important for the designer to know whether a material is isotropic or not. Suppose you are establishing a laboratory for testing structural materials. Present a plan of experiments, and describe the needed instruments.

8.10 A single crystal is usually anisotropic, but polycrystalline material may be isotropic. A single long-chain molecule is anisotropic, but polymer fluids and solids may be isotropic. Explain this from the point of view of our definition of a continuum discussed in Sec. 1.5, with a specified minimum defining dimension, and from statistical characteristics of the structure of the materials.

From a theoretical point of view, is it possible to formulate some rules about isotropy based on the structure of the material? In polycrystalline metals, the grain boundary material is usually amorphous, and dislocations and twinning should be considered. In polymer materials, the molecular structure should be considered.

8.11 An isotropic material can be made anisotropical by mechanical means, e.g., by rolling, hammering, shot peening, explosive forming, stretching, wire drawing. Use statistical arguments to provide an explanation of the changes. Design an experimental facility to test the results.

8.12 Cold work on metal alloys by rolling, pressing, or forging, will deform or crush crystal, create new grain boundaries, causing large movements of dislocations and creating new ones. Subsequent heat treatment can change the shape of crystals, grain boundaries, crystal structure, and solid solution of inclusions (such as carbon in steel). Would these processes change the stress-strain relationship below the proportional limit (in the linear range)? Would the Young's modulus, Poisson's ratio, and the shear modulus be affected? Would the yield point be changed by cold work and heat treatment? Would the ultimate stress at failure be affected?

8.13 For a material which obeys Hooke's law, an experimental program for testing isotropy would include some experiments involving normal stresses and other experiments involving shear stresses; and in addition, checking whether the relationship between the shear modulus, Young's modulus, and Poisson's ratio is satisfied. Now, turn our attention to a material that does not obey Hooke's law, such as the human skin. The stress-strain relationship is nonlinear. What kind of a test program can we follow to ascertain isotropy?

8.14 Isotropy of a material is sometimes confounded by the residual stresses in a test specimen. The residual stresses are stresses in a specimen when there is no external load acting in the specimen. They can be put in the specimen by previous plastic deformation, welding, insertion, or other processes. Hammer a nail into a piece of wood and you put residual stresses in the wood. Bend a wire into a ring, weld the ends, and you have a ring with residual stress. Forge a titanium alloy into a fan blade of a jet engine and you have residual stress in the titanium blade.

If the stress-strain relationship were linear, and the displacements were infinitesimal, then the equations governing the equilibrium or dynamics of the continuum are linear, and the principle of superposition of solutions apply. In this case, the response of a body with residual stress to a given load is the same as that of the same body without residual stress to the same load. In other words, if the material of the body is isotropic, measurements of load-deflection relationship will show isotropy with or without residual stress.

The situation is different if the stress-strain relationship is nonlinear. Explain the nonlinear case. We still wish to know if the basic stress-strain relation is isotropic for the nonlinear material, at least in the neighborhood of the zero-stress state. How should we proceed? Make a plan for the nonlinear case.

8.15 Residual stresses in composite materials, with some components in tension, other components in compression, and all together in equilibrium, is a good way to obtain improved mechanical properties. For example, prestressed-steel-rod reinforced concrete, and high-strength-fiber reinforced metals and plastics are important structural materials. If isotropy or transverse isotropy of the composite material (at a scale much larger than the diameter of the individual fibers) is desired, the fibers should be laid down in some desirable geometric patterns. Design a composite material with high strength and isotropy as objectives.

8.16 Living creatures use cells as basic structure of their bodies. Cell membranes and stress fibers (actin molecules) can harbor tensile residual stress against the pressure (compressive residual stress) in the cell contents. The matrix material in the interstitial space between the cells can be stressed in tension, compression, or shear. The overall mechanical properties of the living tissue (at a scale much larger than the individual cells) depend on the cellular structure of the tissue. Discuss the isotropy or anisotropy of the tissue in relation to the three-dimensional geometric shapes of the cells. Discuss the overall mechanical properties of the tissue (at a scale much larger than the cells) in relation to the intensity of the residual stresses; i.e., in relation to the degree of swelling of the cells.

8.17 Animal cells inside the body rely on blood circulation for access to oxygen. The blood vessels are therefore pervasive: The circulation system supplies blood to within a few microns of distance from every living cell. Consider a piece of tissue whose size is much larger than the diameters of the blood vessels. The blood pressure and the stresses in the blood vessel wall may be regarded as residual stresses in the tissue. The overall mechanical properties of the tissue (at a scale much larger than the diameters of the blood vessels), will depend on how much the vascular system is pressurized; i.e., with how much the vessel walls are stressed. The vascular system is a continuous hollow organ. From this point of view, discuss the relationship between the mechanical properties of the tissue and the blood pressure. Discuss the isotropy of the tissue in relation to the geometry of the blood vessel system.

8.18 Any test of isotropy is a test of a null hypothesis that there is no directional difference. Hence, it must be subjected to statistical principles. What is the principle of statistical design of experiments? How can the principle be applied to the tests planned in the preceding problems?

FURTHER READING

FUNG, Y. C., *Biomechanics: Motion, Flow, Stress, and Growth*. New York: Springer-Verlag (1990), Chapter 11.

JEFFREYS, H., *Cartesian Tensors*. Cambridge: University Press (1957), Chapter 7.

THOMAS, T. Y., *Concepts from Tensor Analysis and Differential Geometry*. New York: Academic Press (1961), pp. 65–69.

9 MECHANICAL PROPERTIES OF REAL FLUIDS AND SOLIDS

In this chapter, we consider real materials in order to see how the idealized constitutive equations of Chaps. 7 and 8 fit into the real world. We begin with gases and liquids from a molecular point of view. Then we consider solids, viscoelastic bodies, and biological materials.

9.1 FLUIDS

Fluids are usually classified as gases or liquids on the basis of the pressure-volume relationship. A typical example of the pressure-volume relationship of carbon dioxide at constant temperatures is shown in Fig. 9.1. The lower curves have horizontal steps at certain values of the pressure. To the left of the step, we have the liquid state, wherein it takes a large increase in pressure to produce a small change in volume. To the right is the vapor or gaseous state. A point on the horizontal step (such as AB in the figure) actually represents a heterogeneous state consisting of a mixture of liquid and vapor. At 31.05°C, the pause in the CO_2 liquid-vapor isotherm is reduced to zero. At temperatures above this critical value, the isotherm passes steadily from high to low pressure with no marked division between gaseous and liquid states. At a higher temperature the equation of state becomes better and better approximated by the "perfect gas" law, Eq. (9.1–1).

From the molecular point of view, studies of gases led Avogadro to propose the hypothesis that equal volumes of gases contain equal numbers of molecules at the same temperature and pressure. A *mole* of molecules (a sample whose weight in grams exactly equals the *molecular weight* of the molecule) contains 6.025×10^{23} particles. This is known as Avogadro's number (N_0). The volume of 1 mole of gas at the normal temperature and pressure (i.e., 0°C and 760 mm mercury pressure) is 22,400 cm^3, which corresponds to an average distance of about 33×10^{-8} cm from one particle to the next. When a vapor is condensed to a liquid or solid, it shrinks to about one-thousandth of its volume at the normal temperature and pressure; i.e., the interparticle spacing is reduced to about 3×10^{-8} cm.

The kinetic theory interprets the pressure in a gas as the reaction of the gas molecules impinging on a surface. From the consideration of changes in momentum

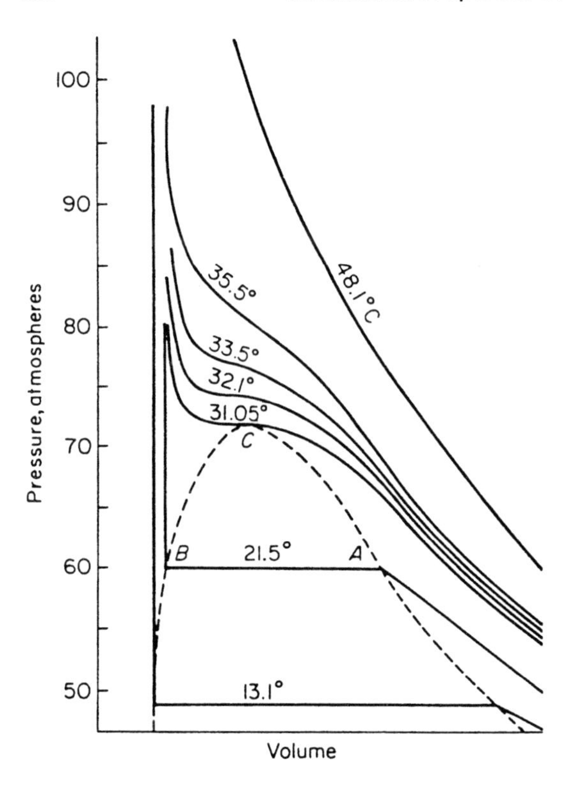

Figure 9.1 Isothermal curves of volume-pressure relationship for CO_2 near the critical point, C. At the liquid-gas critical point of CO_2, the temperature is 304.2°K, the pressure is 72.9 atm, and the volume is 94 cm^3 mole^{-1}

in molecular impacts and rebounds, the kinetic theory derives the equation of state of a perfect gas relating the pressure (p), volume (V), and absolute temperature (T):

$$pV = RT. \tag{9.1–1}$$

Here,

$$R = N_0 k. \tag{9.1–2}$$

For 1 mole of gas, k and R are *universal* constants, the same for all substances. The constant k is Boltzmann's constant $= 1.38 \times 10^{-16}$ erg deg^{-1}, and R is the gas constant $= 8.313 \times 10^7$ erg deg^{-1} mole^{-1} $= 1.986$ cal deg^{-1} mole^{-1}.

For the condensed state, Van der Waals proposed the celebrated equation

$$\left(p + \frac{\alpha}{V^2}\right)(V - \beta) = RT, \tag{9.1–3}$$

in which α/V^2 represents the attractive forces between the gas particles (not accurately, except at low gas densities), while β represents the molecular volume of the particles. Figure 9.2 shows a family of Van der Waals p-V curves. They are like the curves of Fig. 9.1, but the horizontal line AB in Fig. 9.1 has become a continuous curve $AEDFB$ in Fig. 9.2. The parts of the curves below the abscissa,

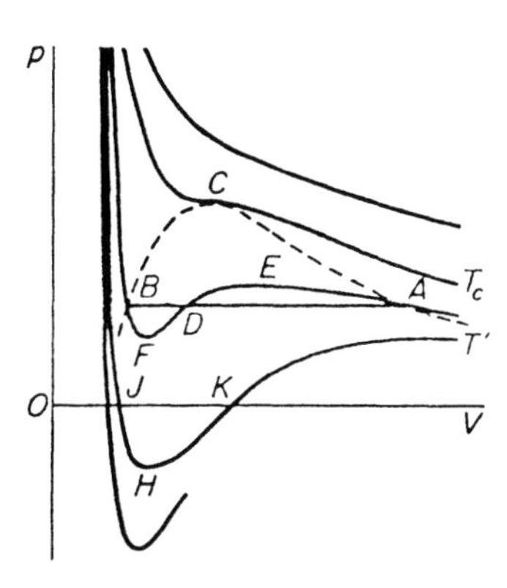

Figure 9.2 A family of Van der Waals isothermals. Here p is pressure, V is volume, T is temperature. C is the critical point, T_c is the critical temperature.

where the pressure is negative, are supposed to represent the tensile strength of the liquid, on account of the short-range attractive forces between the atoms represented by the parameters α/V^2 and β. The minima (e.g., the point H on JHK at temperature T') indicate the *ideal tensile strength* of the liquid. In the case of water, this is $-1{,}168$ atm at 0°C and -875 atm at 50°C. Many authors have devised ingenious methods to measure the tensile strengths of liquids. (See D. E. Gray, ed., *American Institute of Physics Handbook*, New York: McGraw-Hill Book Co. (1957), pp. 2–170.) Experimental values are lower than theoretical values, and the cause is usually thought to be vapor-nucleating agents, small bubbles, and the liquid's tearing away from the walls of the container used in the experiments.

The question of the tensile strength of a liquid is important in problems of cavitation, cavitation damage to ship propellers, water transport in trees, freezing damage to trees, and other problems.

9.2 VISCOSITY

The concept of viscosity in a fluid was given by Newton in terms of a shear flow with a uniform velocity gradient, as shown in Fig. 9.3. Here the coordinates x, y, z represent a rectangular Cartesian frame of reference, and u is the systemic velocity of the fluid (local average of the velocities of the molecules), which points in the direction of the x-axis, and is a function of y only. The shear stress acting on a surface normal to the y-axis is denoted by τ. Newton proposed* the relationship

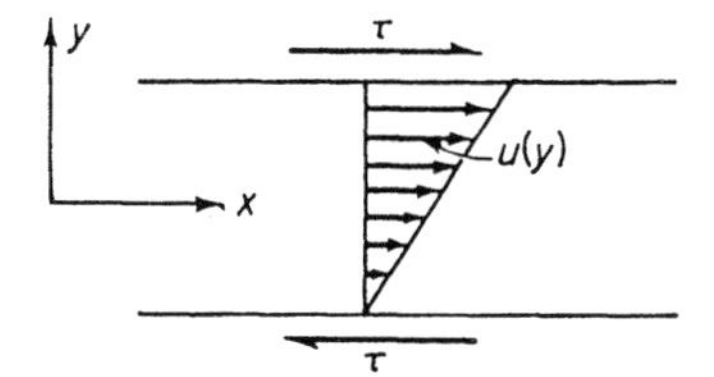

Figure 9.3 Newtonian concept of viscosity.

*In the "Hypothesis" just before Prop. L1, Lib. II, of the *Principia*.

$$\tau = \mu \frac{du}{dy} \tag{9.2–1}$$

for the shear stress τ. The coefficient μ is a constant called the *coefficient of viscosity*. The dimensions of μ are $[ML^{-1}T^{-1}]$. In the centimeter-gram-second system of units, in which the unit of force is the dyne, the unit of μ is called a *poise*, in honor of Poiseuille. In the International units, the unit of viscosity is newton second/meter^2 (Ns/m^2). 1 poise is 0.1 Ns/m^2.

The coefficients of viscosity of air and of water are small—approximately 1.8×10^{-4} poise and 0.01 poise, respectively at atmospheric pressure and 20°C. At the same temperature, the viscosity of glycerine is about 8.7 poises. In liquids, μ diminishes fairly rapidly as the temperature increases; in gases, μ increases as the temperature rises.

An interesting interpretation of the coefficient of viscosity from the kinetic theory of gases was given by Maxwell. Consider a flow with a uniform velocity gradient as shown in Fig. 9.3, and imagine a surface AA normal to the y-axis, as in Fig. 9.4. The shear stress exerted by the gas beneath AA on the gas above has a retarding effect. The shear stress is equal to the rate of loss of ordered momentum

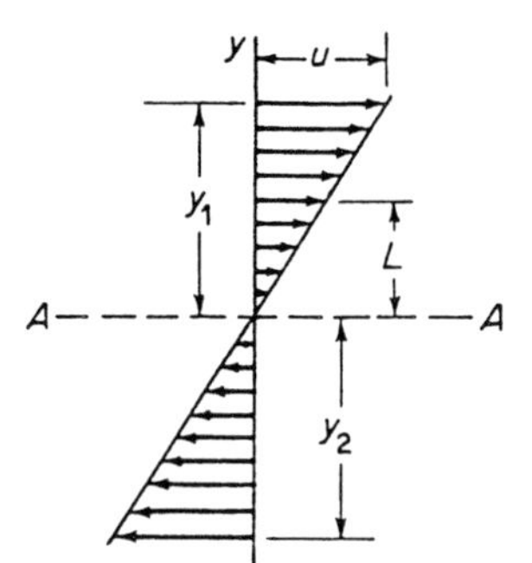

Figure 9.4 Kinetic interpretation of the coefficient of viscosity of gases.

across AA by the random motion of the molecules. A molecule originating at y_1 and moving downward through AA will carry with it a positive momentum $m(du/dy)y_1$, where m is the mass of the molecule, u is the ordered velocity in the x-direction, and du/dy is the vertical velocity gradient, i.e., the shear strain rate. Similarly, a molecule moving upward through AA and originating at y_2 will carry with it a negative momentum $m(du/dy)y_2$. Both of these excursions represent a loss of ordered momentum from the fluid above AA. The sum of such losses that occur in 1 second through unit area AA is equal to the shear stress τ.

Let there be N molecules per unit volume. Suppose that one-third of the molecules are traveling in each of the three coordinate directions. If the average molecular speed is c, and if one-third of the molecules move perpendicularly to AA, $\frac{1}{3}Nc$ molecules will pass through AA each second. Each of these molecules will carry with it a momentum corresponding to the position y at which it originates. Let the average value of the height y be L. Then the shear stress is

$$\tau = \left(\frac{1}{3} Nc\right) m \frac{du}{dy} L.$$

The product Nm is the density ρ. Therefore,

$$\tau = \frac{1}{3} \rho c L \frac{du}{dy}. \tag{9.2–2}$$

A comparison of Eq. (9.2–1) with Eq. (9.2–2) shows that

$$\mu = \tfrac{1}{3}\rho c L. \tag{9.2–3}$$

The effective height L is related to the mean free path l (the average distance a molecule travels before colliding with another molecule), and more accurate calculations by David Enskog and Sydney Chapman* show that

$$\mu = 0.499\, \rho c l. \tag{9.2–4}$$

As the density of a gas decreases, the mean free path increases in such a manner that the product ρl almost remains constant. Then μ is proportional to c, which in turn is proportional to the square root of the absolute temperature. Thus, the coefficient of viscosity of a gas changes with the temperature but not with the pressure. For air under standard conditions (sea level, 59°F) the mean free path of the molecules is approximately 8.8×10^{-6} cm.

The argument that leads to Eq. (9.2–2) can be used on other transport phenomena. When the molecules cross the plane AA, they carry with them not only the momentum of their ordered motion, but also their mass and their energy. In a gas with a density gradient, the transport of mass corresponds to the phenomenon of diffusion. In a gas with a temperature gradient, the transport of energy corresponds to the phenomenon of the conduction of heat. Thus, in the simplest theory, the mechanisms of the transport of a component of ordered momentum, of heat energy, and of mass are identical; and as a result, it is found that the coefficient of heat conduction k is equal to the product of the viscosity μ and the specific heat at constant volume C_v, while the coefficient of self-diffusion D is equal to the viscosity μ divided by the density ρ. Experiments and more accurate calculations give

$$k = 1.91\, \mu C_v, \qquad D = 1.2 \frac{\mu}{\rho}. \tag{9.2–5}$$

The atomic interpretation of the viscosity of liquids and solids is different from that of the viscosity of gases. Solids in the crystal form have *long-range ordered structures*. The atoms are arranged in order by the long-range interaction. On the

*See Chapman, S. and Cowling, T. G., *The Mathematical Theory of NonUniform Gases*. Cambridge, University Press, 3rd ed., 1970.

other hand, gas atoms or molecules interact only when they come "into contact," and the interaction depends on the *short-range* attractive force between two atoms or molecules. The liquid state is an interpolation between the gas and the crystal. Generally speaking, other than those properties such as X-ray diffraction, anisotropy, etc., the structure and properties of a liquid, just above the melting point, are fairly similar to those of its crystal. Metals expand only 3 to 5% on melting (bismuth, like ice, contracts), so that the packing of the atoms cannot be too different. It is as if 3 to 5% of the crystal sites became vacant and their free volume were taken up by neighboring particles in such a way as to destroy the long-range order of the structure. A picture of the cause of viscosity for a simple liquid is proposed by Cottrell (1964), as shown in Fig. 9.5. Here two atoms which gained some free volume are shown as enclosed in a "cage" of other atoms. The figure shows how a relative motion of the two atoms will allow the cage to have a shear

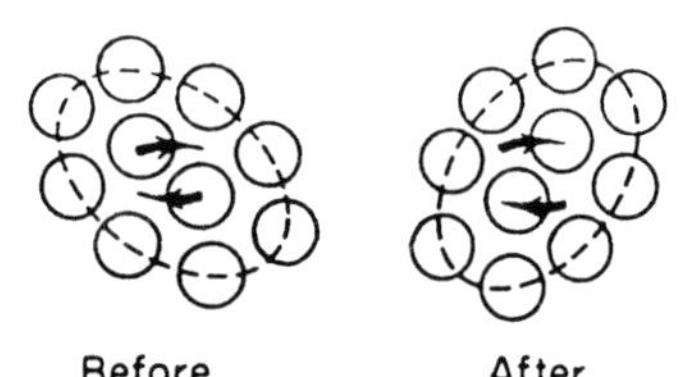

Figure 9.5 Shear due to internal movement in a liquid, suggested by Cottrell as a mechanism for viscosity in the fluid. After A. H. Cottrell, *The Mechanical Properties of Matter*, New York: John Wiley, 1964.

deformation while the relative positions of the rest of the atoms of the cage remain essentially unchanged. The movement of the two atoms distorts the atomic "cage." The surrounding liquid offers elastic (shear) resistance to such a distortion, but this resistance relaxes as similar movements occur nearby.

On the other hand, the atoms in a crystal are arranged in space lattices. The much greater elastic modulus and viscosity of a crystal as compared with its liquid phase is due to the fact that the atomic sites in a crystal are defined exactly by the lattices.

Mixtures, colloidal solutions, suspensions, polycrystalline solids, amorphous solids or glass, etc., can have many other relaxation mechanisms that reveal their viscosity. In many cases, it is not easy to say whether a body behaves as a fluid or as a solid. Silicone putty can be poured slowly from a cup or bounced quickly like a rubber ball. Conventionally, the distinction between fluids and solids is drawn at a low-stress viscosity of 10^{15} poises. A material with viscosity less than that value is called a fluid, while one whose viscosity is greater is called a solid.

9.3 PLASTICITY OF METALS

If a rod of a ductile metal is pulled in a testing machine at room temperature, the load applied on the test specimen may be plotted against the elongation

$$\epsilon = \frac{l - l_0}{l_0}, \tag{9.3-1}$$

where l_0 is the original length of the rod and l is the length under load. Numerous experiments reveal typical load-elongation relationships, as indicated in Fig. 9.6. When ϵ is very small, the load-elongation relationship is usually a straight line. Mild steel shows an upper yield point and a flat yield region that is caused by many

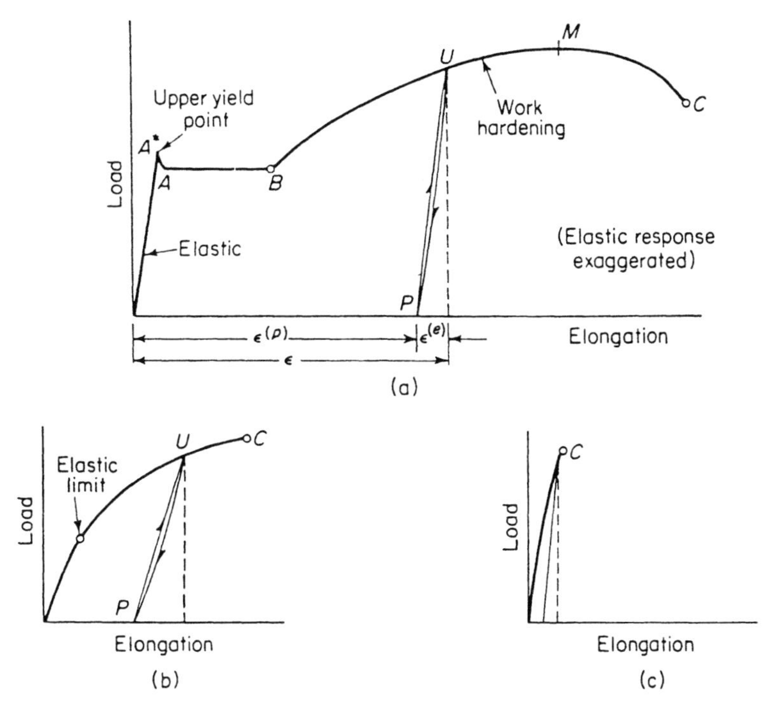

Figure 9.6 Typical load-elongation curves in simple tension tests of metals: (a) of mild steel or structural steel; (b) of aluminum alloys or copper; (c) of brittle materials such as cast iron.

microscopic discontinuous small steps of slip along slip planes of the crystals [Fig. 9.6(a)]. Most other metals do not have such a flat yield region [Fig. 9.6(b)].

Upon unloading at any stage in the deformation, the strain does not retrace the loading curve, but is reduced along an elastic unloading line such as the curve *UP* in Figs. 9.6(a) and (b). Reloading retraces the unloading curve with relatively minor deviations and then produces further plastic deformation when approximately the previous maximum stress is exceeded. The test specimen may "neck" at a certain strain, so that its cross-sectional area is reduced in a small region. When *necking* occurs under continued elongation, the load reaches a maximum and then drops down, although the actual average stress in the neck region (the load divided by the true area of the neck) continues to increase. The maximum M is the ultimate load. Beyond the ultimate load, the metal flows. At the point C in the curves of Fig. 9.6, the specimen breaks.

Materials like cast iron, titanium carbide, beryllium, concrete, rocks, and most ceramics allow minimal plastic deformation before reaching the breaking point, and are called brittle materials. The load-strain curve for a brittle material will appear as that in Fig. 9.6(c). The point C is the breaking point.

A fact of great importance for geology is that brittle materials such as rocks tend to become ductile when subjected to large hydrostatic pressure (large negative mean stress). This was demonstrated by Theodore von Kármán (1881–1963) in his classical experiments on marbles.

Tests of specimens subjected to simple compression or simple shear lead to load-strain diagrams similar to those of Fig. 9.6.

It is well known that, whereas the elastic moduli of all steels are nearly the same, the yield stress and the ultimate strength vary a great deal, depending on the crystal structure (including imperfections, dislocations, vacancies, grain boundaries, twinning, etc.), which can be influenced by small changes in chemical composition, alloying, heat treatment, cold work, and so on. In other words, whereas the elastic moduli are "structurally insensitive," the strengths are "structurally sensitive." For materials without a marked yield point, it is an engineering practice to quote a *yield strength* as the stress at the *proportional limit*, which is defined as the point where a tensile strain of 0.2% is reached. Most engineering structures use materials within the proportional limit—hence, the strain is truly quite small. For this reason, the linearized theory of elasticity is useful in engineering practice.

9.4 MATERIALS WITH NONLINEAR ELASTICITY

Rubber, the material most qualified to be called elastic, cannot be described by Hooke's law. The stress-strain curve obtained when a rubber band is stretched uniaxially in a testing machine is shown in Fig. 9.7. It is nonlinear. A linear, Hookean approximation is applicable only in a range of strain much smaller than

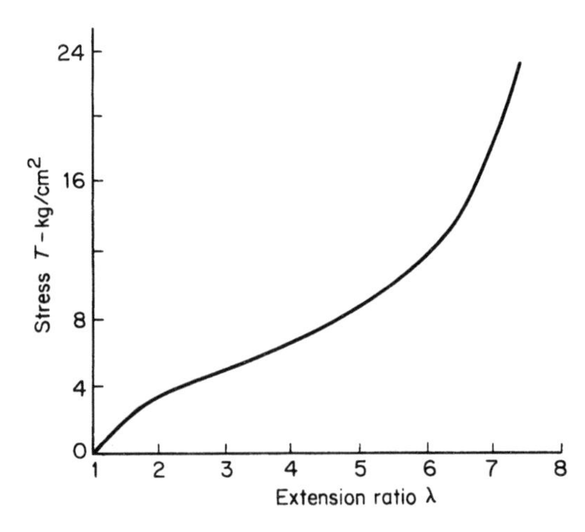

Figure 9.7 A typical force–extension curve for vulcanized rubber.

rubber bands normally undergo. Soft rubber is practically incompressible, in the sense that its volumetric modulus of elasticity is 10^4 to 10^6 times larger than the incremental Young's modulus. Strictly speaking, the stress-strain relationship of rubber is not unique. Features of viscoelasticity exist. Stress relaxation at constant strain, creep at constant stress, and hysteresis in periodic oscillation also exist. Hence, rubber is a nonlinearly viscoelastic material.

The living soft tissues of humans and animals are also nonlinearly viscoelastic in the strain range in which they function normally. Human and animal tissues are generally called elastic if one wishes to emphasize their ability to return to a unique configuration when all the external load is removed. If one looks at their visco-elasticity, one finds some special features also. When a living tissue is subjected to a periodic loading and unloading, it develops a steady-state stress-strain loop that is not very sensitive to strain rate. The hysteresis (the difference between the loading and unloading curves in the stress-strain loop) is largely unaffected by the frequency of the periodic loading. The stress-strain loop is repeatable, i.e., the loading and unloading legs have a certain degree of uniqueness. This feature may be described by the term *pseudoelasticity* (Fung, 1971). On the other hand, the "memory function" that links the present stress with the past stress seems to be linear, even though the stress-strain relationship is nonlinear; hence, the term *quasi-linear viscoelasticity* (Fung, 1971) was introduced.

As an example, let us consider typical connective tissue in an animal: the mesentery of the rabbit. The mesentery is a thin membrane that connects the rabbit's intestines. Nearly transparent to the naked eye, it has good, uniform thickness (about 6×10^{-3} cm) and is a favorite of physiologists because its two-dimensional array of small blood vessels is ideal for observation and experimentation. To obtain the gross mechanical property, a strip of uniform width was cut from the mesentery, tied at both ends with fine silk, and tested in simple tension while immersed in a saline solution, at room temperature, pH 7.4, bubbled with a gas of 95% O_2, 5% CO_2, with concentrations of Ca and other ions similar to those in blood plasma.

After a few cycles of loading and unloading, a repeatable stress-strain loop for each strain rate was obtained, as shown in Fig. 9.8. Note the difference between the shapes of the curves in Figs. 9.7 and 9.8, showing that the constitutive equations of rubber and the rabbit's mesentery are very different.

Since the stress-strain loop of the tissue is repeatable, we can treat loading and unloading separately for this tissue and consider it as one elastic material in loading and another elastic material in unloading, i.e., as two *pseudoelastic materials*.

Of the two typical hysteresis loops shown in Fig. 9.8, the one marked "high" was produced at a strain rate 10 times faster than that marked "low." It is seen that the hysteresis loop did not depend very much on the rate of strain. The abscissa of this figure is extension of the specimen from an arbitrary length. This is done because if the origin of extension were taken at the zero-stress state, the origin would be far to the left and the scale would be too small. For this specimen, the

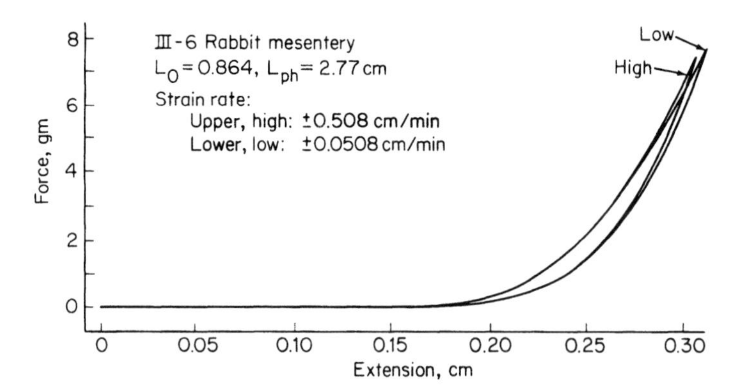

Figure 9.8 Hysteresis curves of rabbit mesentery obtained at two strain rates. The high rate was 10 times that of the low rate. Only a slight change in hysteresis curves was obtained. From Y.C. Fung, "Elasticity of Soft Tissues in Simple Elongation." *American J. of Physiology* **213**(6): 1532–1544, 1957.

length at zero stress was 0.865 cm, the length in the physiological state was 2.77 cm, and the initial cross-sectional area was 1.92×10^{-2} cm^2.

In reducing the experimental data to a stress-strain relationship, we use the Lagrangian stress T (obtained by dividing the force by the original cross-sectional area of the specimen at zero stress) and the extension ratio λ (the deformed length divided by the zero-stress length). The most striking feature of the stress-strain relationship is revealed when $dT/d\lambda$, the slope of the stress-stretch curve, is plotted against T. Figure 9.9 shows such a plot of the slope of the loading curves shown in Fig. 9.8 (the upper curve of the hysteresis loop obtained at the "high" strain rate of 0.508 cm/min). As a first approximation, we may fit the experimental data to a straight line,

$$\frac{dT}{d\lambda} = \alpha(T + \beta) \qquad (\text{for } \lambda < \lambda_y), \tag{9.4–1}$$

where α and β are constants and λ_y is an upper limit of the applicability of this equation (about 3.2 in the case of the rabbit's mesentery).

A simple integration of Eq. (9.4–1), together with the condition that the stress is equal to T* when $\lambda = \lambda_y$, yields

$$T + \beta = (T^* + \beta)e^{a(\lambda - 1)}, \qquad (\text{for } \lambda < \lambda_y). \tag{9.4–2}$$

Several other types of soft tissue, such as the skin, the muscles, the ureter, and the lung tissue, are found to follow similar relationships. Thus, it appears that the exponential type of constitutive equation is common to biological tissues. Further experiments on these tissues over wide ranges of strain rates has led to the experience that, within a 10^4- to 10^6-fold variation in strain rate, the stress at a strain in a loading curve in the physiological range does not differ by more than a

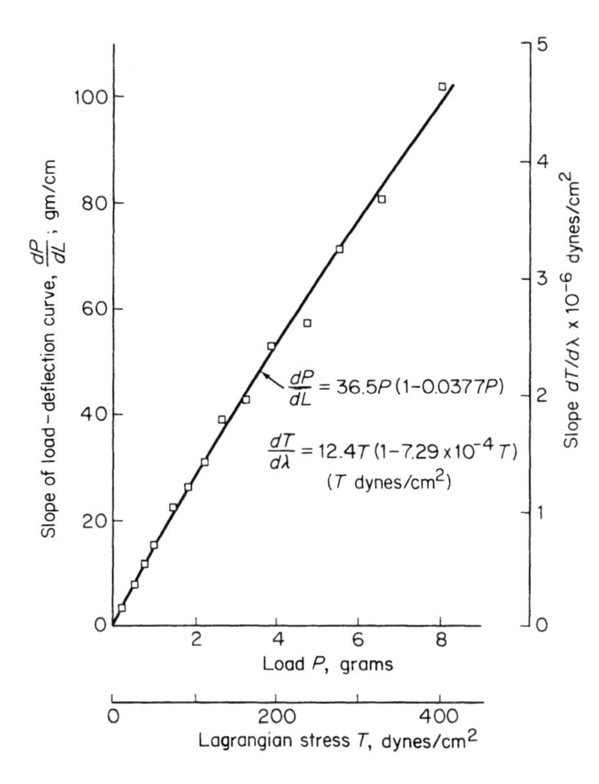

Figure 9.9 The tangential elastic modulus (the slope of the stress–stretch ratio curve) of rabbit mesentery is plotted as a function of the tensile stress T based on the cross sectional area of the specimen at the zero-stress state. Note that the elastic modulus is not a constant as the linear Hooke's law assumes. For this figure, L_0 = 0.865 cm, $A_0 = 1.93 \times 10^{-2}\text{cm}^2$, $\lambda_{ph} = 3.21$.

factor of 2 or 3. Hence, roughly speaking, biological tissues are not sensitive to strain rate. The precise manner in which the strain rate affects the stress-strain curve of a living tissue, however, is very difficult to say, because it varies from one specimen to another (i.e., it is sensitive to local and incidental variations) at all strain rates (i.e., without asymptotic behavior at large or small rates). But an overall insensitivity is a justifiable description.

9.5 NONLINEAR STRESS-STRAIN RELATIONSHIPS OF RUBBER AND BIOLOGICAL TISSUES

In the preceding section, we examined the uniaxial stress-strain relationship of a number of soft tissues. Naturally, for a three-dimensional organ, we need a three-dimensional stress-strain law. No general constitutive equation has been identified for living tissues. But *if a pseudoelastic strain energy function exists*, then the stress-strain relationship can be obtained by a differentiation. The pseudoelastic strain energy function, denoted by $\rho_0 W$, is a function of Green's strain components E_{ij} and is symmetric with respect to E_{ij} and E_{ji}. Taking the partial derivative of $\rho_0 W$ with respect to E_{ij} gives the corresponding Kirchhoff stress components S_{ij}. W is defined for a unit mass of the material, and ρ_0 is the density of the material in the initial state; hence, $\rho_0 W$ is the strain energy per unit volume of the material at the zero-stress state. Thus,

$$S_{ij} = \frac{\partial \rho_0 W}{\partial E_{ij}} \qquad (i, j = 1, 2, 3). \tag{9.5–1}$$

[See Fung (1965, Sec. 16.7) for theoretical details.] If the material is incompressible, then it can take on a *pressure* that is independent of the deformation of the body. In that case, a pressure term should be added to the right-hand side of Eq. (9.5–1). The value of the pressure can vary from point to point, and it can be determined only when the equations of motion, continuity, and boundary conditions are all satisfied. Hence, the pressure in an incompressible fluid is determined by boundary conditions and equations of motion.

A material is recognized by its specific $\rho_0 W$. The determination of $\rho_0 W$ can be helped by theoretical considerations. For example, Green and Adkins (1960) studied the symmetry conditions in all forms of crystals and determined what kind of terms each crystal should have if their $\rho_0 W$'s were polynomials of the strain components. One conclusion reached is that if the material is isotropic, then $\rho_0 W$ must be a function of the strain invariants I_1, I_2, I_3 (see Sec. 5.7); i.e.,

$$\rho_0 W(I_1, I_2, I_3). \tag{9.5–2}$$

If the material is incompressible, then $I_3 = 1$. Soft rubber is incompressible, and the linear form

$$\rho_0 W = C_1(I_1 - 3) + C_2(I_2 - 3) \tag{9.5–3}$$

where C_1 and C_2 are constants, has been found valuable in the study of large deformations of rubber.

Most biological tissues (e.g., skin, muscle, blood vessel wall) are not isotropic. Some (e.g., the lung tissue) are not incompressible. The linear form of strain energy, Eq. (9.5–3), does not fit experimental data of biological tissues.

Based on known experimental data, several strain energy functions have been proposed for the blood vessel wall. If the blood vessel wall is treated as an elastic shell without torsion, then only the average circumferential and longitudinal stresses and strains are of interest. Hence, the vessel wall can be treated as two dimensional and the strain energy is a function of only two strains: the circumferential strain $E_{\theta\theta}$ and the longitudinal strain E_{zz}. Patel and Vaishnav (1972) have used polynomials of $E_{\theta\theta}$, E_{zz} for $\rho_0 W$. Hayashi et al. (1971) have used logarithmic functions. Fung (1973) used an exponential function. A detailed comparison of the polynomial and exponential strain energy functions is given in Fung et al. (1979). In the physiological range, the form

$$\rho_0 W^{(2)} = \frac{c}{2} \exp Q, \tag{9.5–4}$$

where

$$Q = a_1 E_{\theta\theta}^2 + a_2 E_{zz}^2 + a_4 E_{\theta\theta} E_{zz}, \tag{9.5–5}$$

in which c, a_1, a_2, and a_4 are constants, has been shown to work well. The number of material constants in Eq. (9.5–4) is four. The minimum number of material constants in the third-order polynomial expression is seven.

Studies of skin, muscle, ligaments, etc., have shown that the exponential form applies equally well. In the neighborhood of the zero-stress state, it is found that the experimental data can be fitted to a linear stress-strain law or a strain energy function of a second-order polynomial. Hence, for the full range of strain from the zero-stress state to in vivo values, the strain energy function

$$\rho_0 W^{(2)} = P + \frac{C}{2} \exp Q \tag{9.5–6}$$

where

$$P = b_1\, E_{\theta\theta}^2 + b_2\, E_{zz}^2 + b_3\, E_{\theta\theta}\, E_{zz} \tag{9.5–7}$$

in which Q is the same as in Eq. (9.5–5) and b_1, b_2, and b_3 are additional constants, gives a higher accuracy.

9.6 LINEAR VISCOELASTIC BODIES

The features of hysteresis, relaxation, and creep are common to many materials. Collectively, they are called the features of *viscoelasticity*.

Mechanical models are often used to discuss the viscoelastic behavior of materials. In Fig. 9.10 are shown three mechanical models of material behavior, namely, the Maxwell model, the Voigt model, and the "standard linear" model,

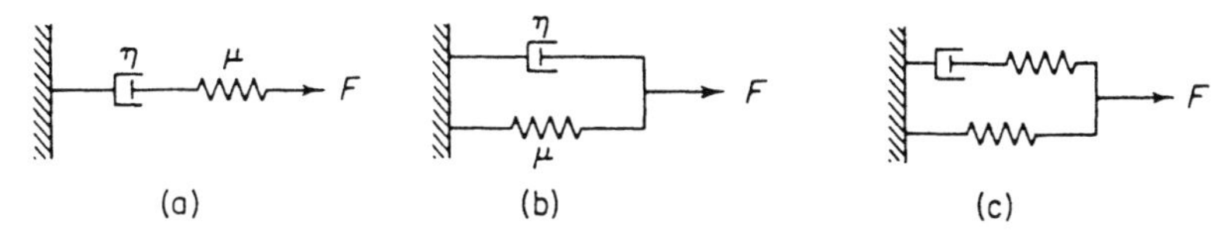

Figure 9.10 Models of linear viscoelasticity. (a) Maxwell, (b) Voigt, (c) standard linear solid.

all of which are composed of combinations of linear springs with spring constant μ and dashpots with coefficient of viscosity η. A *linear spring* is supposed to produce instantaneously a deformation proportional to the load. A *dashpot* is supposed to produce a velocity proportional to the load at any instant. The relationships between the load F and the deflection u at the point of loading are,

Maxwell model:

$$\dot{u} = \frac{\dot{F}}{\mu} + \frac{F}{\eta}, \qquad u(0) = \frac{F(0)}{\mu}, \tag{9.6–1}$$

Voigt model:

$$F = \mu u + \eta \dot{u}, \qquad u(0) = 0, \tag{9.6–2}$$

Standard linear model:

$$F + \tau_\epsilon \dot{F} = E_R(u + \tau_\sigma \dot{u}), \qquad \tau_\epsilon F(0) = E_R \tau_\sigma u(0), \tag{9.6–3}$$

where τ_ϵ, τ_σ are constants, and a dot above F or u indicates a differentiation with respect to time. The initial conditions at $t = 0$ are indicated.

If we solve Eqs. (9.6–1) through (9.6–3) for $u(t)$ when $F(t)$ is a unit-step function $\mathbf{1}(t)$, the result is called the *creep function*, which represents the elongation produced by a sudden application at $t = 0$ of a constant force of magnitude unity. The creep functions for these equations are, respectively,

Maxwell model:

$$c(t) = \left(\frac{1}{\mu} + \frac{1}{\eta} t\right)\mathbf{1}(t), \tag{9.6–4}$$

Voigt model:

$$c(t) = \frac{1}{\mu}(1 - e^{-(\mu/\eta)t})\mathbf{1}(t), \tag{9.6–5}$$

Standard linear model:

$$c(t) = \frac{1}{E_R}\left[1 - \left(1 - \frac{\tau_\epsilon}{\tau_\sigma}\right)e^{-t/\tau_\sigma}\right]\mathbf{1}(t), \tag{9.6–6}$$

where the *unit-step function* $\mathbf{1}(t)$ is defined as

$$\mathbf{1}(t) = \begin{cases} 1 \text{ when } t > 0, \\ \frac{1}{2} \text{ when } t = 0, \\ 0 \text{ when } t < 0. \end{cases} \tag{9.6–7}$$

A body that obeys a load-deflection relation like that given by Maxwell's model is said to be a *Maxwell solid*, and analogously for bodies that obey Voigt and standard linear models. Since a dashpot behaves as a piston moving in a viscous fluid, the preceding models are called models of *viscoelasticity*.

Interchanging the roles of F and u, we obtain the *relaxation function* as a response $F(t) = k(t)$ corresponding to an elongation $u(t) = \mathbf{1}(t)$. The relaxation function $k(t)$ is the force that must be applied in order to produce an elongation, that changes at $t = 0$ from zero to unity and remains unity thereafter. Relaxation functions for Eqs. (9.6–1) through (9.6–3) are, respectively,

Maxwell solid:

$$k(t) = \mu e^{-(\mu/\eta)t}\mathbf{1}(t), \tag{9.6–8}$$

Voigt solid:

$$k(t) = \eta\delta(t) + \mu\mathbf{1}(t), \tag{9.6-9}$$

Standard linear solid:

$$k(t) = E_R\left[1 - \left(1 - \frac{\tau_\sigma}{\tau_\epsilon}\right)e^{-t/\tau_\epsilon}\right]\mathbf{1}(t). \tag{9.6-10}$$

Here, we have used the symbol $\boldsymbol{\delta}(t)$ to indicate the *unit-impulse function*, or *Dirac-delta function*, which is defined as a function with a singularity at the origin, viz.,

$$\begin{aligned} \boldsymbol{\delta}(t) &= 0 \qquad (\text{for } t < 0, \text{ and } t > 0), \\ \int_{-\epsilon}^{\epsilon} f(t)\boldsymbol{\delta}(t)dt &= f(0) \qquad (\epsilon > 0), \end{aligned} \tag{9.6-11}$$

where $f(t)$ is an arbitrary function that is continuous at $t = 0$. The functions $c(t)$ and $k(t)$ are illustrated in Figs. 9.11 and 9.12, respectively.

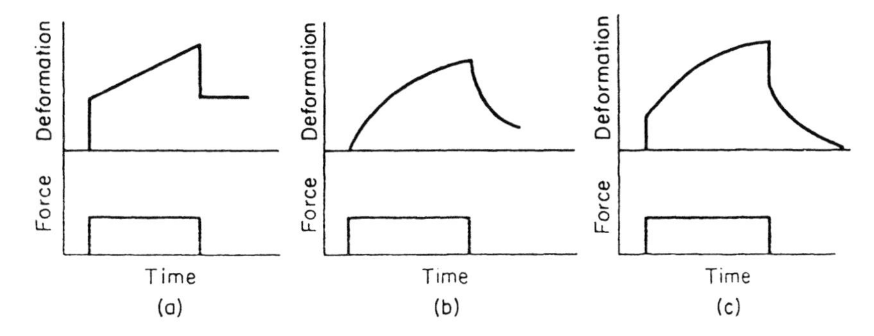

Figure 9.11 Creep function of (a) Maxwell, (b) Voigt, and (c) standard linear solid. A negative phase is superposed at the time of unloading.

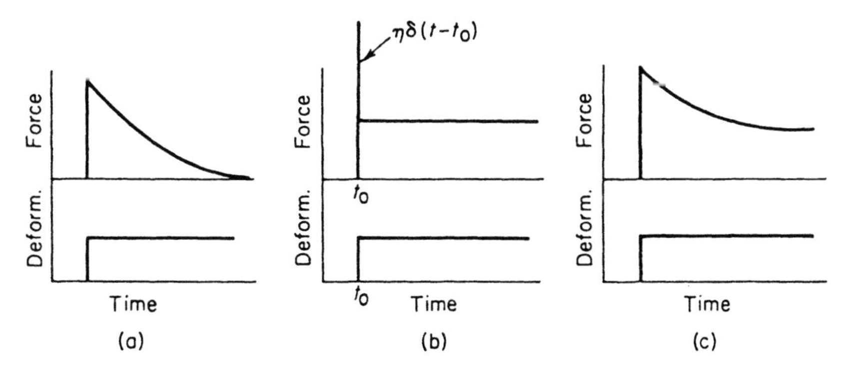

Figure 9.12 Relaxation function of (a) Maxwell, (b) Voigt, and (c) standard linear solid.

For the Maxwell solid, a sudden application of a load induces an immediate deflection by the elastic spring, which is followed by creep of the dashpot. On the other hand, a sudden deformation produces an immediate reaction by the spring, which is followed by stress relaxation according to the exponential law given in Eq. (9.6–8). The factor η/μ, with the dimension of time, may be called a *relaxation time*: It characterizes the rate of decay of the force.

For a Voigt solid, a sudden application of force will produce no immediate deflection, because the dashpot, arranged in parallel with the spring, will not move instantaneously. Instead, as shown by Eq. (9.6–5) and Fig. 9.11(b), a deformation will be gradually built up, while the spring takes a greater and greater share of the load. The displacement of the dashpot relaxes exponentially. Here, the ratio η/μ is again a relaxation time: It characterizes the rate of relaxation of the deflection.

For the standard linear solid, a similar interpretation is applicable. The constant τ_ϵ is the time of relaxation of the load under the condition of constant deflection [see Eq. (9.6–10)], whereas the constant τ_σ is the time of relaxation of deflection under the condition of constant load [see Eq. (9.6–6)]. As $t \to \infty$, the dashpot is completely relaxed, and the load-deflection relation becomes that of the springs, as is characterized by the constant E_R in Eq. (9.6–6) and (9.6–10). Therefore, E_R is called the *relaxed elastic modulus*.

Maxwell introduced the model represented by Eq. (9.6–1), with the idea that all fluids are elastic to some extent. Lord Kelvin showed the inadequacy of the Maxwell and Voigt models in accounting for the rate of dissipation of energy in various materials subjected to cyclic loading. Kelvin's model is commonly called the standard linear model.

More general models may be built by adding more and more elements to the Kelvin model. Equivalently, we may add more and more exponential terms to the creep function or to the relaxation function.

The most general formulation under the assumption of linearity between cause and effect is due to Boltzmann (1844–1906). In the one-dimensional case, we may consider a simple bar subjected to a force $F(t)$ and elongation $u(t)$. The elongation is caused by the total history of the loading up to the time t. If the function $F(t)$ is continuous and differentiable, then in a small time interval $d\tau$ at time τ, the increment of loading is $(dF/d\tau)d\tau$. This increment remains acting on the bar and contributes an element $du(t)$ to the elongation at time t, with a proportionality constant c that depends on the time interval $t - \tau$. Hence, we may write

$$du(t) = c(t - \tau)\,\frac{dF(\tau)}{d\tau}\,d\tau. \tag{9.6–12}$$

Let the origin of time be taken at the beginning of motion and loading. Then, on summing over the entire history, which is permitted under Boltzmann's hypothesis, we obtain

$$u(t) = \int_0^t c(t - \tau)\,\frac{dF(\tau)}{d\tau}\,d\tau. \tag{9.6–13}$$

A similar argument, with the roles of F and u interchanged, gives

$$F(t) = \int_0^t k(t - \tau) \frac{du(\tau)}{d\tau} d\tau. \tag{9.6–14}$$

These laws are linear, since doubling the load doubles the elongation and vice versa. The functions $c(t - \tau)$ and $k(t - \tau)$ are the *creep* and *relaxation* functions, respectively.

The Maxwell, Voigt, and Kelvin models are special examples of the Boltzmann formulation. More generally, we can write the relaxation function in the form

$$k(t) = \sum_{n=0}^{N} \alpha_n e^{-t\nu_n}, \tag{9.6–15}$$

which is a generalization of Eq. (9.6–10). If we plot the amplitude α_n associated with each characteristic frequency ν_n on a frequency axis, we obtain a series of lines that resembles an optical spectrum, Fig. 9–13. Hence, $\alpha_n(\nu_n)$ is called a *spectrum of the relaxation function*. The examples shown in Eqs. (9.6–8)–(9.6–10) are *discrete* spectrums. A generalization to a continuous spectrum is given in the next section.

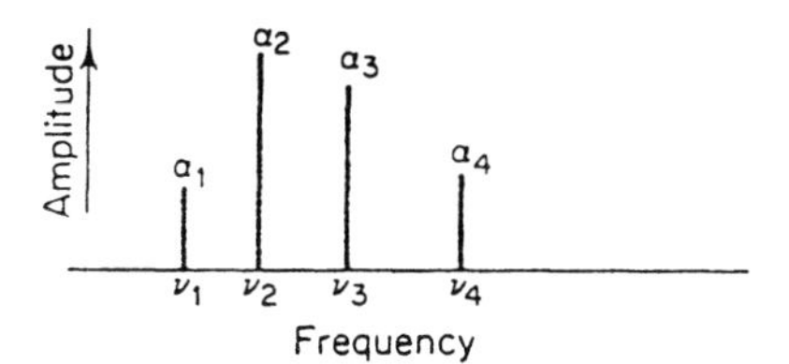

Figure 9.13 A discrete spectrum of a relaxation function.

9.7 QUASI-LINEAR VISCOELASTICITY OF BIOLOGICAL TISSUES

Let us clarify the viscoelastic features of biological soft tissues mentioned in Sec. 9.4. Take the lung tissue as an example. Figure 9.14 shows the stress-strain relationship of the lung tissue in *loading* at different strain rates. Each cycle was done at a constant rate. The period of each is noted in the figure. Over a 360-fold change in strain rate, there was only a minor change in the stress-strain relationship. The *hysteresis* H, defined as the ratio of the area of the hysteresis loop divided by the area under the loading curve, is also noted in the figure. H is seen to be variable, but its variation with strain rate is not large. A similar experience is encountered with other biological tissues. Records of skeletal and cardiac muscles, the ureter, teniae coli, arteries, veins, the pericardium, the mesentery, the bile duct, the skin, tendons, elastin, cartilage, and other tissues show similar characteristics. Typically, in a 1,000-fold change in strain rate, the stress at a given strain in a loading (or unloading) process does not change by more than a factor of 2.

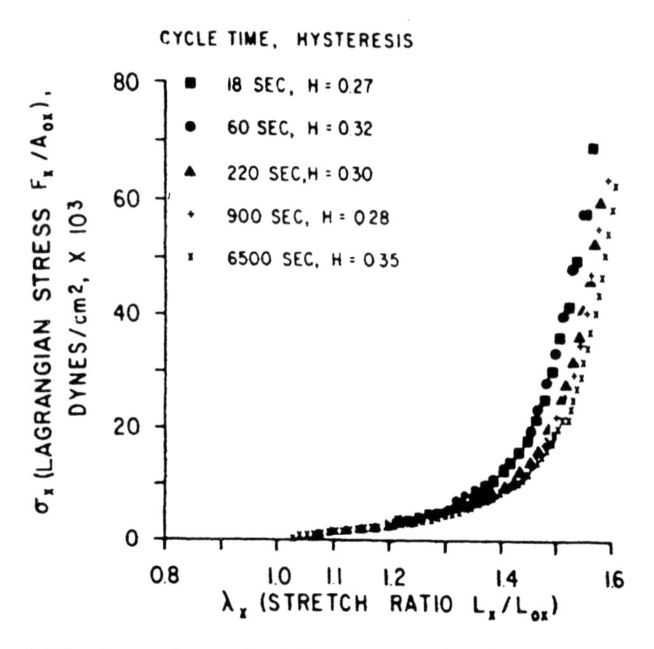

Figure 9.14 A set of records of the stress–strain relationship of the lung tissue subjected to a cyclic biaxial loading are shown in the loading phase, (the unloading phase was not recorded on this graph to improve clarity.) The viscoelasticity of the lung tissue is revealed by the effect of strain rate on the stress–strain relationship. This set of curves cover a range of strain rates varying over 250–fold. It is seen that the stress–strain relationship is not greatly affected by strain rate. The energy dissipated per cycle divided by the work of loading per cycle is called *hysteresis*, and is denoted by H. The values of H at various strain rates (periods of cycles) are shown in the inset. Hysteresis is due to viscoelasticity, and is seen not to vary much with strain rate. For details, see D.L. Vawter, Y.C. Fung, and J.B. West, "Elasticity of Excised Dog Lung Parenchyma," *Journal of Applied Physiology* **45**(2):261–269, 1978.

Figure 9.14 shows two features that cannot be accommodated by the viscoelastic models discussed in Sec. 9.6: the nonlinearity of the stress-strain relationship and the insensitivity of the material to strain rate. The former can be corrected by introducing nonlinear springs. The latter is made clear by reference to Fig. 9.15. The models of Maxwell (a), Voigt (b), and Kelvin (c) are shown, together with their hysteresis characteristics H, (d) through (f), as functions of the frequency of the loading and unloading cycle. Hysteresis decreases with increasing frequency in the Maxwell model because as frequency increases, the dashpot will move less and less. The trend is reversed in Voigt model because here the dashpot takes up more and more of the load. The Kelvin model has a bell-shaped curve of hysteresis vs. the logarithm of the frequency. Each set of constants leads to a characteristic peak. None of the models has the flat hysteresis curve of living tissue.

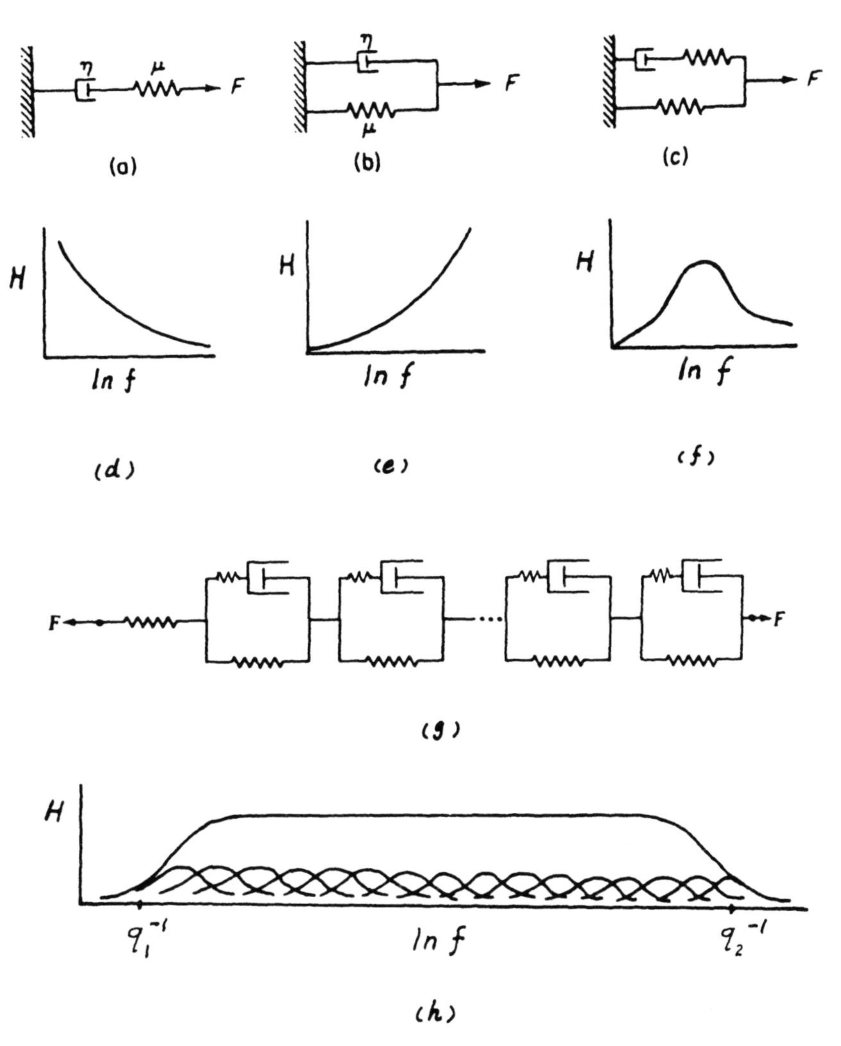

Figure 9.15 Several standard viscoelastic models are shown in the top row, and a mathematical model of the viscoelasticity of biological soft tissues is shown in the third row. Figures in the second row, panels (*d*), (*e*), and (*f*) show the relationships between the hysteresis (*H*) and the logarithm of frequency (*ln f*) of the Maxwell model (*a*), Voight model (*b*), and Kelvin model (*c*), respectively. The figure in the bottom row shows the general hysteresis–log frequency relationship of most living soft tissues, corresponding to the model (*g*) shown in the third row. Living soft tissues usually have a nearly constant hysteresis over a very broad range of frequency. This is modeled in (*g*) by an assemblage of Kelvin models, each of which contributes a small bell-shaped curve; the sum of which is flat over a wide range of frequencies as shown in (*h*).

A model suitable for soft tissue is shown in Fig. 9.15(g). It is composed of a long series of Kelvin bodies whose characteristic times span a broad range. The characteristic curves of hysteresis of these Kelvin bodies are shown by the ripples in the bottom of Fig. 9.15(h). The sum of these ripples is a continuous curve which is flat over a wide frequency range.

To put these observations into mathematical form, we introduce an *Elastic Stress* $\mathbf{T}^{(e)}$ (a tensor), which is a function of the strain $\mathbf{E}$ (a tensor defined with respect to the zero-stress state). If the material is in the zero-stress state until the time $t = 0$, and then suddenly it is strained to $\mathbf{E}$ and maintained constant at that value, the stress developed will be a function of time as well as of $\mathbf{E}$. The history of the stress development may be written as

$$G_{ijmn}(t)T_{mn}^{(e)}(\mathbf{E}), \qquad G_{ijmn}(0) = 1, \tag{9.7–1}$$

in which $G_{ijmn}(t)$, a normalized function of time, is called the *reduced relaxation function*. We then assume that the stress response to an infinitesimal change in a component of strain, δE_{ij}, superposed on a specimen in a state of strain $\mathbf{E}$ at an instant of time τ, is, for $t > \tau$,

$$G_{ijmn}(t - \tau)\,\frac{\partial\, T_{mn}^{(e)}[\mathbf{E}(\tau)]}{\partial E_{ij}}\,\delta E_{ij}(\tau). \tag{9.7–2}$$

Finally, we assume that the superposition principle applies, so that

$$T_{ij}(t) = \int_{-\infty}^{t} G_{ijmn}(t - \tau)\,\frac{\partial\, T_{mn}^{(e)}[\mathbf{E}(\tau)]}{\partial E_{kl}}\,\frac{\partial\, E_{kl}(\tau)}{\partial \tau}\,dt. \tag{9.7–3}$$

That is, the stress at time t is the sum of the contributions of all the past changes, each governed by the same reduced relaxation function. Although $\mathbf{T}^{(e)}(\mathbf{E})$ may be a nonlinear function of strain, the relaxation process is linear. Hence, the theory is called a *quasi-linear viscoelasticity theory*. The lower limit of the integral in Eq. (9.7–3) is written as $-\infty$ to mean the beginning of time.

The reduced relaxation function of a one-dimensional Kelvin model is

$$G(t) = \frac{1}{1 + S}\,[1 + Se^{-t/q}]. \tag{9.7–4}$$

where S and q are constants. If we put an infinite number of Kelvin models in series, we can get a reduced relaxation function in the form

$$G(t) = \left[1 + \int_0^\infty S(q)e^{-t/q}\,dq\right]\left[1 + \int_0^\infty S(q)\,dq\right]^{-1}. \tag{9.7–5}$$

$S(q)$ is called a *relaxation spectrum*, and $1/q$ is a frequency. It has been shown that a specific spectrum, with constants c, q_1, and q_2, namely,

$$S(q) = \begin{cases} \dfrac{c}{q} & \text{for } q_1 \le q \le q_2, \\ \\ 0 & \text{for } q < q_1,\ q > q_2, \end{cases} \tag{9.7–6}$$

fits the data for the skin, arteries, ureter, and teniae coli.

9.8 NON-NEWTONIAN FLUIDS

Newton's law of viscosity describes the behavior of water very well, but there are many other fluids that behave differently. Read the advertisements of some paints: "No drip [will not flow on the brush], spreads easily [offers little resistance to flow], leaves no brush marks [flows to smooth off the surface]." These desirable features for household paints are not Newtonian. Most paints, enamels, and varnishes are non-Newtonian, as are most polymer solutions.

Let us illustrate the subject with a fluid that is most important to our lives—blood. The viscosity of blood depends on the strain rate. Figure 9.16 shows

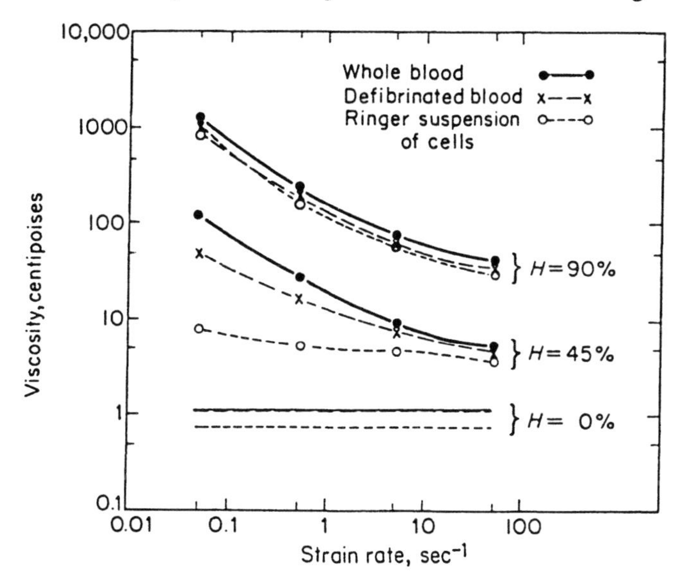

Figure 9.16 The variation in the coefficient of viscosity with the strain rate in human blood, showing data for whole blood, defibrinated blood, and washed red blood cells resuspended in a Ringer solution at 45 and 90% hematocrit H (red cell concentrations by volume). From S. Chien, S. Usami, H. M. Taylor, J. L. Lundberg, and M. I. Gregersen, *J. Appl. Physiol.*, 21 (1966), p. 81, and M. I. Gregersen, "Factors Regulating Blood Viscosity: Relation to Problems of the Microcirculation," *Les Concepts de Claude Bernard sur le Milieu intérieur* (Paris: Masson, 1967).

the variation in the coefficient of viscosity of blood with the strain rate, as measured with a Couette viscometer by Chien, Usami, Gregersen, et al. The coefficient of viscosity increases as the strain rate decreases below about 100 $\sec^{-1}$. At very low strain rate the blood has a finite "yield" stress; i.e., it is visco-plastic. Other visco-plastic materials are discussed in the next section.

The world of non-Newtonian fluids is so much larger than that of Newtonian fluids that the landscape is yet largely unexplored.

9.9 VISCOPLASTIC MATERIALS

A material obeying Newton's law of viscosity must flow under the slightest shear stress (more precisely, under a nonvanishing stress deviation). Materials such as sourdough, paste, and molding clay do not follow such a rule. Bingham, who invented the word "rheology" to describe the science of flow (Greek, ρεος flow), formulated a law for a class of materials known as viscoplastic, to which sourdough seems to belong. A viscoplastic material is often called a Bingham plastic.

A viscoplastic material can sustain stresses with a nonvanishing stress deviation when in a state of rest. See Fig. 9.17. Consider first a body subjected to simple shear, i.e., a state in which all components of the tensors of stress and strain rate

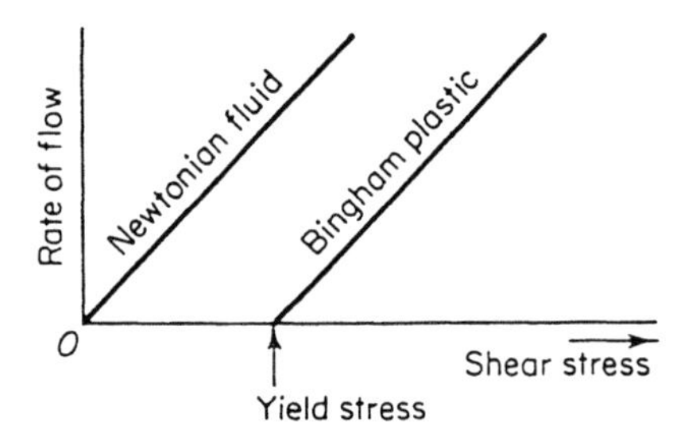

Figure 9.17 Comparison of the flow rate and stress relationship of a viscoplastic material with that of a Newtonian fluid.

vanish, except the shear stress $\sigma_{12} = \sigma_{21} = \tau$ and the shear strain rate $V_{12} = V_{21} = \dot{e}$. As long as the absolute value of the shear stress τ is smaller than a certain constant K, called the *yield stress*, the material remains rigid, so that $\dot{e} = 0$. As soon as $|\tau|$ exceeds K, however, the material flows, with a strain rate $\dot{e}$ having the same sign as τ and an absolute value proportional to $|\tau| - K$. Thus,

$$2\mu\dot{e} = \begin{cases} 0 & \text{if } |\tau| < K, \\ \left(1 - \dfrac{K}{|\tau|}\right)\tau & \text{if } |\tau| > K, \end{cases} \tag{9.9–1}$$

where μ is a coefficient of viscosity. This formulation may be written slightly differently with the introduction of a *yield function* F defined as

$$F = 1 - \frac{K}{|\tau|}. \tag{9.9–2}$$

Then, a viscoplastic material in a state of simple shear is defined by Bingham (1922)* with the relations

$$2\mu\dot{e} = \begin{cases} 0 & \text{if } F < 0, \\ F\tau & \text{if } F \geq 0. \end{cases} \tag{9.9–3}$$

Hohenemser and Prager (1932)† generalized Bingham's definition to arbitrary states of stress in the form

$$2\mu V_{ij} = \begin{cases} 0 & \text{for } F < 0, \\ F\sigma'_{ij} & \text{for } F \geq 0, \end{cases} \tag{9.9–4}$$

with

$$F = 1 - \frac{K}{\sqrt{J_2}}, \tag{9.9–5}$$

where

$$\begin{aligned} \mu &= \text{coefficient of viscosity}, \\ V_{ij} &= \text{strain-rate tensor (see Sec. 6.1)}, \\ \sigma'_{ij} &= \text{stress-deviator tensor} = \sigma_{ij} - \tfrac{1}{3}\sigma_{\alpha\alpha}\delta_{ij}, \\ K &= \text{yield stress}, \\ J_2 &= \text{second invariant of the stress deviator} \\ &= \tfrac{1}{6}[(\sigma_{11} - \sigma_{22})^2 + (\sigma_{22} - \sigma_{33})^2 - (\sigma_{33} - \sigma_{11})^2] + \sigma_{12}^2 + \sigma_{23}^2 + \sigma_{31}^2. \end{aligned}$$

For simple shear, Eqs. (9.9–4) and (9.9–5) reduce to Eqs. (9.9–3) and (9.9–2), respectively.

According to Eq. (9.9–4), the rate-of-deformation tensor of a viscoplastic material is a deviator; i.e., the material is incompressible. When the yield function is negative, the material is rigid. Flow occurs when the yield function has a positive value. The state of stress for which $F = 0$ forms the *yield limit* at which viscoplastic flow sets in or ceases, depending on the sense of direction in which the yield limit is crossed.

Further generalizations of Bingham's equation (9.9–3) are possible. For example, compressibility may be introduced, or other yield criteria may be proposed instead of Eq. (9.9–5).

*Bingham, E. C., *Fluidity and Plasticity*, New York, McGraw-Hill, 1922, p. 215.

†Hohenemser, K., and Prager, W., "Über die Ansätze der Mechanik isotroper Kontinua." *Zeitschrift f. angen Math u. Mech.* 12: 216–226, 1932.

9.10 SOL-GEL TRANSFORMATION AND THIXOTROPY

A colloidal solution may possess rigidity (subjected to shear stress without flow) and be called a *gel*, or it may behave as a fluid without rigidity and be called a *sol*. A gel contains a dispersed component and a dispersion medium, both of which extend continuously throughout the system. The elastic property of the gel may change with its age. The dispersed component of a gel is usually interpreted as forming a network held together by bonds or junction points formed by primary valence bonds, long-range attractive forces, or secondary valence bonds that cause an association between segments of polymer chains or the formation of submicroscopic crystalline regions. Each of the junctions is a mechanism for relaxation under stress. The statistics of the totality of all these relaxation mechanisms is described by the viscoelasticity of the material.

Gels often can be converted into sols and vice versa by a change of temperature, by agitation, or by chemical action in a process called *peptization*. If a reversible gel-sol transformation can be induced isothermally by mechanical vibration, then the material is said to be *thixotropic*, according to Freundlich. The gel is transformed into a sol by mechanical agitation, and the sol reverts to a gel when the agitation is discontinued.

Examples of thixotropic substances are paints, printing inks, iron oxide sols, agar, suspensions of kaolin, and carbon black. Thixotropic materials pervail in the biological world. The protoplasm in an amoeba is perhaps the best known example.

Whether a colloidal system is thixotropic or not may depend on small changes in ionic strength. See Scott-Blair, G. W., *An Introduction to Biorheology*, New York, Elsevier, 1974, for many interesting examples.

PROBLEMS

9.1 Show that the velocity of sound in the material of construction of an aircraft is a criterion for its safety against such dynamic problems as clear air turbulence, gust encounter, and flutter. For this purpose, consider two airplanes identical in geometry and construction, but different in material. Simplify the problem to consider only the following four typical parameters: the density of the material σ; Young's modulus E; the density of air ρ; and the velocity of flight of the airplane, U. Use dimensional analysis to construct the similarity parameters. Let σ, E, ρ, and U refer to one plane and σ', E', ρ', and U' refer to the other. Show that for dynamic similarity, we must have

$$\frac{U'}{U} = \sqrt{\frac{E'}{\sigma'}}\Big/\sqrt{\frac{E}{\sigma}}.$$

If U represents the limit of the safe flight speed (e.g., the critical flutter speed), then the preceding formula relates U to the velocity of sound, $\sqrt{E/\sigma}$ (speed of longitudinal waves in a rod).

9.2 The velocity of sound in a solid is an important similarity parameter for comparing the rigidities of flight structures. Suppose you are an airplane designer selecting materials for construction. Using a handbook, list the velocity of sound for the following structural materials: pure aluminum, magnesium, aluminum alloys, magnesium alloys, carbon steels, stainless steels, titanium, titanium carbide, and the rather exotic materials beryllium oxide and pure beryllium. Compare with the plastics lucite and phenolic laminates with the woods spruce, mahogany, balsa, and bamboo, along the grain. Are you not surprised at the rather small differences between the velocities of sound in many of these materials? What is the best material from this point of view?

9.3 Show that the same conclusion as in Prob. 9.1 would be reached if you considered a suspension bridge that may be induced to vibrate dangerously in wind. (The original Tacoma Narrows Bridge on Puget Sound, Washington, spectacularly failed by flutter on Nov. 7, 1940, four months after it was opened to traffic, in a wind speed of 42 mph. On that morning, the frequency of oscillation of the bridge changed suddenly from 37 to 14 cycles/min, probably because of a failure of a small reinforcing tie rod. The motion grew violently in the torsional mode, and failure occurred half an hour later. If there had not been this aerodynamically induced oscillation (flutter), the bridge should have been able to withstand a steady wind of at least 100 mi/hr.)

9.4 The experimental data on the viscosity of blood measured in a Couette flowmeter (Fig. P3.22, p. 86), as shown in Fig. 9.16, can be expressed approximately by Casson's equation

$$\sqrt{\tau} = \sqrt{\tau_y} + b\sqrt{\dot{\gamma}}$$

in which τ is the shear stress, τ_y is a constant that may be identified as a yield stress, and $\dot{\gamma}$ is the shear strain rate ($\sec^{-1}$). Generalize this result to a constitutive equation for blood that is correct from the point of view of dimensional and tensor analyses.

9.5 Put blood between the cone and plate of a cone-plate viscometer (Fig. P9.5). The cone rotates at an angular speed of n revolutions per second, while the plate remains stationary. Derive the relationship between the torque T acting on the cone, the angular speed n, the radius R, the cone angle θ, and the constants τ_y and b in the constitutive equation derived in Prob. 9.4. In Fig. P9.5, the angle θ is exaggerated. In practice, it has to be small. Discuss what kind of complication will occur when θ is large and why.

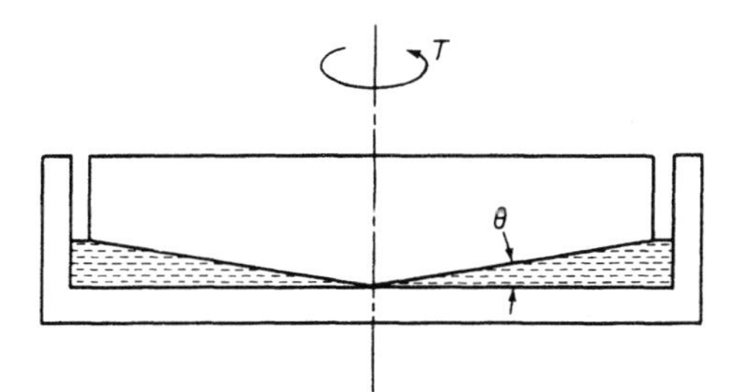

Figure P9.5 A cone-plate viscometer.

9.6 Assume that no material will expand in volume when it is subjected to a hydrostatic pressure. Show that the maximum value of Poisson's ratio ν for any isotropic elastic solid obeying Hooke's law is $\frac{1}{2}$.

9.7 Reinforced concrete is concrete poured over steel rods. A vertical, hollow reinforced concrete column has an internal diameter of 3 ft and thickness of 3 in, with 36 steel rods of 1-sq-in cross-sectional area, uniformly spaced in a circle. The column is subjected to a vertical load, the resultant of which is along the axis of the column. The ratio of Young's modulus of steel to that of concrete is 15. Poisson's ratio for concrete is 0.4, and that for steel is 0.25. Determine the share of the load that is carried by steel at a cross section some distance from the ends of the column.

9.8 Consider a viscoelastic material characterized as Maxwell's model and described by Eq. (9.6–1). Let a sinusoidally varying force $F = a \sin \omega t$ be imposed on the body. What would be the deflection u at steady state?

Answer:

$$u = \frac{A}{\omega}[\sin(\omega t - \alpha) + \sin \alpha],$$

where

$$A = \left[\left(\frac{a\omega}{\mu}\right)^2 + \left(\frac{a}{\eta}\right)^2\right]^{1/2}, \quad \tan \alpha = \frac{\mu}{\eta\omega}.$$

9.9 A liquid flows down a long tube of diameter 1 cm from a reservoir at a rate of 10 cm^3/sec. The streamlines are found to be as shown in Fig. P9.9. The principal feature is that the liquid column expands in diameter as it leaves the tube. Can a Newtonian liquid do this? What kind of stress-strain relationship is suggested? [See A. S. Lodge, *Elastic Liquids*, New York: Academic Press (1964), p. 242.]

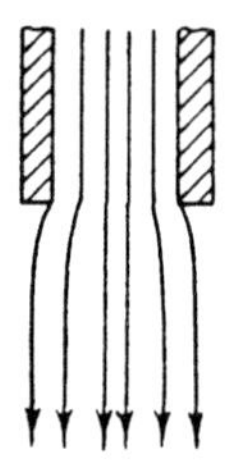

Figure P9.9 A non-Newtonian fluid coming out of a spout.

9.10 When a certain paint was stirred with an electric mixer, it was found that it climbed up the shaft of the mixer. What kind of stress-strain relationship of the paint is revealed by this experiment? (See A. S. Lodge, *ibid.*, p. 232.)

9.11 Take a piece of chalk and twist it to failure. Describe the cleavage surface, and infer the criterion about the strength of the chalk.

Again, break the chalk by bending, and discuss the fracture mechanism.

9.12 Take a piece of nylon thread, pull it to failure, and discuss the failure mechanism of nylon vs. that of the chalk in Prob. 9.11.

9.13 Take a toy rubber balloon. Blow it up. Take a pin. Prick the inflated balloon. Chances are it will explode. Now, without inflating the balloon, stretch the rubber with both hands and ask a friend to prick it with the pin. Chances are that it will not explode. Can you explain that? How would the constitutive equation of the rubber reflect this fact?

9.14 Many engineering and biological structures are made of composite materials consisting of stiffer components embedded in a softer matrix. Consider the following two models:
(**a**) A circular cylindrical tube, in the wall of which are embedded high-strength fibers

of small diameter. Young's moduli for the fiber and the matrix material are E_f and E_0, respectively, with $E_f >> E_0$. The fibers are parallel to the cylinder axis, they are uniformly distributed, and the total cross-sectional area of the fibers is a fraction of the total cross-sectional area of the tube. When the tube is stretched longitudinally, what is Young's modulus for it?

(b) A circular cylindrical tube of inner radius a and outer radius b, embedded with spirally wound fibers. The helical fibers make an angle θ with the cylinder axis. Half of the fibers are wound in the manner of a right-handed screw, and half are left handed. Young's moduli for the fiber and matrix are, again, E_f and E_0, respectively. Compute the effective Young's modulus of the tube in longitudinal tension. Assume that the fibers are perfectly embedded.

(c) When the cylindrical tubes of parts (a) and (b) are subjected to an internal pressure p_i, how is the load resisted by the fibers and matrix?

(d) When a transverse shear load perpendicular to the axis of the cylinder is applied to the cylinder, how is the shear resisted? Analyze the stresses in the fibers and the matrix.

(e) Similarly, analyze the stress distribution when the cylinder has to resist a bending moment.

(f) Torsion resistance is also important. A torque T is applied to the tube. What are the stresses in the fibers and matrix then?

9.15 To measure the tensile strength of water and other liquids, Lyman Briggs (*J. Chem. Physics* 19 (1951), p. 970) used a Z-shaped capillary tube, open at both ends, rotating in the Z-plane about an axis passing through the center of the Z of the tube and perpendicular to the plane. The liquid menisci are located in the bent-back short arms of the Z. The speed of rotation is increased gradually until the liquid in the capillary "breaks." If one uses a straight tube that is open at both ends for this experiment, the fluid will fly away and the experiment will not be possible. The bent-back short arms of the Z provide the stability of the fluid. Examine this stability problem and present a theoretical analysis of the experiment.

REFERENCES

COTTRELL, A. H., *The Mechanical Properties of Matter*. New York: Wiley, 1964.

FUNG, Y. C., *Foundations of Solid Mechanics*. Englewood Cliffs, N.J., Prentice-Hall, 1965.

FUNG, Y. C., "Elasticity of Soft Tissues in Simple Elongation." *Am. J. Physiol.* 213(6): 1532–1544, 1967.

FUNG, Y. C., "Stress-Strain-History Relations of Soft Tissues in Simple Elongation." In *Biomechanics: Its Foundations and Objectives*, Ed. by Y. C. Fung, N. Perrone, and M. Anliker, N.J., Prentice-Hall, 1971.

FUNG, Y. C., "Biorheology of Soft Tissues." *Biorheology* 10: 139–155, 1973.

FUNG, Y. C., FRONEK, K., AND PATITUCCI, P., "Pseudoelasticity of Arteries and the Choice of Its Mathematical Expression." *Am. J. Physiol.* 237: H620–H631, 1979.

FUNG, Y. C., *Biomechanics: Mechanical Properties of Living Tissues*. New York: Springer-Verlag, 1981, 2nd ed., 1993.

FUNG, Y. C., "A Model of the Lung Structure and Its Validation." *J. Appl. Physiol.* 64(5): 2132–2141, 1988.

Fung, Y. C., *Biomechanics: Motion, Flow, Stress, and Growth*. New York: Springer-Verlag, 1990.

Green, A. E., and Adkins, J. E., *Large Elastic Deformations and Nonlinear Continuum Mechanics*. Oxford: Oxford University Press, 1960.

Hayashi, K., Handa, H., Mori, K., and Moritake, K., "Mechanical Behavior of Vascular Walls." *J. Soc. Material Science Japan* 20: 1001–1011, 1971.

Li, Yuan-Hui, "Equation of State of Water and Sea Water." *J. Geophys. Res.* 72(10): 2665–2678, 1967.

Matsuda, M., Fung, Y. C., and Sobin, S. S., "Collagen and Elastin Fibers in Human Pulmonary Alveolar Mouths and Ducts." *J. Appl. Physiol.* 63(3): 1185–1194, 1987.

Patel, D. J., and Vaishnav, R. N., "The Rheology of Large Blood Vessels." In *Cardiovascular Fluid Dynamics*, Vol. 2, edited by D. H. Bergel (pp. 2–64). New York: Academic Press, 1972.

Sobin, S. S., Fung, Y. C., and Tremer, H. M., "Collagen and Elastin Fibers in Human Pulmonary Alveolar Walls." *J. Appl. Physiol.* 64(4): 1659–1675, 1988.

Tait, P. G., "Report on Some of the Physical Properties of Fresh Water and Sea Water." *Report on Scientific Results of Voy. H.M.S., Challenger, Phys. Chem.*, 2, 1–76, 1888.

Tanner, R. I., *Engineering Rheology*. Oxford: Oxford University Press, 1988.

10 DERIVATION OF FIELD EQUATIONS

In the preceding chapters, we have analyzed deformation (strain) and flow (strain rate) and their relationship with the force of interaction (stress) between parts of a material body (continuum). We are now in a position to use this information to derive differential equations describing the motion of the continuum under specific boundary conditions. Our formulation must obey Newton's law of motion, the principle of conservation of mass, and the laws of thermodynamics. This chapter is concerned with expressing these laws in a form suitable for the treatment of a continuum.

One may wonder why there is a need for further elaboration on these well-known laws. The answer may be illustrated in the following example. If we have a single particle, the principle of conservation of mass merely states that the mass of the particle is a constant. However, if we have a large number of particles, such as the water droplets in a cloud, the situation requires some thought. For the cloud, it is no longer practical to identify the individual particles. The most convenient way to describe the cloud is to consider the velocity field, the density distribution, the temperature distribution, etc. It is the description of the classical conservation laws in such a circumstance that will occupy our attention in this chapter.

Our approach is based on the fact that these conservation laws must be applicable to the matter enclosed in a volume bounded by an arbitrary closed surface. In such an approach, we find that some quantities enter naturally in a surface integral, others in a volume integral. A transformation from a surface integral to a volume integral, and vice versa, is often required. This transformation is embodied in Gauss's theorem, which serves as our mathematical starting point.

10.1 GAUSS'S THEOREM

We shall begin with the derivation of Gauss's theorem. Consider a convex region V bounded by a surface S that consists of a finite number of parts whose outer normals form a continuous vector field (e.g., the one shown in Fig. 10.1). Such a

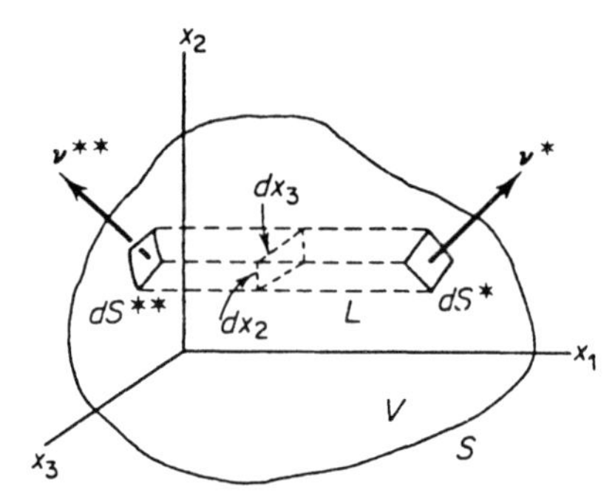

Figure 10.1 Path of integration illustrating the derivation of Gauss's theorem.

region is said to be *regular*. Let a function $A(x_1, x_2, x_3)$ be defined in the volume V and on the surface S. Let A be continuously differentiable in V. Let us consider the volume integral

$$\iiint_V \frac{\partial A}{\partial x_1} dx_1\, dx_2\, dx_3.$$

The integrand is the partial derivative of A with respect to x_1. By integrating with respect to x_1 along a line segment L, we obtain

$$\iiint_V \frac{\partial A}{\partial x_1} dx_1\, dx_2\, dx_3 = \iint_S (A^* - A^{**})\, dx_2\, dx_3 \tag{10.1–1}$$

where A^* and A^{**} are, respectively, the values of A on the surface S at the right and left ends of the line segment L parallel to the x_1-axis. The surface integral on the right-hand side of Eq. (10.1–1) may be written more elegantly. The factors $+dx_2\, dx_3$ and $-dx_2\, dx_3$ are the projections on the x_2x_3-plane of the areas dS^* and dS^{**} at the ends of the line segment L. Let $\boldsymbol{\nu} = (\nu_1, \nu_2, \nu_3)$ be the unit vector along the outer normal to the surface S. For the element shown in Fig. 10.1, we see that $\nu_1^* = \cos(x_1, \boldsymbol{\nu}^*)$ is positive, whereas $\nu_1^{**} = \cos(x_1, \boldsymbol{\nu}^{**})$ is negative. It is easy to see that in this case, $dx_2\, dx_3 = \nu_1^*\, dS^*$ at the right end and $-dx_2\, dx_3 = \nu_1^{**}\, dS^{**}$ at the left end. Therefore, the surface integral in Eq. (10.1–1) can be written as

$$\iint_S (A^*\, dx_2\, dx_3 - A^{**}\, dx_2\, dx_3) = \iint_S (A^*\nu_1^*\, dS^* + A^{**}\nu_1^{**}\, dS^{**}). \tag{10.1–2}$$

The asterisks may be omitted because they merely indicate the appropriate values of A and ν_1 to be taken in a surface integral according to conventional notations. Thus, the right-hand side of Eq. (10.1–1) reduces to $\int_S A\nu_1\, dS$. Now, if we write the volume integral on the left-hand side as $\int_V (\partial A/\partial x_1)\, dV$, then we have

$$\int_V \frac{\partial A}{\partial x_1} dV = \int_S A\nu_1\, dS, \tag{10.1–3}$$

where dV and dS denote the elements of V and S, respectively. A similar argument applies to the volume integral of $\partial A/\partial x_2$ or $\partial A/\partial x_3$. Thus, we obtain Gauss's theorem,

$$\int_V \frac{\partial A}{\partial x_i}\, dV = \int_S A\nu_i\, dS, \qquad (i = 1, 2, 3). \qquad \blacktriangle \quad (10.1\text{–}4)$$

This formula holds for any convex regular region or for any region that can be decomposed into a finite number of convex regular regions.

Now let us consider a tensor field $A_{jkl\ldots}$. Let the region V with boundary surface S be within the region of definition of $A_{jkl\ldots}$. Let every component of $A_{jkl\ldots}$ be continuously differentiable in V. Then Eq. (10.1–4) is applicable to every component of the tensor, and we obtain the general result

$$\int_V \frac{\partial}{\partial x_i} A_{jkl\ldots}\, dV = \int_S \nu_i A_{jkl\ldots}\, dS, \qquad (10.1\text{–}5)$$

which is one of the most useful theorems in applied mathematics.

This theorem was given in various forms by Lagrange (1762), Gauss (1813), Green (1828), and Ostrogradsky (1831). It is best known in this country as *Green's theorem* or *Gauss's theorem.*

Example 1

Let v_i represent a vector. Then, according to Eq. (10.1–5), we have, on identifying $A_i = v_i$, and n_i as the normal vector to the surface S,

$$\int_V \frac{\partial v_i}{\partial x_i}\, dV = \int_S v_i n_i\, dS. \qquad (10.1\text{–}6)$$

If we write the coordinates x_1, x_2, x_3 as x, y, z; the components v_1, v_2, v_3 as u, v, w; and the direction cosines n_1, n_2, n_3 of the outer normal to the surface S as l, m, n, then

$$\iiint_V \left(\frac{\partial u}{\partial x} + \frac{\partial v}{\partial y} + \frac{\partial w}{\partial z}\right) dx\, dy\, dz = \iint_S (lu + mv + nw)\, dS. \qquad (10.1\text{–}7)$$

In another popular notation, we denote the vector by $\mathbf{v}$ and the scalar product $v_i n_i$ by $\mathbf{v}\cdot\mathbf{n}$ and define

$$\operatorname{div} \mathbf{v} = \frac{\partial u}{\partial x} + \frac{\partial v}{\partial y} + \frac{\partial w}{\partial z}. \qquad (10.1\text{–}8)$$

Then Eq. (10.1–7) becomes

$$\int_V \operatorname{div} \mathbf{v}\, dV = \int_S \mathbf{v}\cdot\mathbf{n}\, dS. \qquad \blacktriangle \quad (10.1\text{–}9)$$

Equations (10.1–6), (10.1–7), and (10.1–9) are the best known forms of Gauss's theorem.

Example 2

If A is identified with a potential function ϕ, then Eq. (10.1–3) is usually written in the vector form

$$\int_V \operatorname{grad} \phi \, dV = \int_S \mathbf{n}\phi \, dS.$$

Example 3

Let e_{ijk} be the permutation tensor. Then

$$\int e_{ijk}u_{k,j}dV = e_{ijk}\int u_{k,j}dV = e_{ijk}\int u_k n_j dS = \int e_{ijk}u_k n_j dS;$$

i.e.,

$$\int \operatorname{curl} \mathbf{u} \, dV = \int \mathbf{n} \times \mathbf{u} \, dS.$$

10.2 MATERIAL DESCRIPTION OF THE MOTION OF A CONTINUUM

Let a fixed frame of reference $O\text{-}x_1x_2x_3$ be chosen. Let the location of a material particle be $x_1 = a_1$, $x_2 = a_2$, $x_3 = a_3$ when time $t = t_0$. We shall use (a_1, a_2, a_3) as the label for that particle. As time goes on, the particle moves. Its location has the history

$$x_1 = x_1(a_1, a_2, a_3, t), \qquad x_2 = x_2(a_1, a_2, a_3, t), \qquad x_3 = x_3(a_1, a_2, a_3, t)$$

referred to the same coordinate system or, in short,

$$x_i = x_i(a_1, a_2, a_3, t), \qquad (i = 1, 2, 3). \tag{10.2–1}$$

If such an equation is known for every particle in the body, then we know the history of motion of the entire body. Mathematically, Eq. (10.2–1) defines the *transformation*, or *mapping*, of a domain $D(a_1, a_2, a_3)$ into a domain $D'(x_1, x_2, x_3)$, with t as a parameter. An example is shown in Fig. 10.2. If the mapping is continuous and one to one—i.e., for every point (a_1, a_2, a_3), there is one and only one point (x_1, x_2, x_3) and vice versa—and neighboring points in $D(a_1, a_2, a_3)$ are mapped into neighboring points in $D'(x_1, x_2, x_3)$, then the functions $x_i(a_1, a_2, a_3, t)$ must be single valued, continuous, and continuously differentiable, and the Jacobian must not vanish in the domain D.

The mapping given by Eq. (10.2–1) is said to be a *material description* of the motion of the body. In a material description, the velocity and acceleration of the particle at (a_1, a_2, a_3) are, respectively,

$$v_i(a_1, a_2, a_3, t) = \left.\frac{\partial x_i}{\partial t}\right|_{(a_1,a_2,a_3)}, \tag{10.2–2}$$

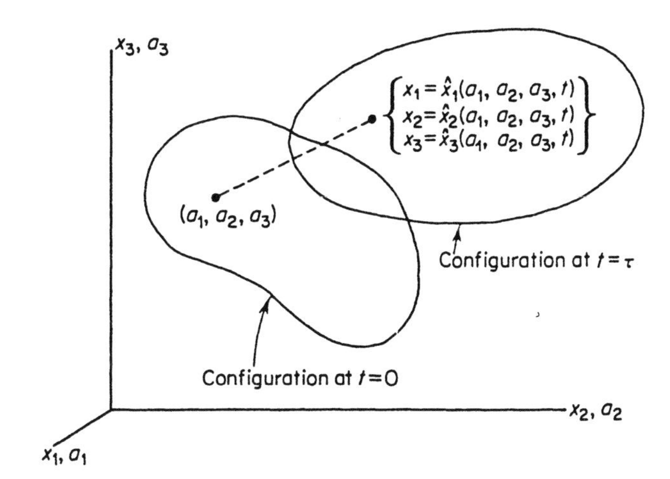

Figure 10.2 Labeling of particles.

and

$$\alpha_i(a_1, a_2, a_3, t) = \left.\frac{\partial v_i}{\partial t}\right|_{(a_1,a_2,a_3)} = \left.\frac{\partial^2 x_i}{\partial t^2}\right|_{(a_1,a_2,a_3)}. \tag{10.2–3}$$

Conservation of mass may be expressed as follows. Let $\rho(\mathbf{x})$ be the density of the material at location $\mathbf{x}$, where the symbol $\mathbf{x}$ stands for (x_1, x_2, x_3). Let $\rho_0(\mathbf{a})$ be the density at the point (a_1, a_2, a_3) when $t = 0$. Then the mass of the material enclosed in a volume V is $\int_D \rho_0(\mathbf{a})\, da_1\, da_2\, da_3$ at $t = 0$ and is $\int_{D'} \rho(\mathbf{x})\, dx_1\, dx_2\, dx_3$ at time t. Thus, conservation of mass is expressed by the formula

$$\int_{D'} \rho(\mathbf{x})\, dx_1\, dx_2\, dx_3 = \int_D \rho_0(\mathbf{a})\, da_1\, da_2\, da_3, \tag{10.2–4}$$

where the integrals extend over the same particles. But

$$\int_{D'} \rho(\mathbf{x})\, dx_1\, dx_2\, dx_3 = \int_D \rho(\mathbf{x}) \det \left|\frac{\partial x_i}{\partial a_j}\right| da_1\, da_2\, da_3, \tag{10.2–5}$$

where $|\partial x_i/\partial a_j|$ is the Jacobian of the transformation, i.e., the determinant of the matrix $(\partial x_i/\partial a_j)$:

$$\det \left|\frac{\partial x_i}{\partial a_j}\right| = \begin{vmatrix} \partial x_1/\partial a_1 & \partial x_1/\partial a_2 & \partial x_1/\partial a_3 \\ \partial x_2/\partial a_1 & \partial x_2/\partial a_2 & \partial x_2/\partial a_3 \\ \partial x_3/\partial a_1 & \partial x_3/\partial a_2 & \partial x_3/\partial a_3 \end{vmatrix}. \tag{10.2–6}$$

Identifying the right-hand sides of Eqs. (10.2–4) and (10.2–5) and realizing that the result must hold for any arbitrary domain D, we see that the integrands must be equal:

$$\rho_0(\mathbf{a}) = \rho(\mathbf{x}) \det \left|\frac{\partial x_i}{\partial a_j}\right|. \tag{10.2–7}$$

Similarly,

$$\rho(\mathbf{x}) = \rho_0(\mathbf{a}) \det \left|\frac{\partial a_i}{\partial x_j}\right|. \tag{10.2–8}$$

These equations relate the density in different configurations of the body to the transformation that leads from one configuration to another.

Thus, the material description of a continuum follows the method used in particle mechanics.

10.3 SPATIAL DESCRIPTION OF THE MOTION OF A CONTINUUM

In the material description, every particle is identified by its coordinates at a given instant of time t_0. This is not always convenient. When we describe the flow of water in a river, we do not desire to identify the location from which every particle of water comes. Instead, we are generally interested in the instantaneous velocity field and its evolution with time. This leads to the *spatial description* traditionally used in hydrodynamics. The location (x_1, x_2, x_3) and the time t are taken as independent variables. It is natural for hydrodynamics because measurements are more easily made and directly interpreted in terms of what happens at a certain place, rather than following the particles.

In a spatial description, the instantaneous motion of the continuum is described by the velocity vector field $v_i(x_1, x_2, x_3, t)$, which, of course, is the velocity of a particle instantaneously located at (x_1, x_2, x_3) at time t. We shall show that the instantaneous acceleration of the particle is given by the formula

$$\dot{v}_i(\mathbf{x}, t) = \frac{\partial v_i}{\partial t}(\mathbf{x}, t) + v_j \frac{\partial v_i}{\partial x_j}(\mathbf{x}, t), \qquad \blacktriangle \quad (10.3\text{–}1)$$

where $\mathbf{x}$ again stands for the variables x_1, x_2, x_3, and every quantity in the formula is evaluated at $(\mathbf{x}, t)$. The proof follows from the fact that a particle located at (x_1, x_2, x_3) at time t is moved to a point with coordinates $x_i + v_i\, dt$ at the time $t + dt$ and that, according to Taylor's theorem, and by omitting the higher-order infinitesimal terms as $dt \to 0$,

$$\begin{aligned}\dot{v}_i(\mathbf{x}, t)\, dt &= v_i(x_j + v_j\, dt,\, t + dt) - v_i(x_j, t) \\ &= v_i + \frac{\partial v_i(\mathbf{x}, t)}{\partial t}\, dt + \frac{\partial v_i(\mathbf{x}, t)}{\partial x_j}\, v_j\, dt - v_i,\end{aligned}$$

which reduces to Eq. (10.3–1). The first term in Eq. (10.3–1) may be interpreted as arising from the dependence of the velocity field on time, the second term as the contribution of the motion of the particle in the nonhomogeneous velocity field. Accordingly, these terms are called the *local* and the *convective* parts of the acceleration, respectively.

The reasoning that leads to Eq. (10.3–1) is applicable to any function $F(x_1, x_2, x_3, t)$ that is attributable to the moving particles, such as the temperature. A convenient terminology is the *material derivative*, which is denoted by a dot or the symbol D/Dt. Thus, the material derivative of F is

$$\dot{F} = \frac{DF}{Dt} \equiv \left(\frac{\partial F}{\partial t}\right)_{\mathbf{x}=\text{const.}} + v_1 \frac{\partial F}{\partial x_1} + v_2 \frac{\partial F}{\partial x^2} + v_3 \frac{\partial F}{\partial x_3}. \quad \blacktriangle \quad (10.3\text{–}2)$$

On the other hand, if $F(x_1, x_2, x_3, t)$ is transformed into $F(a_1, a_2, a_3, t)$ through the transformation given by Eq. (10.2–1), then $F(a_1, a_2, a_3, t)$ is indeed the value of F attached to the particle (a_1, a_2, a_3). Hence, the material derivative F does mean the rate of change of the property F of the particle (a_1, a_2, a_3). Formally,

$$\dot{F} = \left.\frac{\partial F(a_1, a_2, a_3, t)}{\partial t}\right|_{\mathbf{a}}. \quad (10.3\text{–}3)$$

On regarding $F(x_1, x_2, x_3, t)$ as an implicit function of a_1, a_2, a_3, t, we have

$$\dot{F} = \left.\frac{\partial F}{\partial t}\right|_{\mathbf{x}} + \left.\frac{\partial F}{\partial x_1}\right|_t \left.\frac{\partial x_1}{\partial t}\right|_{\mathbf{a}} + \left.\frac{\partial F}{\partial x_2}\right|_t \left.\frac{\partial x_2}{\partial t}\right|_{\mathbf{a}} + \left.\frac{\partial F}{\partial x_3}\right|_t \left.\frac{\partial x_3}{\partial t}\right|_{\mathbf{a}}, \quad (10.3\text{–}4)$$

which reduces to Eq. (10.3–2) by virtue of Eq. (10.2–2).

10.4 THE MATERIAL DERIVATIVE OF A VOLUME INTEGRAL

Let $I(t)$ be a volume integral of a continuously differentiable function $A(\mathbf{x}, t)$ defined over a spatial domain $V(x_1, x_2, x_3, t)$ *occupied by a given set of material particles*:

$$I(t) = \iiint_V A(\mathbf{x}, t)\, dx_1\, dx_2\, dx_3. \quad (10.4\text{–}1)$$

Here again, we write $\mathbf{x}$ for x_1, x_2, x_3. The function $I(t)$ is a function of the time t because both the integrand $A(\mathbf{x}, t)$ and the domain $V(\mathbf{x}, t)$ depend on the parameter t. As t varies, $I(t)$ varies also, and we ask: What is the rate of change of $I(t)$ with respect to t? This rate, denoted by DI/Dt and called the *material derivative* of I, is defined for a given set of material particles.

The phrase "for a given set of particles" is of primary importance. The question is how fast the material body itself "sees" the value of I changing. To evaluate this rate, note that the boundary S of the body at the instant t will have moved at time $t + dt$ to a neighboring surface S', which bounds the domain V' (Fig. 10.3). The material derivative of I is defined as

$$\frac{DI}{Dt} = \lim_{dt \to 0} \frac{1}{dt}\left[\int_{V'} A(\mathbf{x}, t + dt)\, dV - \int_V A(\mathbf{x}, t)\, dV\right]. \quad (10.4\text{–}2)$$

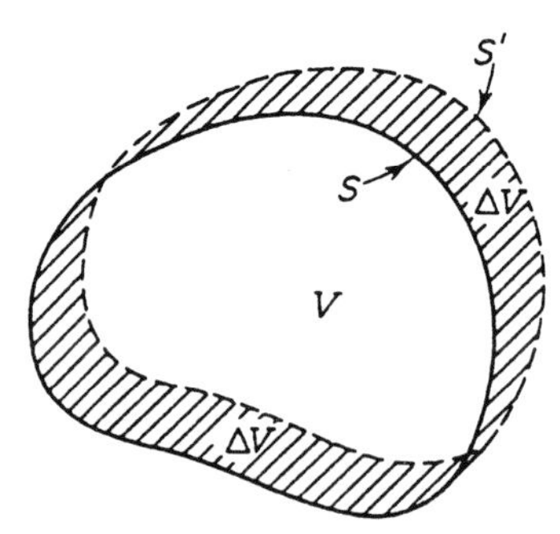

Figure 10.3 Continuous change of the boundary of a region.

Attention is drawn to the difference in the domains V' and V. Let ΔV be the domain $V' - V$. We note that ΔV is swept out by the motion of the surface S in the small time interval dt. Since $V' = V + \Delta V$, we can write Eq. (10.4–2) as

$$\begin{aligned}\frac{DI}{Dt} &= \lim_{dt\to 0}\frac{1}{dt}\left[\int_V A(\mathbf{x}, t + dt)\, dV + \int_{\Delta V} A(\mathbf{x}, t + dt)\, dV \right. \\ &\qquad \left. - \int_V A(\mathbf{x}, t)\, dV\right] \\ &= \lim_{dt\to 0}\left\{\frac{1}{dt}\int_V [A(\mathbf{x}, t + dt) - A(\mathbf{x}, t)]\, dV \right. \\ &\qquad \left. + \frac{1}{dt}\int_{\Delta V} A(\mathbf{x}, t + dt)\, dV\right\}.\end{aligned} \tag{10.4–3}$$

For a continuously differentiable function $A(\mathbf{x}, t)$, the first term on the right-hand side contributes the value $\int_V \partial A/\partial t\, dV$ to DI/Dt. The last term may be evaluated by noting that for an infinitesimal dt, the integrand may be taken to be $A(\mathbf{x}, t)$ on the boundary surface S [because of the assumed continuity of $A(\mathbf{x}, t)$] and that the integral is equal to the sum of $A(\mathbf{x}, t)$ multiplied by the volume swept out by the particles situated on the boundary S in the time interval dt. If n_i is the unit vector along the outer normal of S, then, since the displacement of a particle on the boundary is $v_i\, dt$, the volume swept out by particles occupying an element of area dS on the boundary S is $dV = n_i v_i \mathrm{dS}\cdot\mathrm{dt}$. On ignoring infinitesimal quantities of the second or higher order, we see the contribution of this element to DI/Dt is $Av_i n_i dS$. The total contribution is obtained by an integration over S. Therefore,

$$\frac{D}{Dt}\int_V A\, dV = \int_V \frac{\partial A}{\partial t}\, dV + \int_S A v_j n_j dS. \qquad \blacktriangle \tag{10.4–4}$$

Transforming the last integral by Gauss's theorem and using Eq. (10.3–2), we have

$$\frac{D}{Dt}\int_V A\,dV = \int_V \frac{\partial A}{\partial t}\,dV + \int_V \frac{\partial}{\partial x_j}(Av_j)\,dV$$

$$= \int_V \left(\frac{\partial A}{\partial t} + v_j\frac{\partial A}{\partial x_j} + A\frac{\partial v_j}{\partial x_j}\right) dV \qquad \blacktriangle \quad (10.4\text{–}5)$$

$$= \int_V \left(\frac{DA}{Dt} + A\frac{\partial v_i}{\partial x_j}\right) dV.$$

This important formula will be used repeatedly in the sections that follow. It should be noted that according to Eq. (10.4–5), *the operation of forming the material derivative and that of spatial integration are noncommutative in general.*

10.5 THE EQUATION OF CONTINUITY

The law of conservation of mass was discussed in Sec. 10.2. With the results of Sec. 10.4, we can now give some alternative forms.

The mass contained in a domain V at a time t is

$$m = \int_V \rho\,dV, \qquad (10.5\text{–}1)$$

where $\rho = \rho(\mathbf{x}, t)$ is the density of the continuum at location $\mathbf{x}$ at time t. Conservation of mass requires that $Dm/Dt = 0$. The derivative Dm/Dt is given by Eq. (10.4–4) or Eq. (10.4–5) if A is replaced by ρ. Since the result must hold for an arbitrary domain V, the integrand must vanish. Hence, we obtain the following forms of the law of conservation of mass enclosed in a surface S with outer normal $\mathbf{n}$:

$$\int_V \frac{\partial \rho}{\partial t}\,dV + \int_S \rho v_j n_j\,dS = 0. \qquad \blacktriangle \quad (10.5\text{–}2)$$

$$\frac{\partial \rho}{\partial t} + \frac{\partial\, \rho v_j}{\partial x_j} = 0. \qquad \blacktriangle \quad (10.5\text{–}3)$$

$$\frac{D\rho}{Dt} + \rho\frac{\partial v_j}{\partial x_j} = 0. \qquad \blacktriangle \quad (10.5\text{–}4)$$

These are called the *equations of continuity*. The integral form, Eq. (10.5–2), is useful when the differentiability of ρv_j cannot be assumed.

In problems of statics, these equations are satisfied identically. Then the conservation of mass must be expressed by Eq. (10.2–7) or Eq. (10.2–8).

10.6 THE EQUATIONS OF MOTION

Newton's laws of motion state that in an inertial frame of reference, the material rate of change of the linear momentum of a body is equal to the resultant of the forces applied to the body.

At an instant of time t, the linear momentum of all the particles contained in a domain V is

$$\mathcal{P}_i = \int_V \rho v_i \, dV. \tag{10.6–1}$$

If the body is subjected to surface tractions $\overset{\nu}{T}_i$ and a body force per unit volume X_i, the resultant force is

$$\mathcal{F}_i = \int_S \overset{\nu}{T}_i \, dS + \int_V X_i \, dV. \tag{10.6–2}$$

According to Cauchy's formula, Eq. (3.3–2), the surface traction may be expressed in terms of the stress field σ_{ij}, so that $\overset{\nu}{T}_i = \sigma_{ji}\nu_j$, where ν_j is the unit vector along the outer normal to the boundary surface S of the domain V. On substituting $\sigma_{ij}\nu_j$ for $\overset{\nu}{T}_i$ into Eq. (10.6–2) and transforming the surface integral into a volume integral by Gauss's theorem, we have

$$\mathcal{F}_i = \int_V \left(\frac{\partial \sigma_{ij}}{\partial x_j} + X_i\right) dV. \tag{10.6–3}$$

Newton's law states that

$$\frac{D}{Dt}\mathcal{P}_i = \mathcal{F}_i. \tag{10.6–4}$$

Hence, according to Eq. (10.4–5), with A identified with ρv_i, we have

$$\int_V \left[\frac{\partial \rho v_i}{\partial t} + \frac{\partial}{\partial x_j}(\rho v_i v_j)\right] dV = \int_V \left(\frac{\partial \sigma_{ij}}{\partial x_j} + X_i\right) dV. \tag{10.6–5}$$

Since this equation must hold for an arbitrary domain V, the integrands on the two sides must be equal. Thus,

$$\frac{\partial \rho v_i}{\partial t} + \frac{\partial}{\partial x_j}(\rho v_i v_j) = \frac{\partial \sigma_{ij}}{\partial x_j} + X_i. \tag{10.6–6}$$

The left-hand side of Eq. (10.6–6) is equal to

$$v_i\left(\frac{\partial \rho}{\partial t} + \frac{\partial \rho v_j}{\partial x_j}\right) + \rho\left(\frac{\partial v_i}{\partial t} + v_j \frac{\partial v_i}{\partial x_j}\right).$$

The quantity in the first set of parentheses vanishes according to the equation of continuity, Eq. (10.5–3), while that in the second set of parentheses is the acceleration Dv_i/Dt. Hence, we obtain the celebrated *Eulerian equation of motion of a continuum*:

$$\rho \frac{Dv_i}{Dt} = \frac{\partial \sigma_{ij}}{\partial x_j} + X_i. \qquad \blacktriangle \quad (10.6\text{–}7)$$

The equation of equilibrium discussed in Sec. 3.4 is a special case that can be obtained by setting all velocity components v_i equal to zero.

10.7 MOMENT OF MOMENTUM

An application of the law of balance of angular momentum to the particular case of *static equilibrium* leads to the conclusion that stress tensors are symmetric tensors (see Sec. 3.4.) We shall now show that no additional restriction to the motion of a continuum is introduced in dynamics by the angular momentum postulate, which states that the material rate of change of the moment of momentum with respect to an origin is equal to the resultant moment of all the applied forces about the same origin.

At an instant of time t, a body occupying a regular region V of space with boundary S has the moment of momentum [See Eq. (3.2–2)]

$$\mathcal{H}_i = \int_V e_{ijk} x_j \rho v_k \, dV \qquad (10.7\text{–}1)$$

with respect to the origin of coordinates. If the body is subjected to surface traction $\overset{\nu}{T}_i$ and a body force per unit volume X_i, the resultant moment about the origin is

$$\mathcal{L}_i = \int_V e_{ijk} x_j X_k \, dV + \int_S e_{ijk} x_j \overset{\nu}{T}_k \, dS. \qquad (10.7\text{–}2)$$

Introducing Cauchy's formula, $\overset{\nu}{T}_k = \sigma_{lk} n_l$, into the last integral, and transforming the result into a volume integral by Gauss's theorem, we obtain

$$\mathcal{L}_i = \int_V e_{ijk} x_j X_k \, dV + \int_V (e_{ijk} x_j \sigma_{lk}),_l \, dV. \qquad (10.7\text{–}3)$$

Euler's law states that, for any region V,

$$\frac{D}{Dt} \mathcal{H}_i = \mathcal{L}_i. \qquad (10.7\text{–}4)$$

Evaluating the material derivative of $\mathcal{H}_i$ according to Eq. (10.4–5) and using Eq. (10.7–3), we obtain

$$e_{ijk} x_j \frac{\partial}{\partial t} (\rho v_k) + \frac{\partial}{\partial x_l} (e_{ijk} x_j \rho v_k v_l) = e_{ijk} x_j X_k + e_{ijk} (x_j \sigma_{lk}),_l. \qquad (10.7\text{–}5)$$

The second term in Eq. (10.7–5) can be written as

$$e_{ijk}\rho v_j v_k + e_{ijk}x_j \frac{\partial}{\partial x_l}(\rho v_k v_l) = 0 + e_{ijk}x_j \frac{\partial}{\partial x_l}(\rho v_k v_l)$$

because e_{ijk} is antisymmetric and $v_j v_k$ is symmetric with respect to j, k. The last term in Eq. (10.7–5) can be written as $e_{ijk}\sigma_{jk} + e_{ijk}x_j\sigma_{lk,l}$. Hence, Eq. (10.7–5) becomes

$$e_{ijk}x_j\left[\frac{\partial}{\partial t}(\rho v_k) + \frac{\partial}{\partial x_l}(\rho v_k v_l) - X_k - \sigma_{lk,l}\right] - e_{ijk}\sigma_{jk} = 0. \tag{10.7–6}$$

By the equation of motion Eq. (10.6–6), the sum in the square brackets vanishes. Hence, Eq. (10.7–6) is reduced to

$$e_{ijk}\sigma_{jk} = 0; \tag{10.7–7}$$

i.e., $\sigma_{jk} = \sigma_{kj}$. Thus, if the stress tensor is symmetric, the law of balance of moment of momentum is satisfied identically.

10.8 THE BALANCE OF ENERGY

The motion of a continuum must be governed further by the law of conservation of energy. If mechanical energy alone is of interest in a problem, then the energy equation is merely the first integral of the equation of motion. If a thermal process is significant, then the equation of energy becomes an independent equation to be satisfied.

The law of conservation of energy is the first law of thermodynamics. Its expression for a continuum can be derived as soon as all forms of energy and work are listed. Let us consider a continuum for which there are three forms of energy: the kinetic energy K, the gravitational energy G, and the internal energy E. We have

$$\text{Energy} = K + G + E. \tag{10.8–1}$$

The *kinetic energy* contained in a regular domain V at a time t is

$$K = \int \frac{1}{2}\rho v_i v_i \, dV, \tag{10.8–2}$$

where v_i are the components of the velocity vector of a particle occupying an element of volume dV and ρ is the density of the material. The *gravitational* energy depends on the distribution of mass and may be written as

$$G = \int \rho\phi(x)\, dV, \tag{10.8–3}$$

where ϕ is the gravitational potential per unit mass. In the important special case of a uniform gravitational field, we have

$$G = \int \rho g z \, dV, \qquad (10.8\text{–}4)$$

where g is the gravitational acceleration and z is a distance measured from a certain plane in a direction opposite to the gravitational field. The *internal energy* is written in the form

$$\mathscr{E} = \int \rho E \, dV, \qquad (10.8\text{–}5)$$

where E is the *internal energy per unit mass*. The first law of thermodynamics states that the energy of a system can be changed by absorption of heat Q and by work W done on the system:

$$\Delta \text{ energy} = Q + W. \qquad (10.8\text{–}6)$$

Expressing this in terms of rates, we have

$$\frac{D}{Dt}(K + G + \mathscr{E}) = \dot{Q} + \dot{W}, \qquad (10.8\text{–}7)$$

where $\dot{Q}$ and $\dot{W}$ are the rates of change of Q and W per unit time.

Now, the heat input into the body must be imparted through the boundary. To describe the heat flow, a *heat flux* vector $\mathbf{h}$ (with components h_1, h_2, h_3) is defined as follows. Let dS be a surface element in the body, with unit outer normal n_i. Then the rate at which heat is transmitted across the surface dS in the direction of ν_i is assumed to be representable as $h_i n_i \, dS$. If the medium is moving, we insist that the surface element dS be composed of the same particles. The rate of heat input is, therefore,

$$\dot{Q} = -\int_S h_i n_i \, dS = -\int_V \frac{\partial h_i}{\partial x_i} dV. \qquad (10.8\text{–}8)$$

The rate at which work is done on the body by the body force per unit volume F_i in V and the surface tractions $\overset{\nu}{T}_i$ in S is the *power*

$$\begin{aligned} \dot{W} &= \int F_i v_i \, dV + \int \overset{\nu}{T}_i n_i \, dS \\ &= \int F_i v_i \, dV + \int \sigma_{ij} n_j n_i \, dS \qquad (10.8\text{–}9) \\ &= \int F_i v_i \, dV + \int (\sigma_{ij} n_i)_{,j} \, dV. \end{aligned}$$

Since, in Eq. (10.8–7), the gravitational energy is included in the term G, the power W must be evaluated with the gravitational force excluded from the body force F_i. Substituting Eqs. (10.8–2), (10.8–3), (10.8–5), (10.8–8), and (10.8–9) into the first law of thermodynamics, Eq. (10.8–7), and using Eq. (10.4–5) to compute the material derivatives, we obtain the following result after some calculation:

$$\frac{1}{2}\rho\frac{Dv^2}{Dt} + \frac{v^2}{2}\frac{D\rho}{Dt} + \frac{v^2}{2}\rho \operatorname{div} \mathbf{v} + \rho\frac{DE}{Dt} + E\frac{D\rho}{Dt} \tag{10.8–10}$$

$$+ E\rho \operatorname{div} \mathbf{v} + \rho\frac{D\phi}{Dt} + \phi\frac{D\rho}{Dt} + \phi\rho \operatorname{div} \mathbf{v}$$

$$= -\frac{\partial h_i}{\partial x_i} + F_i v_i + \sigma_{ij,j}v_i + \sigma_{ij}v_{i,j}.$$

This equation can be simplified greatly if we make use of the equations of continuity and motion:

$$\frac{D\rho}{Dt} + \rho \operatorname{div} \mathbf{v} = 0, \qquad \rho\frac{Dv_i}{Dt} = X_i + \sigma_{ij,j}. \tag{10.8–11}$$

Here, X_i is the total body force per unit mass. The difference between X_i and F_i is the gravitational force and is, by definition,

$$X_i - F_i = -\rho\frac{\partial\phi}{\partial x_i}. \tag{10.8–12}$$

Since

$$\frac{D\phi}{Dt} = \frac{\partial\phi}{\partial t} + v_i\frac{\partial\phi}{\partial x_i},$$

and $\partial\phi/\partial t = 0$ for a gravitational field that is independent of time, we have, for such a field, and with Eqs. (10.8–11) and (10.8–12),

$$\frac{1}{2}\rho\frac{Dv^2}{Dt} + \rho\frac{DE}{Dt} = -\frac{\partial h_i}{\partial x_i} + \rho v_i\frac{Dv_i}{Dt} + \sigma_{ij}v_{i,j}. \tag{10.8–13}$$

But

$$\rho v_i\frac{Dv_i}{Dt} = \frac{1}{2}\rho\frac{Dv^2}{Dt}, \tag{10.8–14}$$

and

$$\sigma_{ij}v_{i,j} = \sigma_{ij}[\tfrac{1}{2}(v_{i,j} + v_{j,i}) + \tfrac{1}{2}(v_{i,j} - v_{j,i})] = \sigma_{ij}V_{ij} + 0, \tag{10.8–15}$$

where

$$V_{ij} = \tfrac{1}{2}(v_{i,j} + v_{j,i}) \tag{10.8–16}$$

is the *strain-rate tensor*. The last term in Eq. (10.8–15) vanishes because it is the contraction of the product of a symmetric tensor σ_{ij} with an antisymmetric one. Hence, Eq. (10.8–13) can be simplified, and we obtain the final form of the energy equation:

$$\rho \frac{DE}{Dt} = -\frac{\partial h_i}{\partial x_i} + \sigma_{ij} V_{ij}. \tag{10.8–17}$$

Specialization

(A) If all the nonmechanical transfer of energy consists of heat conduction, which obeys Fourier's law,

$$h_i = -J \lambda \frac{\partial T}{\partial x_i}, \tag{10.8–18}$$

where J is the mechanical equivalent of heat, λ is the conductivity, and T is the absolute temperature, then the energy equation becomes

$$\rho \frac{DE}{Dt} = J \frac{\partial}{\partial x_i} \left(\lambda \frac{\partial T}{\partial x_i} \right) + \sigma_{ij} V_{ij}. \tag{10.8–19}$$

(B) The usual equation of heat conduction in a continuum at rest is obtained by deleting the terms involving ϕ, v_i, and V_{ij} and setting

$$E = JcT, \tag{10.8–20}$$

where c is the specific heat for the vanishing rate of deformation. Then Eq. (10.8–19) becomes

$$\rho c \frac{\partial T}{\partial t} = \frac{\partial}{\partial x_i} \left(\lambda \frac{\partial T}{\partial x_i} \right). \tag{10.8–21}$$

10.9 THE EQUATIONS OF MOTION AND CONTINUITY IN POLAR COORDINATES

In Secs. 3.6 and 5.8, we considered the stress and strain components, respectively, in polar coordinates. The corresponding equations of motion and continuity can be derived in the same manner: by the method of general tensor analysis, by transformation from the Cartesian coordinates, or by direct ad hoc derivation from first principles. Illustrations of the last two approaches follow.

The basic equations for transformation between Cartesian coordinates x, y, z and polar coordinates r, θ, z are given in Sec. 5.8. If we substitute Eq. (3.6–5) into the equation of equilibrium,

$$\frac{\partial \sigma_{ij}}{\partial x_j} = 0, \tag{10.9–1}$$

i.e.,

$$\frac{\partial \sigma_{xx}}{\partial x} + \frac{\partial \sigma_{xy}}{\partial y} + \frac{\partial \sigma_{xz}}{\partial z} = 0,$$

etc., and use Eq. (5.8–3) to transform the derivatives, we obtain

$$\left(\frac{\partial\sigma_{rr}}{\partial r} + \frac{1}{r}\frac{\partial\sigma_{r\theta}}{\partial\theta} + \frac{\sigma_{rr} - \sigma_{\theta\theta}}{r} + \frac{\partial\sigma_{rz}}{\partial z}\right)\cos\theta$$
$$-\left(\frac{1}{r}\frac{\partial\sigma_{\theta\theta}}{\partial\theta} + \frac{\partial\sigma_{r\theta}}{\partial r} + 2\frac{\sigma_{\theta}}{r} + \frac{\partial\sigma_{\theta z}}{\partial z}\right)\sin\theta = 0. \tag{10.9–2}$$

Since this equation must hold for all values of θ, we must have, at $\theta = 0$ and at $\theta = \pi/2$, respectively,

$$\frac{\partial\sigma_{rr}}{\partial r} + \frac{1}{r}\frac{\partial\sigma_{r\theta}}{\partial\theta} + \frac{\sigma_{rr} - \sigma_{\theta\theta}}{r} + \frac{\partial\sigma_{rz}}{\partial z} = 0,$$
$$\frac{1}{r}\frac{\partial\sigma_{\theta\theta}}{\partial\theta} + \frac{\partial\sigma_{r\theta}}{\partial r} + \frac{2\sigma_{r\theta}}{r} + \frac{\partial\sigma_{z\theta}}{\partial z} = 0. \tag{10.9–3}$$

But the choice of the x-direction is arbitrary, so Eq. (10.9–3) must be valid for all values of θ. Similarly, from Eq. (10.9–1) with $i = 3$, we obtain the third equation of equilibrium,

$$\frac{\partial\sigma_{zz}}{\partial z} + \frac{1}{r}\frac{\partial\sigma_{z\theta}}{\partial\theta} + \frac{\partial\sigma_{zr}}{\partial r} + \frac{\sigma_{rz}}{r} = 0. \tag{10.9–4}$$

If the continuum is subjected to an acceleration and a body force, then the equation of motion, Eq. (10.6–7), is

$$\frac{\partial\sigma_{ij}}{\partial x_j} + X_i = \rho\frac{Dv_i}{Dt} = \rho\, a_i. \tag{10.9–5}$$

The body force per unit volume may be resolved into components F_r, F_θ, F_z along the r-, θ-, and z-directions, respectively. The acceleration $Dv_i/Dt = a_i$ must be considered carefully. The component of acceleration in the x-direction in rectangular coordinates is

$$a_x = \frac{\partial v_x}{\partial t} + v_x\frac{\partial v_x}{\partial x} + v_y\frac{\partial v_x}{\partial y} + v_z\frac{\partial v_x}{\partial z}. \tag{10.9–6}$$

The components of acceleration a_x, a_y, a_z and of velocity v_x, v_y, v_z are related to the components a_r, a_θ, a_z and v_r, v_θ, v_z in polar coordinates by the same Eqs. (5.8–4) that relate the displacements, provided that u is replaced by a and v, respectively. Hence, by substitution of Eqs. (5.8–3) and (5.8–4) into Eq. (10.9–6), we obtain

$$a_x = \frac{\partial}{\partial t}(v_r\cos\theta - v_\theta\sin\theta)$$
$$+ (v_r\cos\theta + v_\theta\sin\theta)\left(\cos\theta\frac{\partial}{\partial r} - \frac{\sin\theta}{r}\frac{\partial}{\partial\theta}\right)(v_r\cos\theta - v_\theta\sin\theta)$$

$$+ (v_r \sin\theta + v_\theta \cos\theta)\left(\sin\theta \frac{\partial}{\partial r} + \frac{\cos\theta}{r}\frac{\partial}{\partial\theta}\right)(v_r \cos\theta - v_\theta \sin\theta)$$

$$+ v_z \frac{\partial}{\partial z}(v_r \cos\theta - v_\theta \sin\theta)$$

$$= \cos\theta\left(\frac{\partial v_r}{\partial t} + v_r \frac{\partial v_r}{\partial r} + \frac{v_\theta}{r}\frac{\partial v_r}{\partial\theta} - \frac{v_\theta^2}{r} + v_z \frac{\partial v_r}{\partial z}\right)$$

$$- \sin\theta\left(\frac{\partial v_\theta}{\partial t} + v_r \frac{\partial v_\theta}{\partial r} + \frac{v_\theta}{r}\frac{\partial v_\theta}{\partial\theta} + \frac{v_r v_\theta}{r} + v_z \frac{\partial v_\theta}{\partial z}\right). \tag{10.9–7}$$

Comparing Eq. (10.9–7) with the equation

$$a_x = a_r \cos\theta - a_\theta \sin\theta, \tag{10.9–8}$$

we obtain the components of acceleration:

$$\begin{aligned} a_r &= \frac{\partial v_r}{\partial t} + v_r \frac{\partial v_r}{\partial r} + \frac{v_\theta}{r}\frac{\partial v_r}{\partial\theta} - \frac{v_\theta^2}{r} + v_z \frac{\partial v_r}{\partial z}, \\ a_\theta &= \frac{\partial v_\theta}{\partial t} + v_r \frac{\partial v_\theta}{\partial r} + \frac{v_\theta}{r}\frac{\partial v_\theta}{\partial\theta} + \frac{v_r v_\theta}{r} + v_z \frac{\partial v_\theta}{\partial z}. \end{aligned} \tag{10.9–9}$$

Similarly,

$$a_z = \frac{\partial v_z}{\partial t} + v_r \frac{\partial v_z}{\partial r} + \frac{v_\theta}{r}\frac{\partial v_z}{\partial\theta} + v_z \frac{\partial v_z}{\partial z}. \tag{10.9–10}$$

The full equations of motion are

$$\begin{aligned} \rho a_r &= \frac{\partial\sigma_{rr}}{\partial r} + \frac{1}{r}\frac{\partial\sigma_{r\theta}}{\partial\theta} + \frac{\sigma_{rr} - \sigma_{\theta\theta}}{r} + \frac{\partial\sigma_{rz}}{\partial z} + F_r, \\ \rho a_\theta &- \frac{1}{r}\frac{\partial\sigma_{\theta\theta}}{\partial\theta} + \frac{\partial\sigma_{r\theta}}{\partial r} + \frac{2\sigma_{r\theta}}{\partial r} + \frac{\partial\sigma_{r\theta}}{\partial z} + F_\theta, \\ \rho a_z &= \frac{\partial\sigma_{zz}}{\partial z} + \frac{1}{r}\frac{\partial\sigma_{z\theta}}{\partial\theta} + \frac{\partial\sigma_{zr}}{\partial r} + \frac{\sigma_{rz}}{r} + F_z. \end{aligned} \tag{10.9–11}$$

These derivations are again straightforward, but not very instructive from the physical point of view. A second derivation based on an examination of the balance of forces acting on an element may supply further insight into the equations. Figure 10.4 shows the free-body diagram for an isolated element with the stress pattern

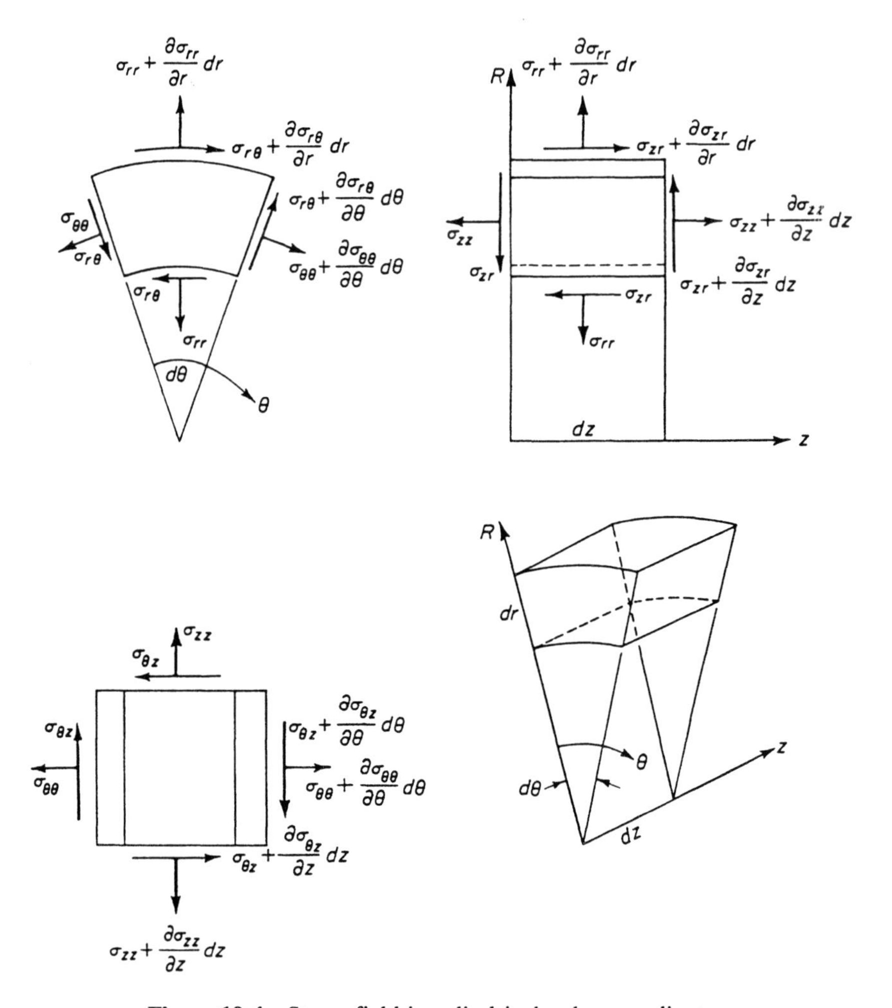

Figure 10.4 Stress field in cylindrical polar coordinates.

indicated. The equation of motion indicates that the acceleration in the radial direction is equal to the sum of all the forces acting in the radial direction. Thus,

$$\rho a_r \, dr \, dz \left[\frac{r \, d\theta + (r + dr)d\theta}{2}\right] = F_r \, dr \, dz \left[\frac{r \, d\theta + (r + dr)d\theta}{2}\right]$$

$$+ \left(\sigma_{rr} + \frac{\partial \sigma_{rr}}{\partial r} dr\right)(r + dr)d\theta \, dz - \sigma_{rr} r \, d\theta \, dz$$

$$- \sigma_{\theta\theta} \, dr \, dz \sin\frac{d\theta}{2} - \left(\sigma_{\theta\theta} + \frac{\partial \sigma_{\theta\theta}}{\partial \theta} d\theta\right) dr \, dz \sin\frac{d\theta}{2}$$

$$+ \left(\sigma_{r\theta} + \frac{\partial \sigma_{r\theta}}{\partial \theta} d\theta\right) dr\, dz - \sigma_{r\theta}\, dr\, dz$$

$$+ \left(\sigma_{rz} + \frac{\partial \sigma_{rz}}{\partial z} dz - \sigma_{rz}\right)\left[\frac{r\, d\theta + (r + dr)\, d\theta}{2}\right] dr. \qquad (10.9\text{–}12)$$

Expanding, dropping higher-order infinitesimal quantities, and dividing through by r, we obtain the first equation of Eq. (10.9–11). The other equations can be obtained in a similar manner. Note that in the equation for radial equilibrium, the term $-\sigma_{\theta\theta}/r$ is a radial pressure in the nature of hoop stress; the term σ_{rr}/r is the contribution due to the larger area of the outer surface at $r + dr$ than that at radius r. The term σ_{rz}/r in the equation for axial equilibrium is present for the same reason. The term $2\sigma_{r\theta}/r$ in the tangential equation has two origins: One is for the same reason as before, viz., that the outer surface is larger; the other arises from the fact that the radial surfaces at θ and $\theta + d\theta$ are not parallel, but make an angle $d\theta$.

A similar graphical interpretation can be made of the individual terms in the expressions for acceleration. The term $-v_\theta^2/r$ in a_r is of the nature of centripetal acceleration. The term $v_\theta v_r/r$ in a_θ arises from the rotation of the radial velocity vector v_r, thus contributing a tangential component of acceleration.

A similar treatment can be used to transform the equation of continuity, Eq. (10.5–3), into polar coordinates. But here it is perhaps most instructive to study the balance of mass flow in an element, as shown in Fig. 10.5. With the area through which the mass flow takes place accounted for properly, we obtain

$$\frac{1}{r}\frac{\partial}{\partial r}(\rho r v_r) + \frac{1}{r}\frac{\partial\, \rho v_\theta}{\partial \theta} + \frac{\partial\, \rho v_z}{\partial z} + \frac{\partial\, \rho}{\partial t} = 0. \qquad (10.9\text{–}13)$$

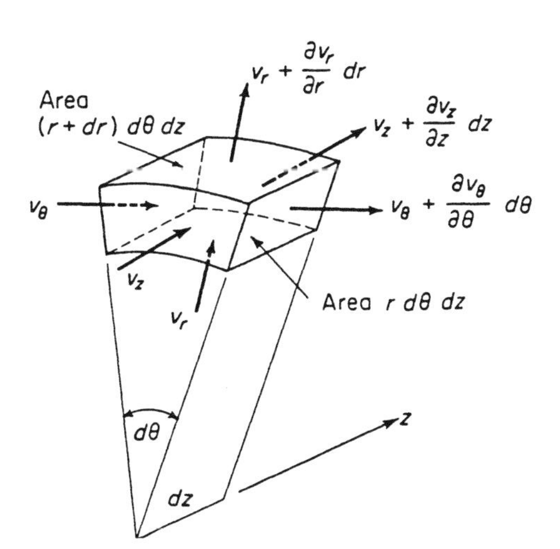

Figure 10.5 Conservation of mass in cylindrical polar coordinates.

PROBLEMS

10.1 State the definitions of (a) a line integral, (b) a surface integral, and (c) a volume integral.

10.2 State the mathematical conditions under which Eqs. (10.1–4), (10.1–5), (10.4–4), and (10.4–5) are valid.

10.3 Evaluate the line integral

$$\oint_C y^2\,dx + x^2\,dy,$$

where C is a triangle with vertices (1,0), (1,1), (0,0). (See Fig. P10.3.)

Answer: 1/3.

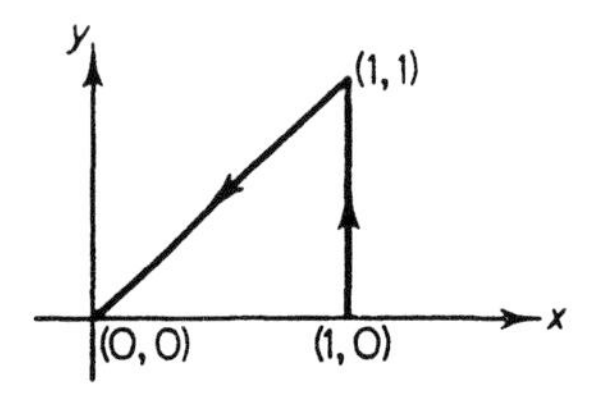

Figure P10.3 Path of integration.

10.4 Evaluate $\oint_C (x^2 - y^2)\,ds$, where C is the circle $x^2 + y^2 = 4$.

10.5 Derive Green's theorem: Let D be a domain of the xy-plane, and let C be a piecewise smooth simple closed curve in D whose interior is also in D. Let $P(x, y)$ and $Q(x, y)$ be functions that are defined and continuous in D and that have continuous first partial derivatives in D. Then

$$\oint_C P\,dx + Q\,dy = \iint_R \left(\frac{\partial Q}{\partial x} - \frac{\partial P}{\partial y}\right) dx\,dy,$$

where R is the closed region bounded by C.

10.6 Interpret Green's theorem vectorially to derive the following theorems:

$$\text{(a)}\ \oint_C u_T\,ds = \iint_R \text{curl}_z\,\mathbf{u}\,dx\,dy,$$

$$\text{(b)}\ \oint_C v_n\,ds = \iint_R \text{div}\,\mathbf{v}\,dx\,dy,$$

where $\mathbf{u}$, $\mathbf{v}$ are vector fields, u_T is the tangential component of $\mathbf{u}$ (tangent to the curve C), ds is the arc length, and v_n is the normal component of $\mathbf{v}$ on C. Equation (a) is a special case of Stokes's theorem. Equation (b) is the two-dimensional form of Gauss's theorem.

10.7 A rubber spherical balloon is quickly blown up in an angry sea by a ditched pilot. Let a particle on the balloon be located at

$$x = x(t), \qquad y = y(t), \qquad z = z(t).$$

Let the surface of the balloon be described by the equation

$$F(t) = (x - \lambda)^2 + (y - \mu)^2 + (z - \nu)^2 - a^2 = 0,$$

where $\lambda(t)$, $\mu(t)$, and $\nu(t)$, which define the center of the sphere, and $a(t)$, the radius, are functions of time. (See Fig. P10.7.)

Show that $DF/Dt = 0$.

Derive the boundary conditions for the air and water moving about the balloon.

Solution: The equation $F(t) = 0$ representing the surface of the balloon is true at all times. Therefore, its derivative with respect to t must vanish. Since x, y, z are coordinates of the particles, and $F(t)$ is associated with the balloon at all times, the time derivative is the material derivative, i.e., DF/Dt, which is zero.

Conversely, from the equation $DF/Dt = 0$, we conclude that $F =$ const. for a given set of particles. In particular, if the set of particles is defined by the equation $F = 0$, it remains the same set. If $F = 0$ defines the balloon at $t = 0$, it defines the balloon at any t.

The equation becomes more significant if we consider the fluid (air and water) around the balloon. Fluid particles once in contact with the balloon remain in contact with it (the so-called no-slip condition of a viscous fluid in contact with a solid body). Hence, the boundary conditions of the flow field are $F = 0$ and $DF/Dt = 0$.

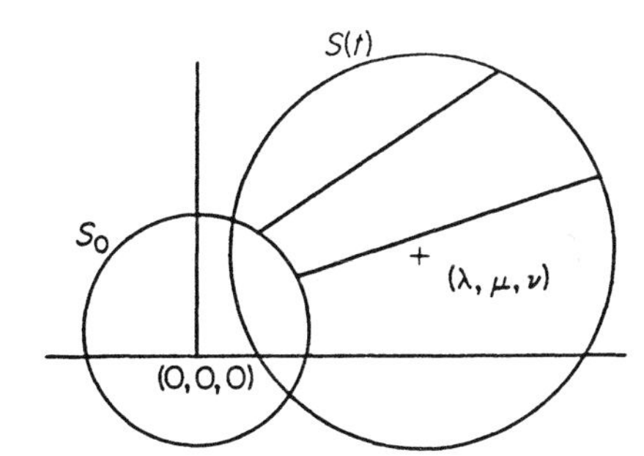

Figure P10.7 Expanding balloon.

10.8 The surface of a flag fluttering in the wind is described by the equation

$$F(x, y, z, t) = 0.$$

Write down analytically the constraints imposed by the flag on the airstream. In other words, given the shape of the boundary surface $F = 0$, derive the boundary condition for the flow. For this problem, consider the air a nonviscous fluid.

What difference would it make if the air were taken to be a viscous fluid?

Solution: As in Prob. 10.7, the boundary condition of the airstream on the flag surface $F = 0$ is

$$\frac{\partial F}{\partial t} + u_x \frac{\partial F}{\partial x} + u_y \frac{\partial F}{\partial y} + u_z \frac{\partial F}{\partial z} = 0 \tag{1}$$

where $\mathbf{u}$ (u_x, u_y, u_z) is the velocity vector. For the surface $F(x, y, z, t) = 0$, the vector $\mathbf{n}$ with components

$$\frac{\partial F}{\partial x}, \frac{\partial F}{\partial y}, \frac{\partial F}{\partial z}$$

is normal to the surface. Hence, Eq. (1) may be written

$$\frac{\partial F}{\partial t} + \mathbf{u}\cdot\mathbf{n} = 0. \tag{2}$$

This means that the normal velocity must be equal to $-\partial F/\partial t$ on the flag surface.

For a viscous fluid, the *no-slip* condition requires, in addition, that $F = 0$. (See the discussion in Sec. 11.2, p. 233.)

10.9 Two components of the velocity field of a fluid are known in the region $-2 \leq x, y, z \leq 2$:

$$u = (1 - y^2)(a + bx + cx^2), \qquad w = 0.$$

The fluid is incompressible. What is the velocity component v in the direction of the y-axis?

10.10 Let the temperature field of the fluid described in Prob. 10.9 be

$$T = T_0 e^{-kt} \sin \alpha x \cos \beta y.$$

Find the material rate of change of the temperature of a particle located at the origin $x = y = z = 0$. Find the same for a particle at $x = y = z = 1$.

10.11 For an isotropic Newtonian viscous fluid, derive an equation of motion expressed in terms of the velocity components.

10.12 The entropy of a moving continuum is $s(x_1, x_2, x_3, t)$ *per unit mass* of the medium. The mass density of the medium is $\rho(x_1, x_2, x_3, t)$. The velocity field is $v_i(x_1, x_2, x_3, t)$. Consider the total amount of entropy in a certain volume of the medium at a certain time. Express the rate of change of the total entropy of the material enclosed in this volume in the form of a volume integral.

11 FIELD EQUATIONS AND BOUNDARY CONDITIONS IN FLUID MECHANICS

We have acquired enough basic equations to deal with a broad range of problems. Most objects on a scale that we can see are continua. Their motion follows the laws of conservation of mass, momentum, and energy. With the proper constitutive equations and boundary conditions, we can describe many physical problems mathematically. In this chapter, we illustrate the formulation of some problems on the flow of fluids.

11.1 THE NAVIER-STOKES EQUATIONS

Let us derive the basic equations governing the flow of a Newtonian viscous fluid. Let x_1, x_2, x_3 or x, y, z be rectangular Cartesian coordinates. Let the velocity components along the x-, y-, z-axis directions be denoted by v_1, v_2, v_3 or u, v, w, respectively. Let p denote pressure; σ_{ij} or σ_{xx}, σ_{xy}, etc., be the stress components; and μ be the coefficient of viscosity. Here, and hereinafter, all Latin indices range over 1, 2, 3. Then, the stress–strain-rate relationship is given by Eq. (7.3–3):

$$\sigma_{ij} = -p\delta_{ij} + \lambda V_{kk}\delta_{ij} + 2\mu V_{ij} = -p\delta_{ij} + \lambda \frac{\partial v_k}{\partial x_k}\delta_{ij} + \mu\left(\frac{\partial v_i}{\partial x_j} + \frac{\partial v_j}{\partial x_i}\right); \qquad (11.1\text{–}1)$$

i.e.,

$$\begin{aligned}
\sigma_{xx} &= -p + 2\mu\frac{\partial u}{\partial x} + \lambda\left(\frac{\partial u}{\partial x} + \frac{\partial v}{\partial y} + \frac{\partial w}{\partial z}\right),\\
\sigma_{yy} &= -p + 2\mu\frac{\partial v}{\partial y} + \lambda\left(\frac{\partial u}{\partial x} + \frac{\partial v}{\partial y} + \frac{\partial w}{\partial z}\right),\\
\sigma_{zz} &= -p + 2\mu\frac{\partial w}{\partial z} + \lambda\left(\frac{\partial u}{\partial x} + \frac{\partial v}{\partial y} + \frac{\partial w}{\partial z}\right), \qquad (11.1\text{–}1a)\\
\sigma_{xy} &= \mu\left(\frac{\partial u}{\partial y} + \frac{\partial v}{\partial x}\right), \qquad \sigma_{yz} = \mu\left(\frac{\partial v}{\partial z} + \frac{\partial w}{\partial y}\right),\\
\sigma_{zx} &= \mu\left(\frac{\partial w}{\partial x} + \frac{\partial u}{\partial z}\right).
\end{aligned}$$

Substituting these into the equation of motion, Eq. (10.6–7), we obtain the Navier-Stokes equations,

$$\rho \frac{Dv_i}{Dt} = \rho X_i - \frac{\partial p}{\partial x_i} + \frac{\partial}{\partial x_i}\left(\lambda \frac{\partial v_k}{\partial x_k}\right) + \frac{\partial}{\partial x_k}\left(\mu \frac{\partial v_k}{\partial x_i}\right) + \frac{\partial}{\partial x_k}\left(\mu \frac{\partial v_i}{\partial x_k}\right), \quad (11.1\text{–}2)$$

where X_i stands for the body force per unit mass.

The velocity components must satisfy the equation of continuity, Eq. (10.5–3), derived from the conservation of mass:

$$\frac{\partial \rho}{\partial t} + \frac{\partial (\rho v_k)}{\partial x_k} = 0. \quad (11.1\text{–}3)$$

These equations are to be supplemented by the equations of thermal state, balance of energy, and heat flow.

If the fluid is *incompressible*, then

$$\rho = \text{const.}, \quad (11.1\text{–}4)$$

and no thermodynamic considerations need be introduced explicitly. Limiting ourselves to an incompressible homogeneous fluid, we see that the equation of continuity becomes

$$\frac{\partial v_k}{\partial x_k} = 0, \quad \text{or} \quad \frac{\partial u}{\partial x} + \frac{\partial v}{\partial y} + \frac{\partial w}{\partial z} = 0, \quad (11.1\text{–}5)$$

and the Navier-Stokes equation is simplified to

$$\rho \frac{Dv_i}{Dt} = \rho X_i - \frac{\partial p}{\partial x_i} + \mu \frac{\partial^2 v_i}{\partial x_k \partial x_k}. \quad (11.1\text{–}6)$$

Written out *in extenso*, these are

$$\begin{aligned} \frac{Du}{Dt} &= X - \frac{1}{\rho}\frac{\partial p}{\partial x} + \nu \nabla^2 u, \\ \frac{Dv}{Dt} &= Y - \frac{1}{\rho}\frac{\partial p}{\partial y} + \nu \nabla^2 v, \\ \frac{Dw}{Dt} &= Z - \frac{1}{\rho}\frac{\partial p}{\partial z} + \nu \nabla^2 w, \end{aligned} \quad (11.1\text{–}7)$$

where $\nu = \mu/\rho$ is the *kinematic viscosity* and

$$\nabla^2 = \frac{\partial^2}{\partial x^2} + \frac{\partial^2}{\partial y^2} + \frac{\partial^2}{\partial z^2}. \quad (11.1\text{–}8)$$

is the *Laplacian operator*. Equations (11.1–5) and (11.1–7) comprise four equations for the four variables u, v, w, and p occurring in an incompressible viscous flow.

The solution of the Navier-Stokes equation is the central problem in fluid mechanics. This equation embraces a tremendous range of physical phenomena and has many applications to science and engineering. The equation is nonlinear and is, in general, very difficult to solve.

To complete the formulation of a problem, we must specify the boundary conditions. In Sec. 11.2, we consider the no-slip condition on a solid-fluid interface. In Sec. 11.3, the condition at a "free," or fluid-fluid, interface is considered, where surface tension plays an important role. Then a dimensional analysis is presented to illustrate the significance of the Reynolds number. We shall then consider the laminar flow in a channel or a tube as an example of a simplified solution when the nonlinear terms can be ignored. As a warning that turbulences may intervene, we discuss the classical experiments of Reynolds in Sec. 11.5.

In some instances, the viscosity of a fluid may be ignored completely, and we deal with the idealized world of "perfect fluids." In association with this idealization, the boundary conditions must be changed: The order of the differential equation would be too low to permit the satisfaction of all the boundary conditions of a viscous fluid. We relinquish the no-slip condition at the solid-fluid interface and ignore any shear gradient requirement at a free surface. As a consequence, sometimes the resulting simpler mathematical problems lead to difficulties in physical interpretations.

11.2 BOUNDARY CONDITIONS AT A SOLID-FLUID INTERFACE

One of the boundary conditions that must be satisfied at a solid-fluid interface is that the fluid must not penetrate the solid if it is impermeable to the fluid. Most containers of fluids are of this nature. Mathematically, this requires that the relative velocity component of the fluid *normal* to the solid surface must vanish.

The specification of the tangential component of velocity of the fluid relative to the solid requires much greater care. It is customary to assume that the *no-slip condition* prevails at an interface between a viscous fluid and a solid boundary. In other words, on the solid-fluid boundary, the velocities of the fluid and the solid are exactly equal. This conviction was realized only after a long historical development by comparing theoretical and experimental results.

If the solid boundary is stationary, the no-slip condition requires that the velocity change continuously from zero at the surface to the free-stream value some distance away. This boundary condition is in drastic contrast to that which is required of a nonviscous fluid, for which we can specify only that no fluid shall penetrate the solid surface; but the fluid must be permitted to slide over the solid so that their tangential velocities can be different. This is a penalty for the idealization of complete absence of viscosity. Figure 11.1 illustrates the difference. In Fig. 11.1(a), the flow of a nonviscous fluid over a stationary solid object is shown. At the interface, the fluid slips over the solid with a tangential velocity. In Fig. 11.1(b), it is shown that for a viscous fluid, the velocity must vanish on the interface.

Since the no-slip condition must be imposed for all real fluid, no matter how small the viscosity, the illustration in Fig. 11.1(b) must prevail for all real fluids.

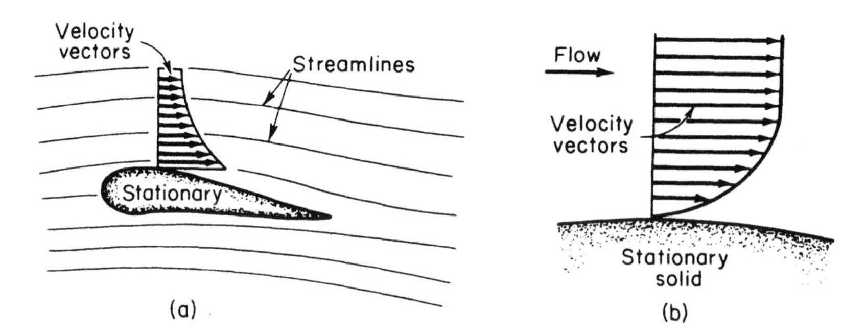

Figure 11.1 The difference in boundary conditions for flows of ideal and real fluids over a solid body. (a) Ideal fluid; (b) Real fluid.

It is known from wind-tunnel measurements that the flow field is well represented by Fig. 11.1(a) for the airfoil shown; i.e., except for the immediate neighborhood of the solid boundary, the flow can be obtained as though air had no viscosity. Yet we know that air has viscosity, even though very small. Therefore, the no-slip condition must prevail. How can we resolve this conflict?

The answer to this question and the resolution of the conflict are a triumph of modern fluid mechanics. The modern view is that the illustration shown in Fig. 11.1(b) is an indication of what happens in the immediate neighborhood of a solid boundary. We should consider that figure as an enlargement of what happens in a very small region of a flow next to an interface. This region is the *boundary layer*. Beyond the boundary layer, the flow is practically nonviscous. The dramatic importance of the boundary layer will be seen at the sharp trailing edge of the airfoil. It dictates the condition that the flow must leave the sharp trailing edge smoothly, with no discontinuity in the velocity field. If we insist on idealized nonviscous flow, the tangential velocity could differ on the top and bottom sides of the trailing edge. In the theory of nonviscous fluids, such a discontinuity can be eliminated either by permitting the flow to round the sharp corner with an infinite velocity gradient or by introducing an exact amount of circulation so that the trailing edge becomes a stagnation point. The latter condition was proposed by the German mathematician Kutta (1902) and the Russian mathematician Joukowski (1907) and is known as *Kutta-Joukowski hypothesis*, which is the basis for our modern theory of flight. Thus, we see that the fluid viscosity, no matter how small, has a profound influence on flow.

But how can we believe the no-slip condition? On what basis is this condition established? The molecular theory of gases does not provide a firm answer. From the molecular hypotheses, Navier deduced (1823) the boundary condition $\beta u = \mu\, \partial u/\partial n$ for flow over a solid wall, where u is the velocity, $\partial u/\partial n$ is the derivative along the normal away from the wall, β is a constant, and μ is the coefficient of viscosity. The ratio μ/β is a length that is zero if there is no slip. Maxwell (1879) calculated that μ/β is a moderate multiple of the mean free path L of the gas molecule—probably about $2L$. This result is in agreement with modern experimental evidence. Since the mean free path of the molecules of the air on the surface of the earth at room temperature is about 5×10^{-8} m, we can say that the no-slip condition may be questioned for micromachines with dimensions on the order of 10^{-6} m; and certainly will not apply to nanomachines, whose dimension is in the nanometer range.

Experiments on the flow of liquids and gases at atmospheric pressure over cm-sized bodies support the no-slip condition conclusively. Coulomb (1800) found that the resistance of an oscillating metallic disk in water was scarcely altered when the disk was smeared with grease or when the surface was covered with powdered sandstone, so the nature of the surface had little influence on the resistance. Poiseuille (1841) and Hagen (1839) obtained precise data on water flow in capillary tubes with diameter on the order of 10–20 μm. Stokes showed that the theoretical result based on the no-slip condition agreed with Poiseuille's experimental results. Other experimenters, such as Whetham (1890) and Couette (1890), came to the same conclusion. Fage and Townsend (1932) used an ultramicroscope to examine the flow of water containing small particles and confirmed the no-slip condition. In addition, there is agreement between theory and experiment on Stokes's and Oseen's theories of motion at small Reynolds numbers, as well as on Taylor's calculations and observations on the stability of flow between rotating cylinders. All these experiences, taken together, support the conclusion that for a liquid, the slip, if it takes place on a solid boundary, is too small to be observed or to make any sensible difference in the results of theoretical deductions.

11.3 SURFACE TENSION AND THE BOUNDARY CONDITIONS AT AN INTERFACE BETWEEN TWO FLUIDS

An interface between two fluids may be regarded as a membrane which has a specific chemical composition and mechanical properties. For example, the surface of a soap bubble in air has a layer of surfactants. The surfaces of pulmonary alveoli have a layer of fluid with surfactants that reduce the surface tension between the lung tissue and the lung gas. A cholesterol vesicle may have a single layer of lipid molecules on its surface or a lipid bilayer. Cell membranes are lipid bilayers. Even at the free surface of water in air, the water molecules at the interface are not in the same state as those in the bulk, and the interface can be regarded as a layer of different material. Hence, if one studies the flow of two fluids separated by an interface, the boundary conditions of the fluids at the interface must take the properties of the interface into consideration.

A membrane is a very thin plate. The stresses in a plate have been discussed in Example 4 of Sec. 1.11, see Fig. 1.6. If the membrane is very thin, we are interested more on the resultant force per unit length in the membrane than in the distribution of stress in its thickness. In thin membrane, the product of the average stress in the membrane and the thickness is called a *stress resultant*, or a *surface tension*, which has the units of [force/length].

Consider a soap bubble in the air, as shown in Fig. 11.2. It is a layer of liquid bounded by two air-liquid interfaces which have surface tension. Assume that the

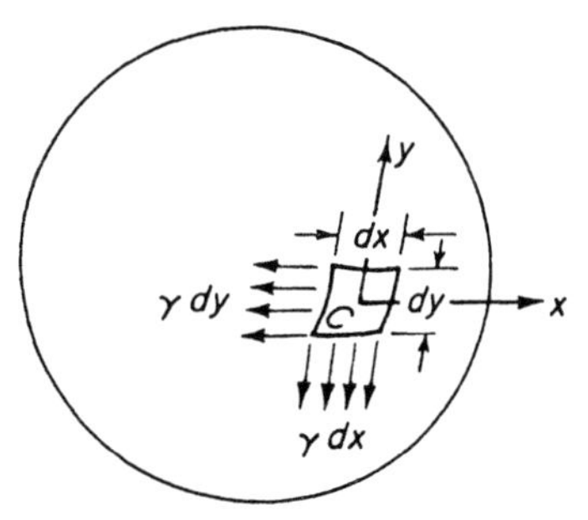

Figure 11.2 A soap bubble.

surface tension is isotropic. Denote the resultant of the surface tensions of the two interfaces by γ. To create the bubble one must blow and create an internal pressure p_i greater than the external pressure p_o, and the force due to the pressure difference must be balanced by the tension in the soap film. Let C be a small, closed rectangular curve of sides dx and dy drawn on the surface of the bubble (Fig. 11.2). The tensions acting on the sides of C are shown in the figure. To compute the pressure required to balance the tensions, let us consider two cross-sectional views: one in the xz-plane (z being normal to the soap film), and another in the yz-plane. The former is shown in Fig. 11.3, where the tensile forces $\gamma\, dy$ act at each end. Since these forces are tangent to the surface, they have a resultant $\gamma\, dy\, d\theta$ normal to

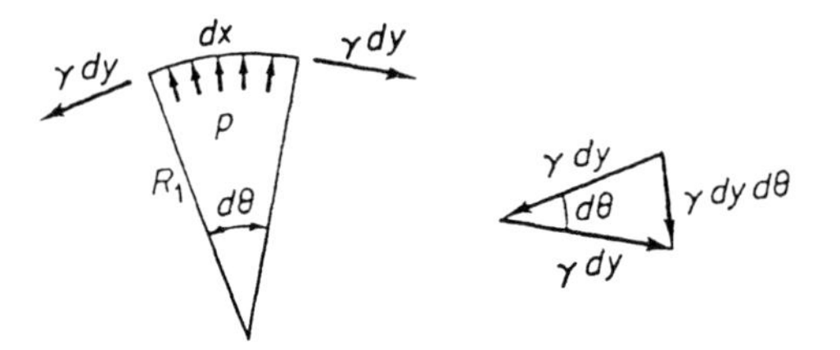

Figure 11.3 Equilibrium of membrane forces acting on an element of the soap bubble.

the surface. But $d\theta = dx/R_1$, where R_1 is the radius of curvature for the soap film. Hence, the normal force is $\gamma\, dx\, dy/R_1$. Similarly, the tensions acting on the other sides of the rectangle contribute a resultant $\gamma\, dx\, dy/R_2$. Since the soap film has two air-liquid interfaces (inside and outside), the total resultant force due to surface tension acting on the curve C is normal to the soap film and is equal to $2\gamma\, dx\, dy/R_1 + 2\gamma\, dx\, dy/R_2$. This force is balanced by the pressure difference multiplied by the area $dx\, dy$. On equating these forces, we obtain, for the soap film, the celebrated equation named after Laplace (1805), although it was actually obtained a year earlier by Thomas Young (1804):

$$2\gamma\left(\frac{1}{R_1} + \frac{1}{R_2}\right) = p_i - p_o \tag{11.3–1}$$

If the soap bubble is spherical, then $R_1 = R_2$. If the bubble is not spherical, we note that the sum

$$\frac{1}{R_1} + \frac{1}{R_2} = \text{mean curvature} \tag{11.3–2}$$

is invariant with respect to the rotation of coordinates on any surface. Hence, the directions chosen for the x- and y-axes are immaterial.

As a particular case, let us consider soap films formed by boundary curves under zero pressure difference. Then the surface is the so-called *minimal surface*, governed by the equation

$$\frac{1}{R_1} + \frac{1}{R_2} = 0. \tag{11.3–3}$$

Equation (11.3–1) indicates that the pressure difference required to balance the surface tension becomes very large if the radii R_1 and R_2 become very small. For a constant γ, if $R_1, R_2 \to 0$, the pressure difference tends toward infinity.

If the fluids are moving and the interface is nonstationary, then the no-slip condition must apply in each fluid relative to the interface if the fluid is real (viscous). If one of the fluids is ideal (nonviscous), then there is no no-slip condition for that fluid. If both fluids are ideal, then there is no restriction on slip.

In the most general case for an interface with a specific surface viscosity, surface compressibility, elasticity, and bending rigidity, the equations of motion (or equilibrium) and continuity of the interface are those of thin membranes or thin shells in solid mechanics. The boundary conditions of the fluids in contact with the interface are the nonpenetration and no-slip conditions.

Surface tension is very important in such chemical engineering problems as foaming, in such mechanical engineering problems as the fracture of metals and rocks, and in such biological problems as the opening and collapse of the lung. Surface tension is variable in general. For example, the alveolar surface in our lungs is moist, and the surface tension is modulated by the presence of "surfactants," lipids such as lecithin. The arrangement of these polar molecules on the interface depends on the concentration of the molecules, the rate at which the surface is strained, and the history of strain, so that the surface tension–area relationship has a huge hysteresis loop when the surface is subjected to a periodic strain. Figure 11.4 gives the experimental results obtained by J. A. Clements by means of a surface balance of the Wilhelmy type. Shown are the surface tension–area relationships between air on the one hand and pure water, blood plasma, 1% Tween 20 detergent, and a saline extract of a normal lung on the other. The loops of water and detergent are exaggerated schematically to show the cylic nature of the strain history.

When there is an interface, there is a question of permeability of the fluid moving through it. The permeability will govern the boundary condition with respect to the normal component of velocity. A certain amount of mass transfer, laminar or turbulent mixing, etc., may occur at the interface.

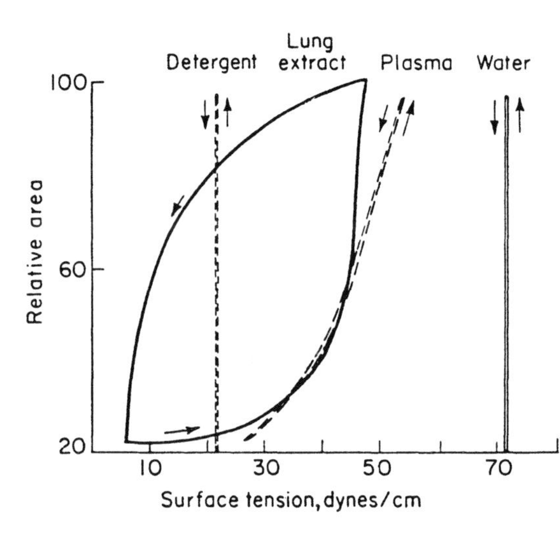

Figure 11.4 The variation in surface tension with strain for several fluids. From J. A. Clements, "Surface Phenomena in Relation to Pulmonary Function," *The Physiologist*, 5(1) (1962), 11–28.

11.4 DYNAMIC SIMILARITY AND REYNOLDS NUMBER

Let us put the Navier-Stokes equation in dimensionless form. For simplicity, we shall consider a homogeneous incompressible fluid. Choose a characteristic velocity V and a characteristic length L. For example, if we investigate the flow of air around an airplane wing, we may take V to be the airplane speed and L to be the wing chord length. If we investigate the flow in a tube, V may be taken as the mean flow speed and L the tube diameter. For a falling sphere, we may take the speed of falling to be V, the diameter of the sphere to be L, and so on. Having chosen these characteristic quantities, we introduce the dimensionless variables

$$x' = \frac{x}{L}, \quad y' = \frac{y}{L}, \quad z' = \frac{z}{L}, \quad u' = \frac{u}{V}, \tag{11.4–1}$$
$$v' = \frac{v}{V}, \quad w' = \frac{w}{V}, \quad p' = \frac{p}{\rho V^2}, \quad t' = \frac{Vt}{L},$$

and the parameter

$$\text{Reynolds number} = R_N = \frac{VL\rho}{\mu} = \frac{VL}{\nu}. \tag{11.4–2}$$

Equation (11.1–7) for an incompressible fluid can then be put into the form

$$\frac{\partial u'}{\partial t'} + u'\frac{\partial u'}{\partial x'} + v'\frac{\partial u'}{\partial y'} + w'\frac{\partial u'}{\partial z} = -\frac{\partial p'}{\partial x'} + \frac{1}{R_N}\left(\frac{\partial^2 u'}{\partial x'^2} + \frac{\partial^2 u'}{\partial y'^2} + \frac{\partial^2 u'}{\partial z'^2}\right) \tag{11.4–3}$$

and two additional equations obtainable from Eq. (11.4–3) by changing u' into v', v' into w', w' into u' and x' into y', y' into z', z' into x'. The equation of continuity, Eq. (11.1–5), can also be put in dimensionless form:

$$\frac{\partial u'}{\partial x'} + \frac{\partial v'}{\partial y'} + \frac{\partial w'}{\partial z'} = 0. \tag{11.4–4}$$

Since Eqs. (11.4–3) and (11.4–4) constitute the complete set of field equations for an incompressible fluid, it is clear that only one physical parameter, the Reynolds number R_N, enters into the field equations of the flow.

Consider two geometrically similar bodies immersed in a moving fluid under identical initial and boundary conditions. One body may be considered a prototype and the other, a model. The bodies are similar (same shape but different size), and the boundary conditions are identical (in the dimensionless variables). The two flows will be identical if the Reynolds numbers for the two bodies are the same, because two geometrically similar bodies having the same Reynolds number will be governed by identical differential equations and boundary conditions (in dimensionless form). Therefore, *flows about geometrically similar bodies at the same Reynolds numbers are completely similar in the sense that the functions* $u'(x', y', z', t')$, $v'(x', y', z', t')$, $w'(x', y', z', t')$, $p'(x', y', z', t')$ *are the same for the various flows*. This kind of similarity of flows is called *dynamic similarity*. Reynolds number governs dynamic similarity of steady flows. For unsteady flows the requirement for the simulation of the differential equation and the initial and boundary conditions may require the simulation of other dimensionless parameters.

The Reynolds number expresses the ratio of the inertial force to the shear stress. In a flow, the inertial force due to convective acceleration arises from terms such as ρu^2, whereas the shear stress arises from terms such as $\mu\, \partial u/\partial y$. The orders of magnitude of these terms are, respectively,

$$\text{inertial force:} \quad \rho V^2,$$

$$\text{shear stress:} \quad \frac{\mu V}{L}.$$

The ratio is

$$\frac{\text{inertial force}}{\text{shear stress}} = \frac{\rho V^2}{\mu V/L} = \frac{\rho V L}{\mu} = \text{Reynolds number}. \tag{11.4–5}$$

A large Reynolds number signals a preponderant inertial effect. A small Reynolds number signals a predominant shear effect.

The wide range of Reynolds numbers that occurs in practical problems is illustrated in the following examples.

PROBLEMS

11.1 Smokestacks are known to sway in the wind if they are not rigid enough. The wind force depends on the Reynolds number of the flow. Let the wind speed be 30 mi/hr (each mi/hr = 0.44704 m/sec) and the smokestack diameter be 20 ft (each ft = 0.3048 m). Compute the Reynolds number of the flow.

Answer: 5.46×10^6.

The coefficient of viscosity of air at 20°C is $\mu = 1.808 \times 10^{-4}$ poise (g/cm sec), and the kinematic viscosity ν is 0.150 Stoke (cm^2/sec).

11.2 Compute the Reynolds number for a submarine periscope of diameter 16 in at 15 knots.

Answer: 2.4×10^6.

For water at 10°C, $\mu = 1.308 \times 10^{-2}$ g/cm sec and $\nu = 1.308 \times 10^{-2}$ cm^2/sec. 1 knot = 1 nautical mile per hour, or 1.852 km/hr.

11.3 Suppose that in a cloud chamber experiment designed to determine the charge of an electron (Robert Millikan's experiment), the water droplet diameter is 5 micra (i.e., 5×10^{-4} cm). The droplet moves in air at 0°C at a speed of 2 mm/sec. What is the Reynolds number?

Answer: 7.6×10^{-4}

For air at 0°C, $\nu = 0.132$ cm^2/sec.

11.4 For blood plasma to flow in a capillary blood vessel of diameter 10 micra (i.e., 10^{-3} cm) at a speed of 2 mm/sec, what is the Reynolds number?

Answer: 1.4×10^{-2}.

For blood plasma at body temperature, μ is about 1.4 centipoises (1.4×10^{-2} g/cm sec).

11.5 Compute the Reynolds number for a large airplane wing with a chord length of 10 ft (3.048 m), flying at 600 mi/hr (268.224 m/s) at an altitude of 7,500 ft (2,286 m), (0°C).

Answer: 6.2×10^7.

11.5 LAMINAR FLOW IN A HORIZONTAL CHANNEL OR TUBE

Navier-Stokes equations are not easy to solve. If, however, one can find a special problem in which the nonlinear terms disappear, then the solution can be obtained easily sometimes. A particularly simple problem of this nature is the steady flow of an incompressible fluid in a horizontal channel of width $2h$ between two parallel planes, as shown in Fig. 11.5.

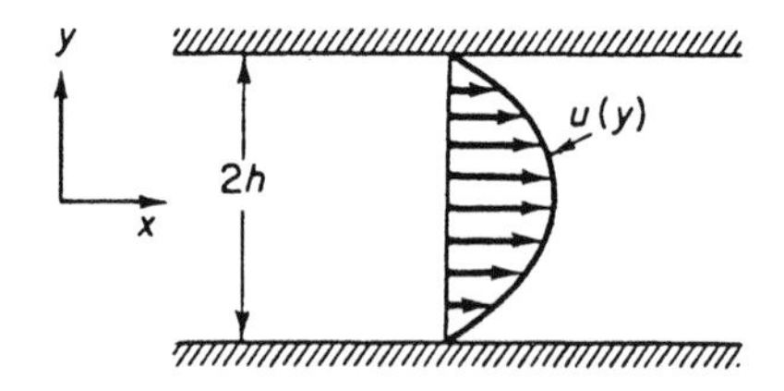

Figure 11.5 Laminar flow in a parallel channel.

We search for a flow

$$u = u(y), \qquad v = 0, \qquad w = 0 \tag{11.5–1}$$

that satisfies the Navier-Stokes equations, the equation of continuity, and the no-slip conditions on the boundaries $y = \pm h$:

$$u(h) = 0, \qquad u(-h) = 0. \tag{11.5–2}$$

Obviously, Eq. (11.5–1) satisfies the equation of continuity, Eq. (11.1–3), exactly, whereas Eq. (11.1–7) becomes

$$0 = -\frac{\partial p}{\partial x} + \mu\frac{d^2u}{dy^2}, \tag{11.5–3}$$

$$0 = \frac{\partial p}{\partial y}, \tag{11.5–4}$$

$$0 = \frac{\partial p}{\partial z}. \tag{11.5–5}$$

Equations (11.5–4) and (11.5–5) show that p is a function of x only. If we differentiate Eq. (11.5–3) with respect to x and use Eq. (11.5–1), we obtain $\partial^2p/\partial x^2 = 0$. Hence, $\partial p/\partial x$ must be a constant, say, $-\alpha$. Equation (11.5–3) then becomes

$$\frac{d^2u}{dy^2} = -\frac{\alpha}{\mu}, \tag{11.5–6}$$

which has a solution

$$u = A + By - \frac{\alpha}{\mu}\frac{y^2}{2}. \tag{11.5–7}$$

The two constants A and B can be determined by the boundary conditions (11.5–2) to yield the final solution,

$$u = \frac{\alpha}{2\mu}(h^2 - y^2). \tag{11.5–8}$$

Thus, the velocity profile is a parabola.

A corresponding problem is the flow through a horizontal circular cylindrical tube of radius a. (See Fig. 11.6.) We search for a solution

$$u = u(y, z), \qquad v = 0, \qquad w = 0.$$

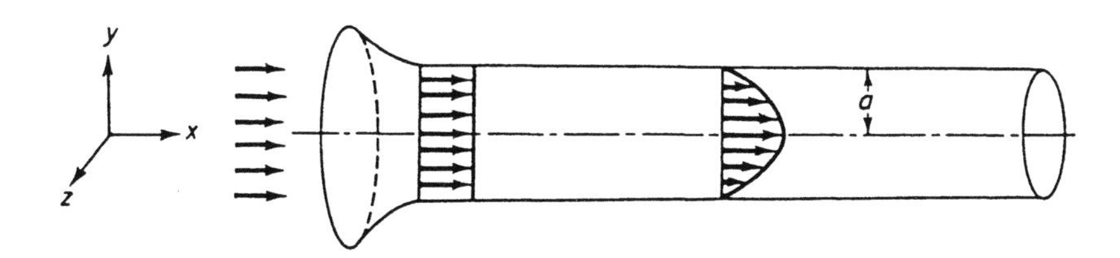

Figure 11.6 Laminar flow in a circular cylindrical tube.

In analogy with Eq. (11.5–6), the Navier-Stokes equation becomes

$$\frac{\partial^2 u}{\partial y^2} + \frac{\partial^2 u}{\partial z^2} = -\frac{\alpha}{\mu}. \tag{11.5–9}$$

It is convenient to transform from the Cartesian coordinates x, y, z to the cylindrical polar coordinates x, r, θ, with $r^2 = y^2 + z^2$. (See Sec. 5.8.) Then Eq. (11.5–9) becomes

$$\frac{\partial^2 u}{\partial y^2} + \frac{\partial^2 u}{\partial z^2} = \frac{1}{r}\frac{\partial}{\partial r}\left(r\frac{\partial u}{\partial r}\right) + \frac{1}{r^2}\frac{\partial^2 u}{\partial \theta^2} = -\frac{\alpha}{\mu}. \tag{11.5–10}$$

Let us assume that the flow is symmetric, so that u is a function of r only; then $\partial^2 u/\partial\theta^2 = 0$, and the equation

$$\frac{1}{r}\frac{d}{dr}\left(r\frac{du}{dr}\right) = -\frac{\alpha}{\mu} \tag{11.5–11}$$

can be integrated immediately to yield

$$u = -\frac{\alpha}{\mu}\frac{r^2}{4} + A\log r + B. \tag{11.5–12}$$

The constants A and B are determined by the conditions of no slip at $r = a$ and symmetry on the centerline, $r = 0$:

$$u = 0 \quad \text{at} \quad r = a. \tag{11.5–13}$$

$$\frac{du}{dr} = 0 \quad \text{at} \quad r = 0. \tag{11.5–14}$$

The final solution is

$$u = \frac{\alpha}{4\mu}(a^2 - r^2). \tag{11.5–15}$$

This is the famous parabolic velocity profile of the Hagen-Poiseuille flow; the theoretical solution was worked out by Stokes.

The classical solution of the Hagen-Poiseuille flow has been subjected to innumerable experimental observations. It is not valid near the entrance to a tube. It is satisfactory at a sufficiently large distance from the entrance, but is again invalid if the tube is too large or if the velocity is too high. The difficulty at the entry region is due to the transitional nature of the flow in that region, so that our assumption that $v = 0$ and $w = 0$ is not valid. The difficulty with too large a Reynolds number, however, is of a different kind: The flow becomes turbulent!

Osborne Reynolds demonstrated the transition from laminar to turbulent flow in a classical experiment in which he examined an outflow through a small tube from a large water tank. At the end of the tube, he used a stopcock to vary the speed of water through the tube. The junction of the tube with the tank was nicely rounded, and a filament of colored ink was introduced at the mouth. When the speed of water was slow, the filament remained distinct through the entire length of the tube. When the speed was increased, the filament broke up at a given point and diffused throughout the cross section (see Fig. 11.7). Reynolds identified the governing parameter $u_m d/\nu$—the Reynolds number—where u_m is the mean velocity, d is the diameter, and ν is the kinematic viscosity. Reynolds found that the transition from laminar to turbulent flow occurred at Reynolds numbers btween 2,000 and 13,000, depending on the smoothness of the entry conditions. When extreme care is taken, the transition can be delayed to Reynolds numbers as high as 40,000. On the other hand, a value of 2,000 appears to be about the lowest value obtainable on a rough entrance.

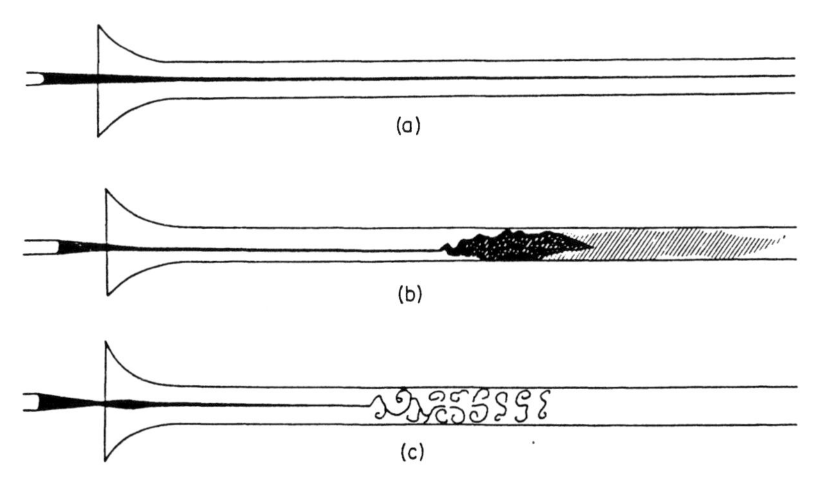

Figure 11.7 Reynolds's turbulence experiment: (a) laminar flow; (b) and (c), transition from laminar to turbulent flow. After Osborne Reynolds, "An Experimental Investigation of the Circumstances which Determine whether the Motion of Water Shall Be Direct or Sinuous, and of the Law of Resistance in Parallel Channels, *Phil. Trans., Roy. Soc.*, 174 (1883), 935–982.

Turbulence is one of the most important and most difficult problems in fluid mechanics. It is technically important not only because turbulence affects skin friction, resistance to flow, heat generation and transfer, diffusion, etc., but also because it is widespread. One might say that the normal mode of fluid flow is turbulent. The water in the ocean, the air above the earth, and the state of motion in the sun are turbulent. The theory of turbulence will greet you wherever you turn when you study fluid mechanics in greater depth.

PROBLEM

11.6 From the basic solution given by Eq. (11.5–15), show that the rate of mass flow through the tube is

$$Q = \frac{\pi a^4 \rho}{8\mu} \alpha, \tag{11.5–16}$$

that the mean velocity is

$$u_m = \frac{a^2}{8\mu}\alpha, \tag{11.5–17}$$

and that the skin friction coefficient is

$$c_f = \frac{\text{shear stress}}{\text{mean dynamic pressure}} = \frac{-\mu(\partial u/\partial r)_{r=a}}{\frac{1}{2}\rho u_m^2} = \frac{16}{R_N}, \tag{11.5–18}$$

where $R_N = 2au_m/\nu$.

11.6 BOUNDARY LAYER

If we let $R_N \to \infty$ in the dimensionless Navier-Stokes equation (11.4–3) for a homogeneous incompressible fluid, namely,

$$\frac{Du_i'}{Dt'} = -\frac{\partial p'}{\partial x_i'} + \frac{1}{R_N}\nabla^2 u_i', \qquad (i = 1, 2, 3), \tag{11.6–1}$$

the last term would drop out unless the second derivatives become very large. In a general flow field in which the velocity and its derivatives are finite, the effect of viscosity would disappear when the Reynolds number tends toward infinity. Near a solid wall, however, a rapid transition takes place for the velocity to vary from that of the free stream to that of the solid, because of the no-slip condition. If this transition layer is very thin, the last term cannot be dropped, even though the Reynolds number is very large.

We shall define the boundary layer as the region of a fluid in which the effect of viscosity is felt, even though the Reynolds number is very large. In the boundary layer, the flow is such that the shear-stress term—the last term in Eq. (11.6–1)—is of the same order of magnitude as the convective force term. Based on the observation that in a high-speed flow the boundary layer is very thin, Prandtl (1904) simplified the Navier-Stokes equation into a much more tractable boundary-layer equation.

To see the nature of the boundary-layer equation, let us consider a two-dimensional flow over a fixed flat plate. (See Fig. 11.8.) We take the x'-axis in the direction of flow along the surface and the y'-axis normal to it. The velocity component w along the z'-axis is assumed to vanish. Then Eq. (11.6–1) becomes

$$\frac{\partial u'}{\partial t'} + u'\frac{\partial u'}{\partial x'} + v'\frac{\partial u'}{\partial y'} = -\frac{\partial p'}{\partial x'} + \frac{1}{R_N}\left(\frac{\partial^2 u'}{\partial x'^2} + \frac{\partial^2 u'}{\partial y'^2}\right), \tag{11.6–2}$$

$$\frac{\partial v'}{\partial t'} + u'\frac{\partial v'}{\partial x'} + v'\frac{\partial v'}{\partial y'} = -\frac{\partial p'}{\partial y'} + \frac{1}{R_N}\left(\frac{\partial^2 v'}{\partial x'^2} + \frac{\partial^2 v'}{\partial y'^2}\right). \tag{11.6–3}$$

If we take the free-stream velocity as the characteristic velocity, then the dimensionless velocity u' is equal to 1 in the free stream (outside the boundary layer).

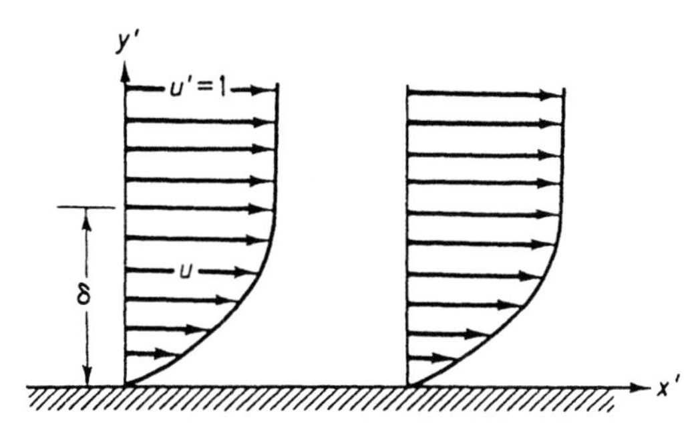

Figure 11.8 A boundary layer of flow.

The velocity u' varies from 0 on the solid surface $y' = 0$ to 1 at $y' = \delta$, where δ denotes the boundary-layer thickness (which is dimensionless and numerically small). We can now estimate the order of magnitude of the terms occurring in Eq. (11.6–2) as follows. We write $u' = O(1)$ to mean that u' is *at most* on the order of unity. We notice that $O(1) + O(1) = O(1)$, $O(1)\cdot O(1) = O(1)$, $O(1) + O(\delta) = O(1)$, and $O(1)\cdot O(\delta) = O(\delta)$. Then, since the variation of u' with respect to t' and x' is finite, we have

$$u' = O(1), \qquad \frac{\partial u'}{\partial x'} = O(1),$$
$$\frac{\partial^2 u'}{\partial x'^2} = O(1), \qquad \frac{\partial u'}{\partial t'} O(1). \tag{11.6–4}$$

By the equation of continuity, Eq. (11.4–4), we have

$$\frac{\partial u'}{\partial x'} = -\frac{\partial v'}{\partial y'} = O(1). \tag{11.6–5}$$

Hence,

$$v' = \int_0^\delta \frac{\partial v'}{\partial y'} dy' \sim \int_0^\delta O(1)dy = O(\delta). \tag{11.6–6}$$

Thus, the vertical velocity is at most on the order of δ, which is numerically small:

$$\delta \ll 1. \tag{11.6–7}$$

Since $v' = O(\delta)$ while $\partial v'/\partial y' = O(1)$ according to Eq. (11.6–5), we see that a differentiation of a quantity with respect to y' in the boundary layer increases the order of magnitude of that quantity by $1/\delta$. Then

$$\frac{\partial u'}{\partial y'} = O\left(\frac{1}{\delta}\right), \quad \frac{\partial^2 u'}{\partial y'^2} = O\left(\frac{1}{\delta^2}\right),$$
$$\frac{\partial v'}{\partial x'} = O(\delta), \quad \frac{\partial^2 v'}{\partial x'^2} = O(\delta), \tag{11.6–8}$$
$$\frac{\partial v'}{\partial t'} = O(\delta), \quad \frac{\partial^2 v'}{\partial y'^2} = O\left(\frac{1}{\delta}\right).$$

Now, by definition, the shear stress term is of the same order of magnitude as the inertial force term in the boundary layer. But the terms on the left-hand side of Eq. (11.6–2) are all $O(1)$; hence, those on the right-hand side must be also $O(1)$; in particular,

$$O(1) = \frac{\partial p'}{\partial x'},$$
$$O(1) = \frac{1}{R_N}\left(\frac{\partial^2 u'}{\partial x'^2} + \frac{\partial^2 u'}{\partial y'^2}\right) = \frac{1}{R_N}\left[O(1) + O\left(\frac{1}{\delta^2}\right)\right]. \tag{11.6–9}$$

Since the first term in the bracket is much smaller than the second term, we have

$$O(1) = \frac{1}{R_N} O\left(\frac{1}{\delta^2}\right).$$

Hence,

$$R_N = O\left(\frac{1}{\delta^2}\right). \tag{11.6–10}$$

Thus, we obtain an estimate of the boundary-layer thickness:

$$\delta = O\left(\frac{1}{\sqrt{R_N}}\right). \tag{11.6–11}$$

Substituting Eqs. (11.6–4), (11.6–8), and (11.6–10) into Eq. (11.6–3), we see that all terms involving v' are $O(\delta)$; hence, the remaining term $\partial p'/\partial y'$ must also be $O(\delta)$. Thus,

$$\frac{\partial p'}{\partial y'} = O(\delta) \sim 0. \tag{11.6–12}$$

In other words, *the pressure is approximately constant through the boundary layer*. By retaining only terms of order 1, the Navier-Stokes equations are reduced to

$$\frac{\partial u'}{\partial t'} + u'\frac{\partial u'}{\partial x'} + v'\frac{\partial u'}{\partial y'} = -\frac{\partial p'}{\partial x'} + \frac{1}{R_N}\frac{\partial^2 u'}{\partial y'^2}, \tag{11.6–13}$$

and Eq. (11.6–12). Equation (11.6–13) is Prandtl's boundary-layer equation; it is subjected to the boundary conditions

$$\begin{aligned} u' &= v' = 0 \qquad &&\text{for } y' = 0, \\ u' &= 1 &&\text{for } y' = \delta. \end{aligned} \tag{11.6–14}$$

PROBLEM

11.7 Estimate the boundary-layer thickness of air flowing over a plate 10 ft (3.048 m) long at 100 ft/sec (30.48 m/s).

Answer: At 20°C, $\delta = O(4.018 \times 10^{-4})$. With a chord length of 3.048 m, the boundary-layer thickness $\doteq$ 0.12 cm.

11.7 LAMINAR BOUNDARY LAYER OVER A FLAT PLATE

To apply Prandtl's boundary-layer theory, let us consider an incompressible fluid flowing over a flat plate, as in Fig. 11.9, in which the vertical scale is magnified to make the picture clearer. The velocity outside the boundary layer is assumed constant, $\bar{u}$. We shall seek a steady-state solution for which $\partial u/\partial t = 0$. An additional assumption will be made, to be justified *a posteriori*, that the pressure gradient $\partial p/\partial x$ is negligible, compared with the other terms in the boundary-layer equation. Then Eq. (11.6–13) becomes

$$u\frac{\partial u}{\partial x} + v\frac{\partial u}{\partial y} = \nu\frac{\partial^2 u}{\partial y^2}. \qquad (11.7\text{–}1)$$

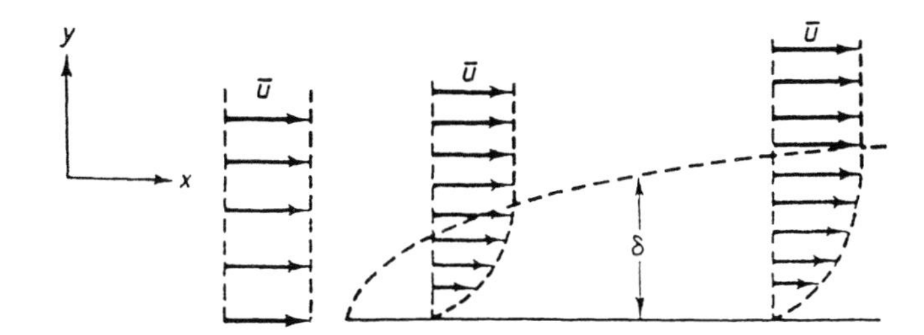

Figure 11.9 Laminar boundary layer over a flat plate, showing the growth in thickness of the boundary layer.

Here we return to the physical quantities and drop the primes. The equation of continuity is

$$\frac{\partial u}{\partial x} + \frac{\partial v}{\partial y} = 0. \qquad (11.7\text{–}2)$$

Equation (11.7–2) is satisfied identically if u, v are derived from a stream function $\psi(x, y)$:

$$u = -\frac{\partial \psi}{\partial y}, \qquad v = \frac{\partial \psi}{\partial x}. \qquad (11.7\text{–}3)$$

Then Eq. (11.7–1) becomes

$$\frac{\partial \psi}{\partial x}\frac{\partial^2 \psi}{\partial y^2} - \frac{\partial \psi}{\partial y}\frac{\partial^2 \psi}{\partial x \partial y} = \nu\frac{\partial^3 \psi}{\partial y^3}. \qquad (11.7\text{–}4)$$

The boundary conditions are (a) no slip on the plate and (b) continuity at the free stream outside the boundary layer; i.e.,

$$u = v = 0 \quad \text{or} \quad \frac{\partial \psi}{\partial x} = \frac{\partial \psi}{\partial y} = 0 \qquad \text{for } y = 0, \qquad (11.7\text{–}5)$$

$$u = \bar{u} \qquad \text{or} \quad -\frac{\partial \psi}{\partial y} = \bar{u} \qquad \text{for } y = \delta. \tag{11.7–6}$$

Following Blasius,* we seek a "similarity" solution. Consider the transformation

$$\bar{x} = \alpha x, \qquad \bar{y} = \beta y, \qquad \bar{\psi} = \gamma \psi, \tag{11.7–7}$$

in which α, β, and γ are constants. A substitution of Eq. (11.7–7) into Eq. (11.7–4) shows that the equation for the function $\bar{\psi}(\bar{x}, \bar{y})$ has the same form as Eq. (11.7–4) if we choose $\gamma = \alpha/\beta$. A similar substitution into Eq. (11.7–6) shows that $-\partial\bar{\psi}/\partial\bar{y} = \bar{u}$ if we choose $\gamma = \beta$. Hence, $\beta = \alpha/\beta$, or $\beta = \sqrt{\alpha}$. With this choice, we have

$$\frac{\bar{y}}{\sqrt{\bar{x}}} = \frac{y}{\sqrt{x}}, \qquad \frac{\bar{\psi}}{\sqrt{\bar{x}}} = \frac{\psi}{\sqrt{x}}. \tag{11.7–8}$$

These relations suggest that there are solutions of the form

$$\psi = -f(\xi)\sqrt{\nu \bar{u} x}, \qquad \xi = \sqrt{\frac{\bar{u}}{\nu}}\frac{y}{\sqrt{x}}. \tag{11.7–9}$$

Substitution of Eq. (11.7–9) into Eq. (11.7–4) yields the ordinary differential equation

$$2f''' + ff'' = 0, \tag{11.7–10}$$

where the primes indicate differentiation with respect to ξ. This equation has been solved numerically to a high degree of accuracy under the boundary conditions

$$f(0) = 0, \qquad f'(0) = 0, \qquad f'(\infty) = 1, \tag{11.7–11}$$

which say that $u = 0$ and $v = 0$ at the plate and $u \to \bar{u}$, the free-stream velocity, outside the boundary layer. From Eq. (11.7–9), it is seen that for fixed x/L, $\xi \to \infty$ means that y/L is large, compared with the boundary-layer thickness $\sqrt{\nu/L\bar{u}}$, or δ. The velocity distribution, yielded by the solution of Eqs. (11.7–10) and (11.7–11), agrees closely with experimental evidence,† as seen in Fig. 11.10, except very near the leading edge of the plate, where the boundary-layer approximation breaks down, and far downstream, where the flow becomes turbulent.

The flow corresponding to the solution given by Eq. (11.7–9), (11.7–10), and (11.7–11) is a laminar flow. At a sufficient distance downstream from the leading edge, the flow becomes turbulent and the Blasius solution fails. The transition occurs when a Reynolds number based on the boundary layer thickness,

$$R = \frac{\bar{u}\delta}{\nu},$$

*H. Blasius, "Grenzschichten in Flüssigkeiten mit kleiner Reibung," *Zeitschrift f. Math. u. Phys.*, **56** (1908), 1.

†J. Nikuradse, *Laminare Reibungsschichten an der längsangeströmten platte*. Monograph, Zentrale f. Wiss. Berichtswesen, Berlin, 1942. See H. Schlichting, *Boundary Layer Theory*, translated by J. Kestin, New York: McGraw-Hill Book Company (1960), p. 124.

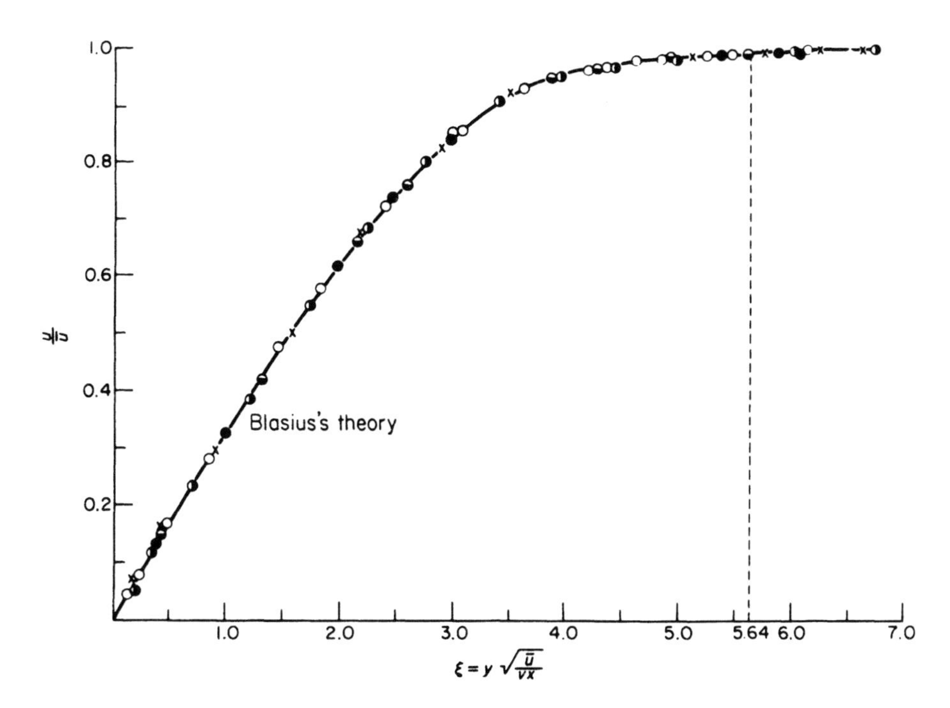

Figure 11.10 Blasius's solution of velocity distribution in a laminar boundary layer on a flat plate at zero incidence and comparison with Nikuradse's measurements.

reaches a critical value. Generally, the value of the critical transitional Reynolds number is on the order of 3,000, but the exact value depends on the surface roughness, curvature, Mach number, etc.

There is a tremendous difference between a laminar boundary layer and a turbulent one with respect to heat transfer, skin friction, heat generation, etc. In our space age, the question of laminar-turbulent transition is of supreme importance for reentry vehicles. As a satellite reenters the atmosphere, the heat generated by skin friction in the boundary layer is tremendous—but a turbulent boundary layer generates much more heat than a laminar one. For most reentry vehicles, survival is possible if the boundary layer over the nose cone is laminar; if the flow became turbulent, the nose cone could be burned out.

11.8 NONVISCOUS FLUID

A great simplification is obtained if the coefficient of viscosity vanishes exactly. Then the stress tensor is isotropic, i.e.,

$$\sigma_{ij} = -p\delta_{ij}, \tag{11.8–1}$$

and the equation of motion can be simplified to

$$\rho \frac{Dv_i}{Dt} = \rho X_i - \frac{\partial p}{\partial x_i}. \tag{11.8–2}$$

Here, ρ is the density of the fluid; p is the pressure; v_1, v_2, v_3 are the velocity components; and X_1, X_2, X_3 are the body force components per unit mass.

If, in addition, the fluid is homogeneous and incompressible, then its density is a constant, and the equation of continuity, Eq. (11.1–3), is reduced to the form

$$\frac{\partial u}{\partial x} + \frac{\partial v}{\partial y} + \frac{\partial w}{\partial z} = 0 \quad \text{or} \quad \frac{\partial u_i}{\partial x_i} = 0. \tag{11.8–3}$$

A vector field satisfying Eq. (11.8–3) is said to be *solenoidal*. According to the general theory of potentials, a solenoidal field can be derived from another vector field. This can be illustrated in the simple case of a *two-dimensional flow field* for which $w = 0$ and u, v are independent of z and for which the equation of continuity is

$$\frac{\partial u}{\partial x} + \frac{\partial v}{\partial y} = 0. \tag{11.8–4}$$

Then it is obvious that if we take an arbitrary function $\psi(x, y)$ and derive u, v according to the rules

$$u = \frac{\partial \psi}{\partial y}, \qquad v = -\frac{\partial \psi}{\partial x}, \tag{11.8–5}$$

Eq. (11.8–4) will be satisfied identically. Such a function ψ is called a *stream function*.

Substituting Eq. (11.8–5) into the equation of motion, Eq. (11.8–2), we obtain the governing equations (for the two-dimensional flow),

$$\begin{aligned} \frac{\partial^2 \psi}{\partial t \partial y} + \frac{\partial \psi}{\partial y}\frac{\partial^2 \psi}{\partial x \partial y} - \frac{\partial \psi}{\partial x}\frac{\partial^2 \psi}{\partial y^2} &= X - \frac{1}{\rho}\frac{\partial p}{\partial x}, \\ -\frac{\partial^2 \psi}{\partial t \partial x} - \frac{\partial \psi}{\partial y}\frac{\partial^2 \psi}{\partial x^2} + \frac{\partial \psi}{\partial x}\frac{\partial^2 \psi}{\partial x \partial y} &= Y - \frac{1}{\rho}\frac{\partial p}{\partial y}. \end{aligned} \tag{11.8–6}$$

If the body force is zero, an elimination of p yields

$$\frac{\partial}{\partial t}\nabla^2\psi + \psi_y \nabla^2 \psi_x - \psi_x \nabla^2 \psi_y = 0, \tag{11.8–7}$$

in which

$$\nabla^2 = \frac{\partial^2}{\partial x^2} + \frac{\partial^2}{\partial y^2},$$

and the subscripts indicate partial differentiation.

PROBLEM

11.8 Show that for a two-dimensional flow of an incompressible viscous fluid, the governing equation for the stream function defined by Eq. (11.8–5) is

$$\frac{\partial}{\partial t}\nabla^2\psi + \psi_y\nabla^2\psi_x - \psi_x\nabla^2\psi_y = \nu\nabla^2\nabla^2\psi + \frac{\partial X}{\partial y} - \frac{\partial Y}{\partial x}. \qquad (11.8\text{–}8)$$

11.9 VORTICITY AND CIRCULATION

The concepts of circulation and vorticity are of great importance in fluid mechanics. The *circulation* $I(\mathscr{C})$ in any closed circuit $\mathscr{C}$ is defined by the line integral

$$I(\mathscr{C}) = \int_{\mathscr{C}} \mathbf{v}\cdot\mathbf{dl} = \int_{\mathscr{C}} v_i dx_i, \qquad (11.9\text{–}1)$$

where $\mathscr{C}$ is any closed curve in the fluid and the integrand is the scalar product of the velocity vector $\mathbf{v}$ and the vector $\mathbf{dl}$, which is tangent to the curve $\mathscr{C}$ and of length dl (Fig. 11.11). Clearly, the circulation is a function of both the velocity field and the chosen curve $\mathscr{C}$.

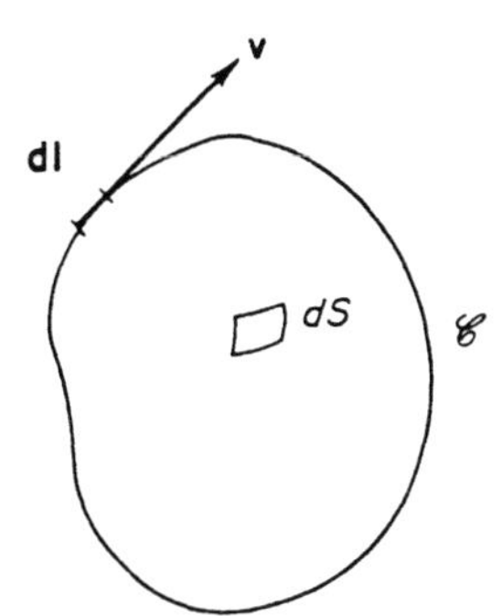

Figure 11.11 Circulation: Notations.

By means of Stokes's theorem, if $\mathscr{C}$ encloses a simply connected region, the line integral can be transformed into a surface integral

$$I(\mathscr{C}) = \int_S (\nabla \times \mathbf{v})_n dS = \int_S (\text{curl } \mathbf{v})_i \nu_i dS, \qquad (11.9\text{–}2)$$

where S is any surface in the fluid bounded by the curve $\mathscr{C}$, ν_i is the unit normal to the surface, and curl $\mathbf{v} = e_{ijk}v_{j,k}$. The curl $\mathbf{v}$ is called the *vorticity* of the velocity field.

The law of change of circulation with time, when the circuit $\mathscr{C}$ is a *fluid line*, i.e., a curve $\mathscr{C}$ formed by the same set of fluid particles as time changes, is given by the *theorem of Lord Kelvin*: If the fluid is nonviscous and the body force is conservative, then

$$\frac{DI}{Dt} = -\int_{\mathscr{C}} \frac{dp}{\rho}. \qquad (11.9\text{–}3)$$

If, in addition to the preceding conditions, the density ρ is a unique function of the pressure, then the fluid is called *barotropic*, and the last integral vanishes because the integral would be single valued and $\mathscr{C}$ is a closed curve. We then have the *Helmholtz theorem* that

$$\frac{DI}{Dt} = 0. \tag{11.9–4}$$

To prove the foregoing theorems, we note that since $\mathscr{C}$ is a fluid line composed always of the same particles, the order of differentiation and integration may be interchanged in the following:

$$\frac{D}{Dt}\int_{\mathscr{C}} v_i\, dx_i = \int_{\mathscr{C}} \frac{D}{Dt}(v_i dx_i) = \int_{\mathscr{C}} \left(\frac{Dv_i}{Dt}\, dx_i + v_i \frac{D\, dx_i}{Dt}\right). \tag{11.9–5}$$

But $D\, dx_i/Dt$ is the rate at which dx_i is increasing as a consequence of the motion of the fluid; hence, it is equal to the difference of the velocities parallel to x_i at the ends of the element, i.e., dv_i. Substituting Dv_i/Dt from the equation of motion, Eq. (11.8–2), and replacing $D\, dx_i/Dt$ by dv_i, we obtain

$$\frac{DI}{Dt} = \int_{\mathscr{C}} \left[\left(-\frac{1}{\rho}\frac{\partial p}{\partial x_i} + X_i\right) dx_i + v_i\, dv_i\right]$$

$$= -\int_{\mathscr{C}} \frac{dp}{\rho} + \int_{\mathscr{C}} X_i\, dx_i + \int_{\mathscr{C}} dv^2. \tag{11.9–6}$$

Of the terms on the right-hand side, the last vanishes because v^2 is single valued in the flow field; the second vanishes if the body force X_i is conservative. Hence, Kelvin's theorem is proved. Helmholtz's theorem follows immediately as a special case because the integral on the right-hand side vanishes if the fluid is barotropic.

In the clear-cut conclusion of Helmholtz's theorem lies its importance. For if we limit our attention to a barotropic fluid, then we have I = const. Hence, if the circulation vanishes at one instant of time, it must vanish for all times. If this is so for any arbitrary fluid lines in a field, then, according to Eq. (11.9–2), the vorticity vanishes in the whole field. This leads to a great simplification, which will be discussed in Sec. 11.10, namely, the irrotational flow. To appreciate the importance of this simplification, one need observe only that a vast majority of the classical literature on fluid mechanics deals with irrotational flows.

Note that the circulation around a fluid line *does not* have to remain constant if the density ρ depends on other variables in addition to pressure. Into this category fall most geophysical problems in which the temperature enters as a parameter affecting both ρ and p. Also, in stratified flows, ρ is a function of location, not necessarily a function of p alone.

The significance of the term *fluid line* in the theorems of Kelvin and Helmholtz may be seen by considering the problem of a thin airfoil moving in the air. The conditions of the Helmholtz theorem are satisfied. Hence, the circulation I about any fluid line never changes with time. Since the motion of the fluid is caused by the motion of the airfoil, and since at the beginning the fluid is at rest and $I = 0$, it follows that I vanishes at all times. Note, however, that the volume occupied by the airfoil is exclusive of the fluid. A fluid line $\mathscr{C}$ enclosing the boundary of the airfoil becomes elongated when the airfoil moves forward, as shown in Fig. 11.12. According to the Helmholtz theorem, the circulation about $\mathscr{C}$ is zero, so that the total vorticity inside $\mathscr{C}$ vanishes, but one cannot conclude that the vorticity actually vanishes everywhere inside $\mathscr{C}$. In the region occupied by the airfoil *and* in the wake behind the airfoil, vorticity does exist. However, the Helmholtz theorem applies to the region outside the airfoil and its wake, and the vanishing of circulation about every possible fluid line shows clearly that the flow is irrotational outside the airfoil *and* its wake.

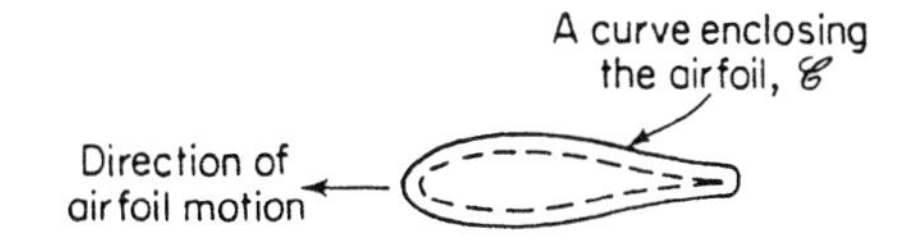

Figure 11.12 Fluid line enclosing an airfoil and its wake.

11.10 IRROTATIONAL FLOW

A flow is said to be *irrotational* if the vorticity vanishes everywhere, i.e., if

$$\nabla \times \mathbf{v} = \operatorname{curl} \mathbf{v} = 0, \tag{11.10–1}$$

or

$$e_{ijk} v_{j,k} = 0.$$

For a two-dimensional irrotational flow, we must have

$$\frac{\partial u}{\partial y} - \frac{\partial v}{\partial x} = 0. \tag{11.10–2}$$

If the fluid is incompressible and a stream function defined by Eq. (11.8–5) is introduced, then a substitution of Eq. (11.8–5) into Eq. (11.10–2) yields the equation

$$\frac{\partial^2 \psi}{\partial x^2} + \frac{\partial^2 \psi}{\partial y^2} = 0. \tag{11.10–3}$$

This is the famous *Laplace equation*, whose solution is the concern of many books on applied mathematics.

We can show that an irrotational flow of an incompressible fluid is governed by a Laplace equation even in the three-dimensional case, because by the definition of irrotationality, the following three equations must hold:

$$\frac{\partial u}{\partial y} - \frac{\partial v}{\partial x} = 0, \qquad \frac{\partial v}{\partial z} - \frac{\partial w}{\partial y} = 0, \qquad \frac{\partial w}{\partial x} - \frac{\partial u}{\partial z} = 0. \tag{11.10–4}$$

These equations can be satisfied identically if the velocities u, v, w are derived from a *potential function* $\Phi(x, y, z)$ according to the rule

$$u = \frac{\partial \Phi}{\partial x}, \qquad \mathrm{v} = \frac{\partial \Phi}{\partial y}, \qquad \mathrm{w} = \frac{\partial \Phi}{\partial z}. \tag{11.10–5}$$

If, in addition, the fluid is *incompressible*, then a substitution of Eq. (11.10–5) into Eq. (11.1–5) yields the *Laplace equation*

$$\frac{\partial^2 \Phi}{\partial x^2} + \frac{\partial^2 \Phi}{\partial y^2} + \frac{\partial^2 \Phi}{\partial z^2} = 0. \tag{11.10–6}$$

Since Φ is a potential function, this equation is also called a *potential equation*.

The incompressible potential flow is governed by the Laplace equation. If a solution can be found that satisfies all the boundary conditions, then the Eulerian equation of motion yields the pressure gradient, and the problem is solved. The nonlinear convective acceleration, which causes the central difficulty of fluid mechanics, does not hinder the solution of potential flows of an incompressible fluid. This is why the potential theory is so simple and so important.

To realize the usefulness of the potential theory, we quote the Helmholtz theorem (see Sec. 11.9): If the motion of any portion of a fluid mass is irrotational at any one instant of time it will continue to be irrotational at all times, provided that the body forces are conservative and that the fluid is *barotropic* (i.e., its density is a function of pressure alone). These conditions are met in many problems. If a solid body is immersed in a fluid and suddenly set in motion, the motion generated in a nonviscous fluid is irrotational.* Hence, a whole class of technologically important problems is irrotational.

11.11 COMPRESSIBLE NONVISCOUS FLUIDS

Basic Equations

If a fluid is compressible, the equation of continuity, Eq. (10.5–3), is

$$\frac{\partial \rho}{\partial t} + \frac{\partial \rho v_j}{\partial x_j} = 0. \tag{11.11–1}$$

*See H. Lamb, *Hydrodynamics*, New York: Dover Publications, 6th ed. (1945), pp. 10, 11.

If the fluid is nonviscous, the Eulerian equation of motion is

$$\frac{\partial v_i}{\partial t} + v_j \frac{\partial v_i}{\partial x_j} = -\frac{1}{\rho}\frac{\partial p}{\partial x_i} + X_i. \tag{11.11–2}$$

The density is uniquely related to the pressure only if the temperature T is explicitly accounted for. Thus, if the temperature is known to be constant (isothermal), we have, for an ideal gas,

$$\frac{p}{\rho} = \text{const.}, \qquad T = \text{const.}, \tag{11.11–3}$$

whereas if the flow is isentropic (adiabatic and reversible), we have

$$\frac{p}{\rho^{\gamma}} = \text{const.}, \qquad \frac{T}{\rho^{\gamma-1}} = \text{const.}, \tag{11.11–4}$$

where γ is the ratio of the specific heats of the gas at constant pressure, C_p, and constant volume, C_v; i.e., $\gamma = C_p/C_v$. Both cases are *barotropic.*

In other cases, it is necessary to introduce the temperature explicitly as a variable. Then we must introduce also the equation of state relating p, ρ, and T and the *caloric* equation of state relating C_p, C_v, and T.

Small Disturbances

Let us consider, as an example, the propagation of small disturbances in a barotropic fluid in the absence of body force. Let us write

$$c^2 = \frac{dp}{d\rho}. \tag{11.11–5}$$

The velocity of flow will be assumed to be so small that the second-order terms may be neglected in comparison with the first-order term. Correspondingly, the disturbances in the density ρ and the pressure p and the derivatives of ρ and p are also first-order infinitesimal quantities. Then, on neglecting the body force X_i and all small quantities of the second or higher order, Eqs. (11.11–1) and (11.11–2) are linearized to

$$\frac{\partial \rho}{\partial t} + \rho \frac{\partial v_j}{\partial x_j} = 0, \tag{11.11–6}$$

$$\frac{\partial v_i}{\partial t} = -\frac{1}{\rho}\frac{\partial p}{\partial x_i} = -\frac{1}{\rho}\frac{dp}{d\rho}\frac{\partial \rho}{\partial x_i} = -\frac{c^2}{\rho}\frac{\partial \rho}{\partial x_i}. \tag{11.11–7}$$

Differentiating Eq. (11.11–6) with respect to t and Eq. (11.11–7) with respect to x_i, again neglecting the second-order terms, and eliminating the sum $\rho\, \partial^2 v_j/\partial t\, \partial x_j$, we obtain

$$\frac{1}{c^2}\frac{\partial^2\rho}{\partial t^2} = \frac{\partial^2\rho}{\partial x_i \partial x_i}; \tag{11.11–8}$$

i.e.,

$$\frac{1}{c^2}\frac{\partial^2\rho}{\partial t^2} = \frac{\partial^2\rho}{\partial x^2} + \frac{\partial^2 p}{\partial y^2} + \frac{\partial^2\rho}{\partial z^2}.$$

This is the *wave equation for the propagation of small disturbances*. It is the basic equation of acoustics.

By the same linearization procedure, and because the change in pressure is proportional to the change in density, $dp = c^2\, d\rho$, we see that the same wave equation governs the pressure disturbance:

$$\frac{1}{c^2}\frac{\partial^2 p}{\partial t^2} = \frac{\partial^2 p}{\partial x_k \partial x_k}. \tag{11.11–9}$$

Further, from Eqs. (11.11–7) and (11.11–8) or (11.11–9), we deduce that

$$\frac{1}{c^2}\frac{\partial^2 v_i}{\partial t^2} = \frac{\partial^2 v_i}{\partial x_k \partial x_k}. \tag{11.11–10}$$

Hence, in the linearized theory, ρ, p, v_1, v_2, and v_3 are governed by the same wave equation.

Propagation of Sound

Let us apply these equations to the problem of a source of disturbance (sound) located at the origin and radiating symmetrically in all directions. We may visualize a spherical siren. Because of the radial symmetry, we have

$$\frac{\partial^2}{\partial x^2} + \frac{\partial^2}{\partial y^2} + \frac{\partial^2}{\partial z^2} = \frac{\partial^2}{\partial r^2} + \frac{2}{r}\frac{\partial}{\partial r}. \tag{11.11–11}$$

Hence, Eq. (11.11–8) becomes

$$\frac{1}{c^2}\frac{\partial^2\rho}{\partial t^2} = \frac{\partial^2\rho}{\partial r^2} + \frac{2}{r}\frac{\partial\rho}{\partial r}. \tag{11.11–12}$$

It can be verified by direct substitution that a general solution of this equation is the sum of two arbitrary functions f and g:

$$\rho = \rho_0 + \frac{1}{r}f(r - ct) + \frac{1}{r}g(r + ct), \tag{11.11–13}$$

Here, ρ_0 is a constant (the undisturbed density of the field), the f term represents a wave radiating out from the origin, and the g term represents a wave converging toward the origin. Perhaps the clearest way to see this is to consider a special case in which the function $f(r - ct)$ is a step function: $f(r - ct) = \epsilon\mathbf{1}(r - ct)$, where ϵ

is small and $\mathbf{1}(r - ct)$ is the unit-step function, which is zero when $r - ct < 0$ and is 1 when $r - ct > 0$. The disturbance is, therefore, a small jump across a line of discontinuity described by the equation $r - ct = 0$. At time $t = 0$, the disturbance is located at the origin. At time t, the line of discontinuity is moved to $r = ct$. Thus, c is the speed of propagation of the disturbance. The general case follows by the principle of superposition. In acoustics, c is called the *velocity of sound*.

The velocity of sound $c = (dp/d\rho)^{1/2}$ depends on the relationship between pressure and density. If we are concerned with an ideal gas and the condition is isentropic, we have, from Eq. (11.11–4),

$$c = \sqrt{\frac{\gamma p}{\rho}}. \tag{11.11–14}$$

In the history of mechanics, there was a long story about the propagation of sound in air. The first theoretical investigation of the velocity of sound was made by Newton (1642–1727), who assumed Eq. (11.11–3) and obtained $c = \sqrt{p/\rho}$ in a publication in 1687. It was found that the value calculated from Newton's formula falls short of the experimental value of the speed of sound by a factor of approximately one-sixth. This discrepancy was not explained until Laplace (1749–1827) pointed out that the rate of compression and expansion in a sound wave is so fast that there is no time for any appreciable interchange of heat by conduction; thus, the process must be considered adiabatic. This argument becomes plausible if we think of the step wave discussed in the preceding paragraph. For a step wave, the sudden changes in ρ and p that take place as the wave front sweeps by must be accomplished at the wave front in an infinitesimal region of space and time. Heat transfer in such a small time interval is negligible. Hence, the gas flows isentropically across the discontinuity. As a general sound wave is a superposition of such step waves, the entire flow is isentropic. Therefore, Eq. (11.11–4) applies and Eq. (11.11–14) results. Experiments have verified that Laplace was right.

Generally, then, the wave equations (11.11–8), et seq., are associated with isentropic flows. To apply these equations, conditions that guarantee isentropy, such as the absence of strong shock waves and small thermal diffusivity, must be observed.

11.12 SUBSONIC AND SUPERSONIC FLOW

Basic Equations in Laboratory Frame of Reference

The basic wave equation (11.11–8) is referred to a frame of reference that is at rest relative to the fluid at infinity. The equation imposes no restriction on where and how the disturbances are generated. The sources of disturbances may be moving or changing with time; the same equation holds. The nature of the sources would appear only in the boundary conditions and initial conditions.

A flying aircraft is a source of disturbances in still air. The disturbances come to us as sound waves governed by the wave equation. As we all know, the nature of the disturbances changes drastically as the aircraft's flight speed changes from subsonic to supersonic. In the latter case, we hear the sonic boom.

It is convenient to study the nature of flow about an aircraft in a wind tunnel. We shall therefore write down the wave-propagation equation for disturbances in the air flowing in a tunnel as they appear to us standing on the ground.

Consider a body of fluid coming from, say, the left, with a uniform velocity U at infinity. If the disturbances are indicated by a prime, we assume the velocity components to be

$$u = U + u', \qquad v = v', \qquad w = w', \qquad U = \text{const.} \tag{11.12–1}$$

and the pressure and density

$$p = p_0 + p', \qquad \rho = \rho_0 + \rho'. \tag{11.12–2}$$

The whole investigation would be simplified if we could assume that the disturbances are infinitesimal quantities of the first order; i.e.,

$$u', v', w' \ll U, \qquad p' \ll p_0, \qquad \rho' \ll \rho_0. \tag{11.12–3}$$

Under these assumptions, the basic equations (11.11–1) and (11.11–4) may be linearized as before. In fact, repeating the relevant steps in Sec. 11.11 with our new assumptions, we obtain the equation of continuity

$$\frac{\partial \rho'}{\partial t} + \rho_0\left(\frac{\partial U}{\partial x} + \frac{\partial u'}{\partial x} + \frac{\partial v'}{\partial y} + \frac{\partial w'}{\partial z}\right) + (U + u')\frac{\partial \rho'}{\partial x} + v'\frac{\partial \rho'}{\partial y} + w'\frac{\partial \rho'}{\partial z} = 0,$$

which is linearized to

$$\frac{\partial \rho'}{\partial t} + \rho_0\left(\frac{\partial u'}{\partial x} + \frac{\partial v'}{\partial y} + \frac{\partial w'}{\partial z}\right) + U\frac{\partial \rho'}{\partial x} = 0. \tag{11.12–4}$$

Similarly, the equations of motion are linearized to

$$\begin{aligned}
\frac{\partial u'}{\partial t} + U\frac{\partial u'}{\partial x} &= -\frac{1}{\rho_0}\frac{\partial p'}{\partial x} = -\frac{c^2}{\rho_0}\frac{\partial \rho'}{\partial x}, \\
\frac{\partial v'}{\partial t} + U\frac{\partial v'}{\partial x} &= -\frac{c^2}{\rho_0}\frac{\partial \rho'}{\partial y}, \\
\frac{\partial w'}{\partial t} + U\frac{\partial w'}{\partial x} &= -\frac{c^2}{\rho_0}\frac{\partial \rho'}{\partial z}.
\end{aligned} \tag{11.12–5}$$

Differentiating the three equations (11.12–5) with respect to x, y, and z, respectively, adding, and again neglecting the second-order terms, we obtain

$$\frac{\partial}{\partial t}\left(\frac{\partial u'}{\partial x} + \frac{\partial v'}{\partial y} + \frac{\partial w'}{\partial z}\right) + U\frac{\partial}{\partial x}\left(\frac{\partial u'}{\partial x} + \frac{\partial v'}{\partial y} + \frac{\partial w'}{\partial z}\right) = -\frac{c^2}{\rho_0}\left(\frac{\partial^2 \rho'}{\partial y^2} + \frac{\partial^2 \rho'}{\partial y^2} + \frac{\partial^2 \rho'}{\partial z^2}\right).$$

Hence, on eliminating $\partial u'/\partial x + \partial v'/\partial y + \partial w'/\partial z$ with Eqs. (11.12–4), we have

$$\frac{\partial^2 \rho'}{\partial t^2} + 2U\frac{\partial^2 \rho'}{\partial x \partial t} + U^2\frac{\partial^2 \rho'}{\partial x^2} = c^2\left(\frac{\partial^2 \rho'}{\partial x^2} + \frac{\partial^2 \rho'}{\partial y^2} + \frac{\partial^2 \rho'}{\partial z^2}\right). \tag{11.12–6}$$

This is the basic equation for compressible flow in aerodynamics.

If, to Eq. (11.12–6) we apply the method used in Sec. 11 to derive Eqs. (11.11–9), (11.11–10) from Eq. (11.11–8), we can show that the pressure p' and velocity components v_i' satisfy the same equation. If the flow is irrotational, then the velocity potential Φ, for which $v_i = \Phi,_i$, also satisfies this equation.

Steady Flow

Let us examine the basic equation (11.12–6) in some simpler cases. Consider a steady flow around a model at rest. Then all derivatives with respect to time t vanish, and the velocity potential Φ is governed by the equation

$$\frac{U^2}{c^2}\frac{\partial^2 \Phi}{\partial x^2} = \frac{\partial^2 \Phi}{\partial x^2} + \frac{\partial^2 \Phi}{\partial y^2} + \frac{\partial^2 \Phi}{\partial z^2}. \tag{11.12–7}$$

This equation now depends on only one dimensionless parameter, U/c, which is called the *Mach number* and is denoted by

$$M = \frac{U}{c}. \tag{11.12–8}$$

The nature of the solution to Eq. (11.12–7) depends on whether M is greater or less than 1. We call a flow *subsonic* if $M < 1$, *supersonic* if $M > 1$. We write, for a subsonic flow,

$$(1 - M^2)\frac{\partial^2 \Phi}{\partial x^2} + \frac{\partial^2 \Phi}{\partial y^2} + \frac{\partial^2 \Phi}{\partial z^2} = 0 \qquad (M < 1), \tag{11.12–9}$$

whereas, for a supersonic flow, we have

$$(M^2 - 1)\frac{\partial^2 \Phi}{\partial x^2} - \frac{\partial^2 \Phi}{\partial y^2} - \frac{\partial^2 \Phi}{\partial z^2} = 0 \qquad (M > 1). \tag{11.12–10}$$

Equation (11.12–9) is a partial differential equation of the *elliptic type*. Equation (11.12–10) is one of the *hyperbolic type*. Let us consider an example showing the difference between these equations.

Example: Steady Flow over a Wavy Plate

Let a very thin plate with a small sinusoidal wavy profile be placed in a steady flow, with the mean chord of the plate parallel to the velocity U at infinity. (See Figs. 11.13 and 11.14.) The waves of the plate are described by the equation

$$z = a \sin\frac{\pi x}{L}. \tag{11.12–11}$$

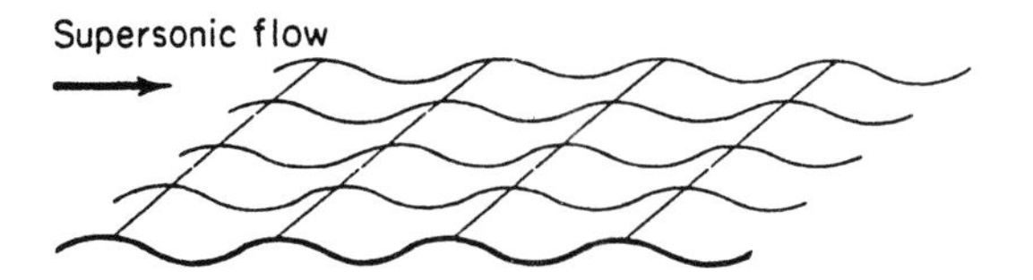

Figure 11.13 A wavy plate in a steady supersonic flow.

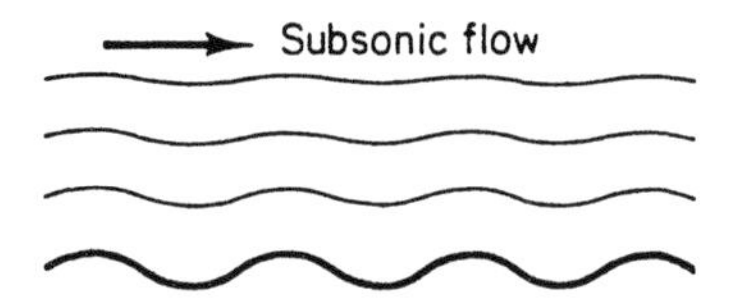

Figure 11.14 A wavy plate in a steady subsonic flow.

The amplitude a is assumed to be small compared with the wave length L:

$$a \ll L. \tag{11.12–12}$$

The fluid, since it is assumed to be perfect, can glide over the plate, but cannot penetrate it. Therefore, the velocity vector of the flow must be tangent to the plate. Now the velocity vector has the components

$$U + u', \; v', \; w' \tag{11.12–13}$$

in the x-, y-, and z-directions, respectively. On the other hand, the normal vector to the surface described by Eq. (11.12–11) has the following components (see Fig. 11.15):

$$-\frac{\partial z}{\partial x}, \quad -\frac{\partial z}{\partial y}, \quad 1. \tag{11.12–14}$$

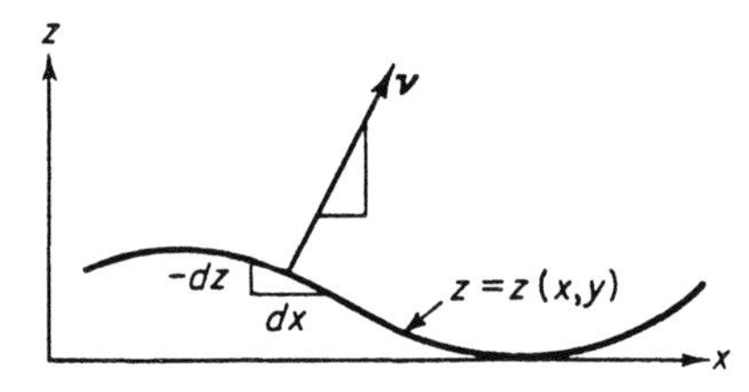

Figure 11.15 Surface normal and the velocity boundary condition.

If the velocity vector, with components given by Eq. (11.12–13), is to be tangent to the surface of the plate, it must be normal to the normal vector given by Eq. (11.12–14). Hence, the condition of nonpenetration can be stated as the orthogonality of the vectors of Eqs. (11.12–13) and (11.12–14), i.e., by the condition that their scalar product vanishes:

$$-(U + u')\frac{\partial z}{\partial x} - v'\frac{\partial z}{\partial y} + w' \cdot 1 = 0.$$

Omitting higher-order terms, we obtain the boundary condition

$$w' = U\frac{\partial z}{\partial x} \tag{11.12–15}$$

on the plate. From Eq. (11.12–11), this is

$$w' = U\frac{a\pi}{L}\cos\frac{\pi x}{L} \qquad \left(\text{when } z = a\sin\frac{\pi z}{L}\right). \tag{11.12–16}$$

Again, counting on the continuity and differentiability of the function $w'(x, y, z)$, we can write

$$w'(x, y, z) = w'(x, y, 0) + z\left(\frac{\partial w'}{\partial z}\right)_{z=0} + \dots . \tag{11.12–17}$$

For small z, all the terms following the first are higher-order terms. Consistently neglecting these terms, we can simplify the boundary condition to

$$w' = U\frac{a\pi}{L}\cos\frac{\pi x}{L} \qquad (\text{when } z = 0). \tag{11.12–18}$$

Condition at Infinity

The boundary condition given in Eq. (11.12–18) is not sufficient to determine the solution to our problem, which is governed by either Eq. (11.12–9) or Eq. (11.12–10), depending on whether the flow is subsonic or supersonic. In addition, the conditions at infinity must be specified. There is a great difference between the elliptic and hyperbolic equations with respect to the appropriate types of boundary conditions that may be specified, and we must consider them in some detail.

Subsonic case. For the elliptic equation (11.12–9), the influence of the disturbances is spread out in all directions, and it is reasonable to assume that, for any finite body, the disturbances tend toward zero at distances infinitely far away from the body. A rigorous argument may be based on the total energy that may be imparted to the fluid. If the fluid velocity is distributed in a certain fashion, and if it does not tend toward zero at a certain rate as the distance from the body increases toward infinity, an infinitely large energy would have to be imparted to the fluid in order to create the motion, which is impossible. (For further details, see texts on partial differential equations or aerodynamics.) Accordingly, we impose the following conditions on our problem:

(a) The flow is two-dimensional and parallel to the xz-plane, and there is no dependence on the y-coordinate.

(b) All disturbances tend toward zero as $z \to \pm\infty$. In particular,

$$u', v', w' \to 0; \quad \text{i.e.,} \quad \Phi \to \text{const. as } z \to \pm\infty. \tag{11.12–19}$$

Supersonic case. Turning now to the hyperbolic equation (11.12–10), we find that the disturbances can be carried away along waves of limited dimension. The argument of decreasing amplitude does not apply. Instead, the boundary condition must be replaced by the *radiating condition*: that the plate is the only source of disturbances and that the disturbances radiate *from* the source, not toward it.

This description of the radiation condition is easy to apply when we are concerned with a single source. For example, of the two solutions on the right-hand side of Eq. (11.11–13), the term $f(r - ct)/r$ represents a wave radiating from the origin; hence, for a source at the origin, it is the only term admissible under the radiation condition. The condition becomes somewhat confounded, however, when applied to two-dimensional steady flow. Perhaps the matter can be clarified by examining some photographs of supersonic flow about stationary models in a wind tunnel, such as those shown in Fig. 11.16. Here, the flows are from left to right. We see that the lines of disturbances, which are contours of density of the fluid as revealed by the Schlieren photographs, incline to the right. This direction of inclination of the strong (shock) and weak (Mach) waves is determined by the radiation condition.

Solution of the Wavy Wall Problem

Now we can return to our problem. It is easily verified by direct substitution that, in the subsonic case, Eq. (11.12–9) can be satisfied by a function of the form

$$\Phi = Ae^{\mu z} \cos \frac{\pi x}{L}. \tag{11.12–20}$$

On substituting Eq. (11.12–20) into Eq. (11.12–9), we obtain

$$-(1 - M^2)\left(\frac{\pi}{L}\right)^2 Ae^{\mu z} \cos \frac{\pi x}{L} + A\mu^2 e^{\mu z} \cos \frac{\pi x}{L} = 0,$$

or

$$\mu = \pm\left(\frac{\pi}{L}\right) \sqrt{1 - M^2}. \tag{11.12–21}$$

If the plus sign is used in Eq. (11.12–21), the function Φ in Eq. (11.12–20) will grow exponentially without limit as $z \to \infty$. On the other hand, if the minus sign is used, Eq. (11.12–19) can be satisfied. Hence, we may try

$$\Phi = Ae^{-(\pi/L)\sqrt{1-M^2}z} \cos \frac{\pi x}{L}. \tag{11.12–22}$$

The vertical velocity w' computed from Φ is

$$w' = \frac{\partial \Phi}{\partial z} = -\frac{\pi}{L} \sqrt{1 - M^2}\, Ae^{-(\pi/L)\sqrt{1-M^2}z} \cos \frac{\pi x}{L}. \tag{11.12–23}$$

On setting $z = 0$ in Eq. (11.12–23) and applying the boundary condition, Eq. (11.12–18), we obtain

$$A = -\frac{U\, a}{\sqrt{1 - M^2}}. \tag{11.12–24}$$

Figure 11.16 (a) Flow past a flat plate with a beveled, sharp leading edge, the top surface being aligned with the free stream of Mach number 8. On the top side of the plate, a laminar boundary layer is revealed by the lighter line. A shock wave is induced by the displacement effect of the boundary layer. Similar features are seen on the lower side. Schlieren system. Flow left to right. *Courtesy of Toshi Kubota, California Institute of Technology*; (b) Scale mode of the Nimbus spacecraft in a 50-in hypersonic tunnel, at Mach number 8 and Reynolds number of 0.42×10^6/ft. Schlieren system. Flow left to right. *Courtesy of Von Karman Gas Dynamics Facility, ARO, Inc.*

Now all the boundary conditions for the subsonic case are satisfied. Hence, the solution for the subsonic case is

$$\Phi = -\frac{U\,a}{\sqrt{1 - M^2}}\, e^{-(\pi/L)\sqrt{1-M^2}z} \cos \frac{\pi x}{L}. \tag{11.12–25}$$

We see that the disturbances decrease exponentially with increasing z. From this solution, we can deduce the velocity field, the pressure field, and the density field. In particular, since

$$U\frac{\partial u'}{\partial x} = -\frac{1}{\rho}\frac{\partial p'}{\partial x'}, \tag{11.12–26}$$

we have

$$p' = -\rho U u' = -\rho U \frac{\partial \Phi}{\partial x}. \tag{11.12–27}$$

The streamlines for such a flow are plotted in Fig. 11.14.

Turning now to the supersonic case, Eq. (11.12–10), we see that it can be satisfied by the function

$$\Phi = f(x - \sqrt{M^2 - 1}\,z) + g(x + \sqrt{M^2 - 1}\,z), \tag{11.12–28}$$

where f and g are arbitrary functions, because if we set

$$\xi = x - \sqrt{M^2 - 1}\,z, \tag{11.12–29}$$

then

$$\frac{\partial f}{\partial x} = \frac{df}{d\xi}, \qquad \frac{\partial f}{\partial z} = -\sqrt{M^2 - 1}\,\frac{df}{d\xi};$$

hence,

$$(M^2 - 1)\frac{\partial^2 f}{\partial x^2} - \frac{\partial^2 f}{\partial z^2} = (M^2 - 1)\frac{d^2 f}{d\xi^2} - (M^2 - 1)\frac{d^2 f}{d\xi^2} = 0,$$

and Eq. (11.12–10) is satisfied. The lines

$$\xi = \text{const.}, \quad \text{i.e.,} \quad x - \sqrt{M^2 - 1}\,z = \text{const.}, \tag{11.12–30}$$

are the Mach waves, along which the disturbances are propagated with undiminished intensity. These lines are inclined in the correct direction, as revealed by the wind-tunnel photographs. On the other hand, the Mach lines for the function $g(x + \sqrt{M^2 - 1}\,z)$ are inclined in the wrong direction. Hence, the function g must be rejected on the basis of the radiation condition. Therefore, we may try

$$\Phi = f(x - \sqrt{M^2 - 1}\,z). \tag{11.12–31}$$

From Eq. (11.12–31), we obtain

$$w' = \frac{\partial \Phi}{\partial z} = -\sqrt{M^2 - 1}\,\frac{df}{d\xi}. \qquad (11.12\text{–}32)$$

Comparing Eq. (11.12–32) with the boundary condition, Eq. (11.12–18), we obtain, when $z = 0$,

$$-\sqrt{M^2 - 1}\left(\frac{df}{d\xi}\right)_{z=0} = \frac{U a\pi}{L}\cos\frac{\pi x}{L} = \frac{U a\pi}{L}\cos\frac{\pi\xi}{L}\bigg|_{z=0}. \qquad (11.12\text{–}33)$$

Hence, on integrating and returning to Eq. (11.12–29), we have

$$\Phi = f = -\frac{U a}{\sqrt{M^2 - 1}}\sin\frac{\pi}{L}(x - \sqrt{M^2 - 1}\,z), \qquad (11.12\text{–}34)$$

which solves the problem. A plot of the streamlines is shown in Fig. 11.13.

The contrast between the two cases is dramatic. Whereas in the subsonic case the pressure disturbance is diminished as the distance from the plate increases, in the supersonic case it is not. This is, of course, the reason why a sonic boom hits us with all its fury from a supersonic aircraft, but not from a subsonic one.

11.13 APPLICATIONS TO BIOLOGY

Fluid mechanics is as relevant to living creatures as to machines and physical objects. The gas in the airway and lung, the urine, and the sap in the xylem of trees are Newtonian fluids to which the Navier-Stokes equation and the no-slip boundary conditions apply. Blood is a non-Newtonian fluid. If the shear strain rate is sufficiently high (e.g., $> 100\ \text{s}^{-1}$), the viscosity of blood is almost constant, i.e., its behavior is almost Newtonian. If the shear strain rate is low, however, the viscosity of blood increases. Saliva, mucus, synovial fluid in the knee joint, and other body fluids are also non-Newtonian. Analysis of the flow of these must take their non-Newtonian behavior into consideration.

Blood can be treated as a homogeneous fluid only when one is considering flow in a blood vessel whose diameter is much larger than the diameter of the red blood cells. Flow in a small blood vessel, such as in the capillaries, whose diameter is about the same as that of the red cells, must treat the cells as individual bodies. The blood is, then, a biphasic fluid. Other body fluids that contain proteins and other suspensions may have to be treated as biphasic or multiphasic if the dimensions of the vessels in which they flow are sufficiently small.

Animals and plants live in gas, water, and earth. Understanding their movements requires fluid mechanics. Body fluids circulate inside of animals and plants. Understanding their movement also requires fluid mechanics. In either case, the boundary conditions are, in general, nonstationary.

The examples considered in this chapter have applications to biology. The analysis of the flow in a channel or tube is relevant to the blood flow problem. The blood vessels, however, are elastic. The diameter of a blood vessel varies with the pressure. The interaction between the flow and the elastic deformation of the wall can produce some very interesting phenomena. In biology, solid mechanics and fluid mechanics are often closely knit together.

The reader may gain some insight into the broad subject of fluid mechanics and biomechanics from the references listed at the end of the chapter.

PROBLEMS

11.9 Derive the Navier-Stokes equation for an incompressible fluid in cylindrical polar coordinates.

Solution: The left-hand side of the Navier-Stokes equation represents acceleration. In polar coordinates, the components are a_r, a_θ, a_z, which are given in Eq. (10.9–9) on p. 225. The right-hand side is the vector divergence of the stress tensor. In polar coordinates, these components are given by Eq. (10.9–11). It remains to write down the stresses in terms of the velocities u, v, w along the radial, circumferential, and axial directions, respectively. On p. 128, we have e_{rr}, $e_{\theta\theta}$, etc., expressed in terms of u_r, u_θ, u_z. The relationship between the strain rates $\dot{e}_{rr}$, $\dot{e}_{\theta\theta}$, etc., to the velocities u, v, w are the same. Hence,

$$\dot{e}_{rr} = \frac{\partial u}{\partial r}, \qquad \dot{e}_{\theta\theta} = \frac{u}{r} + \frac{1}{r}\frac{\partial v}{\partial \theta}, \text{ etc.}$$

Therefore, from Eq. (7.3–6), and for an incompressible fluid, we have

$$\sigma_{rr} = -p + 2\mu\dot{e}_{rr} = -p + 2\mu \frac{\partial u}{\partial r},$$

$$\sigma_{\theta\theta} = -p + 2\mu\dot{e}_{\theta\theta} = -p + 2\mu \left(\frac{u}{r} + \frac{1}{r}\frac{\partial v}{\partial \theta}\right),$$

$$\sigma_{zz} = -p + 2\mu\dot{e}_{zz} = -p + 2\mu \frac{\partial w}{\partial z},$$

$$\sigma_{r\theta} = 2\mu\dot{e}_{r\theta} = \mu \left(r \frac{\partial (v/r)}{\partial r} + \frac{1}{r}\frac{\partial u}{\partial \theta}\right),$$

$$\sigma_{\theta z} = 2\mu\dot{e}_{\theta z} = \mu \left(\frac{1}{r}\frac{\partial w}{\partial \theta} + \frac{\partial v}{\partial z}\right),$$

$$\sigma_{zr} = 2\mu\dot{e}_{zr} = \mu \left(\frac{\partial u}{\partial z} + \frac{\partial w}{\partial r}\right).$$

A substitution into Eq. (10.9–11) yields the Navier-Stokes equations,

$$\frac{\partial u}{\partial t} + u\frac{\partial u}{\partial r} + \frac{v}{r}\frac{\partial u}{\partial \theta} + w\frac{\partial u}{\partial z} - \frac{v^2}{r} = -\frac{1}{\rho}\frac{\partial p}{\partial r} + \nu\left(\nabla^2 u - \frac{u}{r^2} - \frac{2}{r^2}\frac{\partial v}{\partial \theta}\right) + F_r$$

$$\frac{\partial v}{\partial t} + u\frac{\partial v}{\partial r} + \frac{v}{r}\frac{\partial v}{\partial \theta} + w\frac{\partial v}{\partial z} + \frac{uv}{r} = -\frac{1}{\rho}\frac{1}{r}\frac{\partial p}{\partial \theta} + \nu\left(\nabla^2 v + \frac{2}{r^2}\frac{\partial u}{\partial \theta} - \frac{v}{r^2}\right) + F_\theta$$

$$\frac{\partial w}{\partial t} + u\frac{\partial w}{\partial r} + \frac{v}{r}\frac{\partial w}{\partial \theta} + w\frac{\partial w}{\partial z} = -\frac{1}{\rho}\frac{\partial p}{\partial z} + \nu\nabla^2 w + F_z$$

where

$$\nabla^2 \equiv \frac{\partial^2}{\partial r^2} + \frac{1}{r}\frac{\partial}{\partial r} + \frac{1}{r^2}\frac{\partial^2}{\partial \theta^2} + \frac{\partial^2}{\partial z^2}.$$

The equation of continuity is

$$\frac{1}{r}\frac{\partial}{\partial r}(ru) + \frac{1}{r}\frac{\partial v}{\partial \theta} + \frac{\partial w}{\partial z} = 0.$$

11.10 Blood is a non-Newtonian fluid whose viscosity varies with the strain rate. (See Fig. 9.15 and Prob. 9.4.) Derive the equation of motion of blood in a form analogous to the Navier-Stokes equations. Formulate mathematically the problem of blood flow in a living heart.

11.11 If air is truly nonviscous, would an airplane be able to fly? What about birds and insects? Why?

11.12 If water is nonviscous, would fish be able to swim? What are the differences in the arguments for fish in water and birds in the air?

11.13 Formulate the mathematical problem of tides induced on the earth under the influence of the moon. (See Lamb, *Hydrodynamics*, pp. 358–362.)

11.14 Waves are generated in water in a long channel of rectangular cross section. What are the equations with which the wavelength and frequency can be determined?

11.15 Ripples are generated on the surface of water in a deep pond. Does the wave speed depend on the wavelength? Even though the full solution is rather complicated, whether or not the waves are dispersive (i.e., whether the speed depends on wavelength) can be detected when all the basic equations are written down. Take the free surface of the pond to be the xy-plane, let the z-axis point downward, and try a two-dimensional solution with velocity components

$$v \equiv 0, \qquad u = ae^{-kz}\sin kx \sin \omega t, \quad \text{and} \quad w = -ae^{-kz}\cos kx \sin \omega t.$$

11.16 Consider a ground-effect machine, which uses one or more reaction jets and hovers above the ground. Sketch the streamlines of the flow and write the equations and boundary conditions that govern the machine when it is hovering.

11.17 Analyze the motion in a cumulus cloud in a summer thunderstorm. What are the variables relevant to this problem? If temperature is an important consideration, how would it be incorporated into the basic equations? Gravity must not be neglected. Present the basic equations. Make a dimensional analysis to determine fundamental dimensionless parameters.

11.18 Water waves run up a sloping beach and create all the panorama on the seashore: surf, riptides, waves, ripples, and foam. Analyze the phenomenon mathematically. Give an appropriate choice of variables. Write down the differential equations and boundary conditions. Make simplifying assumptions if you think they are appropriate, but state your assumptions clearly.

11.19 On the beach, there are riptides, which are fast-moving narrow streams of water that move toward the ocean in a direction perpendicular to the shoreline and are dangerous to swimmers. Now this is an anomaly: For a two-dimensional sloping beach and a two-dimensional water wave, we obtain a three-dimensional solution. Is there any basic objection to this situation (from the mathematician's point of view, not the swimmer's)? Can you name another example of such a phenomenon in nature?

11.20 When wind blows over (perpendicular to) long cylindrical pipes, vortices are shed in the wake. These vortices induce vibrations in the pipe. A trans-Arabic oil line (the aboveground part) was reported to have suffered severe vibrations due to wind. Smokestacks, large rockets, and the like are subjected to these disturbances. Vortex shedding over a long cylinder is three dimensional; in other words, the shedding is nonuniform along the length of the cylinder, even if the wind and the cylinder are both uniform. Formulate the aerodynamic problem for a fixed, rigid cylinder. Furnish all the differential equations and boundary conditions. Make a dimensional analysis to determine all the dimensionless parameters involved.

11.21 Generalize Prob. 11.20 to take account of the vortex shedding over a flexible, vibrating cylinder.

11.22 Using the equations derived in Prob. 11.9, find the velocity field in a Couette flowmeter (Fig. P3.22, p. 86).

Answer: Let $v = \omega_1 a$ at $r = a$ and $v = \omega_2 b$ at $r = b$. Then

$$v = (a^2 - b^2)^{-1}[(\omega_1 a^2 - \omega_2 b^2)r - a^2 b^2(\omega_2 - \omega_1)/r].$$

11.23 Using the Navier-Stokes equation, find the velocity distribution of a flow in a long cylindrical pipe of rectangular cross section.

11.24 Discuss whether the concept of a boundary layer is important in each of the following problems. Explain briefly how and why boundary-layer theory is used in those problems to which it is applicable.

(a) Blood flow in the aorta. Assume a viscosity coefficient $\mu = 0.04$ poise, radius $r = 3$ mm, density $\rho = 1$, and velocity $v = 50$ cm/sec.

(b) Blood flow in small blood vessels. Assume a coefficient of viscosity $\mu = 0.04$ poise, radius $a = 10^{-3}$ cm, density $\rho = 1$, and mean velocity $v = 0.07$ cm/sec.

Note: Compute the Reynolds number $R_N = 2\,VL/\mu$. In (a), $R_N = 750$. In (b), $R_N = 3.5 \times 10^{-3}$. The boundary layer thickness δ is on the order of $(R_N)^{-1/2}$.

11.25 I have a garden hose curved on the ground. One end is connected to a water faucet. When the valve is opened, the pressure is high, a water jet comes out with good force, and the hose whips like a snake. Why?

Now consider an analagous problem for a pipeline suspended in air above ground. One span L is supported between two pillars. The pipe is a thin-walled circular cylindrical shell, in which flows a fluid. The pipe is straight if there is no load. It is loaded by its own weight, the weight of the fluid, and the pressure of the flowing fluid. To design the pipe and the pillars, what fluid mechanical problems should be considered? Formulate a mathematical theory for an important problem that you identified. Write down the differential equations and the boundary conditions. Outline a method of solution.

FURTHER READING

BATCHELOR, G. K., *An Introduction to Fluid Mechanics*, Cambridge: Cambridge University Press (1967).

FUNG, Y. C., *Biodynamics: Circulation*, New York: Springer-Verlag (1984).

FUNG, Y. C., *Biomechanics: Motion, Flow, Stress, and Growth*, New York: Springer-Verlag (1990).

GOLDSTEIN, S. (ed.), *Modern Development in Fluid Dynamics* (2 vol.), London: Oxford University Press (1938).

LAMB, HORACE, *Hydrodynamics*, 1st ed., 1879, 6th ed., 1932, New York: Dover Publications (1945).

LIEPMANN, H. W., AND A. ROSHKO, *Elements of Gasdynamics*, New York: Wiley (1957).

PRANDTL, L., *The Essentials of Fluid Dynamics*, London: Blackie (1953).

SCHLICHTING, H., *Boundary Layer Theory*, 4th ed., New York: McGraw-Hill (1960).

YIH, CHIA-SHUN, *Fluid Mechanics, a Concise Introduction to the Theory*, West River Press, 3530 W. Huron River Dr., Ann Arbor, MI 48103 (1990).

12 SOME SIMPLE PROBLEMS IN ELASTICITY

Basic equations, elastic waves, torsion of shafts, bending of beams, and some remarks about biomechanics.

12.1 BASIC EQUATIONS OF ELASTICITY FOR HOMOGENEOUS, ISOTROPIC BODIES

In the preceding chapter, we discussed the equations governing the flow of fluids. In this chapter, we shall consider the motion of solids that obey the Hooke's law. A Hookean body has a unique zero-stress state. All strains and particle displacements are measured from this state, in which their values are counted as zero.

The basic equations can be gleaned from the preceding chapters. Let $u_i(x_1, x_2, x_3, t)$, $i = 1, 2, 3$, describe the displacement of a particle located at x_1, x_2, x_3 at time t from its position in the zero-stress state. Various strain measures can be defined for the displacement field. The Green strain tensor is expressed in terms of $u_i(x_1, x_2, x_3, t)$ according to Eq. (5.3–3):

$$e_{ij} = \frac{1}{2}\left[\frac{\partial u_j}{\partial x_i} + \frac{\partial u_i}{\partial x_j} + \frac{\partial u_k}{\partial x_i}\frac{\partial u_k}{\partial x_j}\right]. \qquad (12.1\text{–}1)$$

Here, and hereinafter, all Latin indices range over 1, 2, 3. The particle velocity v_i is given by the material derivative of the displacement,

$$v_i = \frac{\partial u_i}{\partial t} + v_j\frac{\partial u_i}{\partial x_j}. \qquad (12.1\text{–}2)$$

The particle acceleration α_i is given by the material derivative of the velocity, Eq. (10.3–7),

$$\alpha_i = \frac{\partial v_i}{\partial t} + v_j\frac{\partial v_i}{\partial x_j}. \qquad (12.1\text{–}3)$$

The conservation of mass is expressed by the equation of continuity, Eq. (10.5–3),

$$\frac{\partial \rho}{\partial t} + \frac{\partial(\rho v_i)}{\partial x_i} = 0. \tag{12.1–4}$$

The conservation of momentum is expressed by the Eulerian equation of motion, Eq. (10.6–7),

$$\rho \alpha_i = \frac{\partial \sigma_{ij}}{\partial x_j} + X_i. \tag{12.1–5}$$

Hooke's law for a homogeneous, isotropic material is

$$\sigma_{ij} = \lambda e_{kk} \delta_{ij} + 2G e_{ij}, \tag{12.1–6}$$

where λ and G are Lamé constants.

Equations (12.1–1) through (12.1–6) together describe a theory of elasticity. If we compare these equations with the corresponding equations for a viscous fluid, as given in Sec. 11.1, we see that their theoretical structures are similar, except that here we have a nonlinear strain-and-displacement-gradient relation [Eq. (12.1–1)], in contrast to the linear rate-of-deformation-and-velocity-gradient relation [Eq. (6.1–3)], for the fluid. Hence, the theory of elasticity is more deeply nonlinear than the theory of viscous fluids.

The nonlinear problem is so wrought with mathematical complexities that only a few exact solutions are known. For this reason, it is common to simplify the theory by introducing a severe restriction, namely, that *the displacements and velocities are infinitesimal*. In this way, Eqs. (12.1–1) through (12.1–3) can be linearized. One tries to learn as much as possible about the linearized theory and then proceed to discover what features are introduced by the nonlinearities.

We *linearize* the equations by restricting ourselves to values of u_i, v_i so small that the nonlinear terms in Eqs. (12.1–1) through (12.1–3) may be neglected. Thus,

$$e_{ij} = \frac{1}{2}\left(\frac{\partial u_i}{\partial x_j} + \frac{\partial u_j}{\partial x_i}\right), \tag{12.1–7}$$

$$v_i = \frac{\partial u_i}{\partial t}, \qquad \alpha_i = \frac{\partial v_i}{\partial t}. \tag{12.1–8}$$

Equations (12.1–4) through (12.1–8) together are 22 equations for the 22 unknowns ρ, u_i, v_i, α_i, e_{ij}, σ_{ij}; $i, j = 1, 2, 3$. We may eliminate σ_{ij} by substituting Eq. (12.1–6) into Eq. (12.1–5) and using Eq. (12.1–7) to obtain the well-known *Navier's equation*,

$$G\nabla^2 u_i + (\lambda + G)\frac{\partial e}{\partial x_i} + X_i = \rho\frac{\partial^2 u_i}{\partial t^2}, \tag{12.1–9}$$

where e is the divergence of the displacement vector $\mathbf{u}$; i.e.,

$$e = \frac{\partial u_j}{\partial x_j} = \frac{\partial u_1}{\partial x_1} + \frac{\partial u_2}{\partial x_2} + \frac{\partial u_3}{\partial x_3}. \tag{12.1–10}$$

∇^2 is the *Laplace operator.* If we write x, y, z instead of x_1, x_2, x_3, we have

$$\nabla^2 = \frac{\partial^2}{\partial x^2} + \frac{\partial^2}{\partial y^2} + \frac{\partial^2}{\partial z^2}. \tag{12.1–11}$$

If we introduce Poisson's ratio ν, as in Eq. (7.4–8), we can write Navier's equation (12.1–9) as

$$G\left(\nabla^2 u_i + \frac{1}{1 - 2\nu}\frac{\partial e}{\partial x_i}\right) + X_i = \rho\frac{\partial^2 u_i}{\partial t^2}. \tag{12.1–12}$$

This is the basic field equation of the linearized theory of elasticity.

Navier's equation (12.1–9) must be solved for appropriate boundary conditions, which are usually one of two kinds:

(1) *Specified displacements.* The components of displacement u_i are prescribed on the boundary.

(2) *Specified surface tractions.* The components of surface traction $\overset{\nu}{T}_i$ are assigned on the boundary.

In most problems of elasticity, the boundary conditions are such that over one part of the boundary the displacements are specified, whereas over another part the surface tractions are specified. In the latter case, Hooke's law may be used to convert the boundary condition into prescribed values of a certain combination of the first derivatives of u_i.

12.2 PLANE ELASTIC WAVES

To illustrate the use of the linearized equations, let us consider a simple harmonic wave train in an elastic medium. Let us assume that the displacement components u_1, u_2, u_3 (or, in unabridged notation, u, v, w) are infinitesimal and that the body force X_i vanishes. Then it is easy to verify that a solution of Navier's equation (12.1–9) is

$$u = A \sin\frac{2\pi}{l}(x \pm c_L t), \qquad v = w = 0, \tag{12.2–1}$$

where A, l, and c_L are constants, provided that the constant c_L is chosen to be

$$c_L = \sqrt{\frac{\lambda + 2G}{\rho}} = \sqrt{\frac{E(1 - \nu)}{(1 + \nu)(1 - 2\nu)\rho}}. \tag{12.2–2}$$

The pattern of motion expressed by Eq. (12.2–1) is unchanged when $x \pm c_L t$ remains constant. Hence, if the negative sign were taken, the pattern would move to the right with a velocity c_L as the time t increased. The constant c_L is called the *phase velocity* of the wave motion. The constant l is the *wavelength*, as can be seen from the sinusoidal pattern of u as a function of x, at any instant of time. The particle velocity computed from Eq. (12.2–1) is in the same direction as that of the wave propagation (namely, in the direction of the x-axis). Such a motion is said to constitute a train of *longitudinal waves*. Since at any instant of time the wave crests lie in parallel planes, the motion represented by this equation is called a train of *plane waves*.

Next, let us consider the motion

$$u = 0, \qquad v = A \sin \frac{2\pi}{l}(x \pm ct), \qquad w = 0, \tag{12.2–3}$$

which represents a train of plane waves of wavelength l propagating in the direction of the x-axis with a phase velocity c. When Eqs. (12.2–3) are substituted into Eq. (12.1–9), it is seen that c must assume the value

$$c_T = \sqrt{\frac{G}{\rho}}. \tag{12.2–4}$$

The particle velocity (in the y-direction) computed from Eq. (12.2–3) is perpendicular to the direction of wave propagation (the x-direction). Hence, the wave generated is said to be a *transverse wave*. The speeds c_L and c_T are called the characteristic *longitudinal wave speed* and *transverse wave speed*, respectively. They depend on the elastic constants and the density of the material. The ratio c_T/c_L depends on Poisson's ratio only and is given by

$$c_T = c_L\sqrt{\frac{1 - 2\nu}{2(1 - \nu)}}. \tag{12.2–5}$$

If $\nu = 0.25$, then $c_L = \sqrt{3}\, c_T$.

Similar to Eq. (12.2–3), the following equations represent a transverse wave in which the particles move in the direction of the z-axis:

$$u = 0, \qquad v = 0, \qquad w = A \sin \frac{2\pi}{l}(x \pm c_T t). \tag{12.2–6}$$

The plane parallel to which the particles move [such as the xy-plane in Eq. (12.2–3) or the xz-plane in Eq. (12.2–6)] is called the *plane of polarization*.

Plane waves may exist only in an unbounded elastic continuum. In a finite body, a plane wave will be reflected when it hits a boundary. If there is another elastic medium beyond the boundary, refracted waves occur in the second medium. The features of reflection and refraction are similar to those in acoustics and optics; the main difference is that, in elasticity, an incident longitudinal wave will be reflected and refracted in a combination of longitudinal and transverse waves, and an incident transverse wave will also be reflected in a combination of both types of waves. The details can be worked out by the proper combination of these waves so that the boundary conditions are satisfied.

12.3 SIMPLIFICATIONS

Important simplifications to the equation of the linearized theory of elasticity may come from

(1) Homogeneity and isotropy.
(2) The absence of inertial forces.
(3) A high degree of symmetry in geometry.
(4) Plane stress and plane strain.
(5) Thin-walled structures—plates and shells.

Clearly, a simplification is obtained if the number of independent or dependent variables is reduced. Thus, if nothing changes with time, the variable t will be suppressed. Homogeneity of materials makes the coefficients of the differential equations constant. Isotropy reduces the number of independent material constants. High degree of symmetry reduces the number of geometric parameters in a problem. Reduction of the general field equations to two dimensions or one dimension reduces the number of independent and dependent variables.

Example 1. A Plane State of Stress

A *plane-stress* state depending on x, y only may be visualized as a state that exists in a thin membrane stressed in its own plane. Figure 4.1 on p. 89 shows an example of such a case. Analytically, a plane-stress state is defined by the condition that the stress components σ_{zz}, σ_{zx}, σ_{zy} vanish everywhere, i.e.,

$$\sigma_{zz} = \sigma_{zx} = \sigma_{zy} = 0, \qquad (12.3\text{–}1)$$

whereas the stress components σ_{xx}, σ_{xy}, σ_{yy} are independent of the coordinate z.

Example 2. A Plane State of Strain

If the z-component of the displacement w vanishes everywhere, and if the displacements u, v are functions of x, y only, and not of z, the body is said to be in a *plane-strain* state, depending on x, y only. Such a state may be visualized as one that exists in a long cylindrical body loaded uniformly along the axis. With a plane-strain state, we must have

$$\frac{\partial u}{\partial z} = \frac{\partial v}{\partial z} = w = 0. \qquad (12.3\text{–}2)$$

12.4 TORSION OF A CIRCULAR CYLINDRICAL SHAFT

We shall now illustrate an application of the linearized elasticity theory by considering the problem of torsion. To transmit a torque from one place to another, a shaft is employed. The problem is to solve the Navier's equations to obtain the stress distribution in the shaft. The degree of difficulty to solve this problem depends on the geometry of the shaft. If the shaft is a circular cylinder, the solution is simple. If it is a cylinder of a noncircular cross section, or if the shaft has variable cross sections, then it is difficult.

Let us consider the simple problem of the torsion of a cylindrical shaft of circular cross section. (See Fig. 12.1, which shows the notations and the coordinate axes to be used.) Before tackling the problem analytically, let us look at the physical conditions. Under the torque, the shaft twists. Let the cross section at $z = 0$ be fixed. Since the shaft, as well as the loading, is homogeneous along the z-axis, the twist must be uniform along the z-axis. Hence, the deformation must be expressible in terms of twist per unit length α, which is a constant independent of z. The quantity α represents the rotation of a section at $z = 1$ relative to that at $z = 0$.

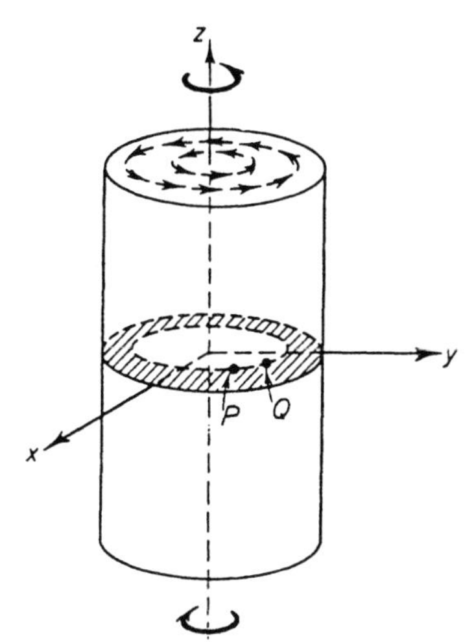

Figure 12.1 Torsion of a circular shaft.

By reason of symmetry, it is obvious that a circular cross section of the shaft remains circular when a torque is applied. But what about the axial displacements of such a section? Consider a plane cross section such as that at $z = 0$ before the torque is applied. When the torque T is applied, the boundary conditions are axisymmetric, hence any axial displacement must be axisymmetric. But the boundary conditions are also preserved by reversal of the z-axis, hence the axial displacement must be zero, and the plane section remains plane.

Summarizing this discussion, we see that the distortion of a circular shaft under a torque must be a relative rotation of the cross sections at a uniform rate of twist. Therefore, the displacement of a particle located at (x, y, z) would appear to be, in polar coordinates,

$$u_r = 0, \qquad u_\theta = \alpha z r, \qquad u_z = 0, \tag{12.4–1}$$

or, in rectangular Cartesian coordinates,

$$u_x = -\alpha zy, \qquad u_y = \alpha zx, \qquad u_z = 0, \tag{12.4–2}$$

as is shown in Fig. 12.2.

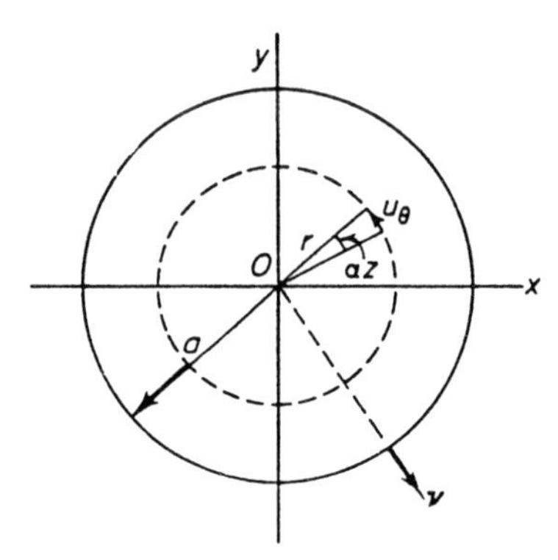

Figure 12.2 Notations.

We shall now show that this is indeed correct. Since the displacements are given, there is no need to check the compatibility conditions. We must, however, check the equation of equilibrium and the boundary conditions.

From Eq. (12.4–2), we have the strain components

$$\begin{aligned} e_{xx} &= 0, \qquad e_{yy} = 0, \qquad e_{zz} = 0, \\ e_{xy} &= \frac{1}{2}\left(\frac{\partial u_x}{\partial y} + \frac{\partial u_y}{\partial x}\right) = \frac{1}{2}(\alpha z - \alpha z) = 0, \\ e_{xz} &= \frac{1}{2}\left(\frac{\partial u_x}{\partial z} + \frac{\partial u_z}{\partial x}\right) = -\frac{1}{2}\alpha y, \\ e_{yz} &= \frac{1}{2}\left(\frac{\partial u_y}{\partial z} + \frac{\partial u_z}{\partial y}\right) = \frac{1}{2}\alpha x. \end{aligned} \tag{12.4–3}$$

The stress-strain relation yields the corresponding stress components. We have

$$\begin{aligned} \sigma_{xx} &= \sigma_{yy} = \sigma_{zz} = \sigma_{xy} = 0, \\ \sigma_{xz} &= -G\alpha y, \\ \sigma_{yz} &= G\alpha x, \end{aligned} \tag{12.4–4}$$

where G is the shear modulus of the shaft material.

The equations of equilibrium are obtained by omitting α_i and X_i in Eq. (12.1–5). We obtain

$$\begin{aligned} \frac{\partial \sigma_{xx}}{\partial x} + \frac{\partial \sigma_{xy}}{\partial y} + \frac{\partial \sigma_{xz}}{\partial z} &= 0, \\ \frac{\partial \sigma_{yx}}{\partial x} + \frac{\partial \sigma_{yy}}{\partial y} + \frac{\partial \sigma_{yz}}{\partial z} &= 0, \end{aligned} \tag{12.4–5}$$

$$\frac{\partial \sigma_{zx}}{\partial x} + \frac{\partial \sigma_{zy}}{\partial y} + \frac{\partial \sigma_{zz}}{\partial z} = 0,$$

which are obviously satisfied by the stress components given in Eq. (12.4–4).

The boundary conditions of our problem consists of the facts that the lateral surfaces are stress free and that the ends are acted on by a torque. Since there is no tension or compression on the ends, we have

$$\sigma_{zz} = 0 \qquad (\text{on } z = -L \text{ and } z = L). \tag{12.4–6}$$

This is satisfied by Eq. (12.4–4).

The stress vector acting on the lateral surface is given by $\overset{\nu}{T}_i$, where $\boldsymbol{\nu}$ denotes the vector normal to the lateral surface. By Cauchy's formula,

$$\overset{\nu}{T}_i = \nu_j \sigma_{ij}. \tag{12.4–7}$$

Setting $i = 1, 2, 3$, we have the three equations,

$$\begin{aligned} \sigma_{xx}\nu_x + \sigma_{xy}\nu_y + \sigma_{xz}\nu_z &= 0, \\ \sigma_{yx}\nu_x + \sigma_{yy}\nu_y + \sigma_{yz}\nu_z &= 0, \\ \sigma_{zx}\nu_x + \sigma_{zy}\nu_y + \sigma_{zz}\nu_z &= 0, \end{aligned} \tag{12.4–8}$$

where ν_x, ν_y, ν_z are the direction cosines of the normal vector to the lateral surface. Now, on the lateral surface, it is evident from Fig. 12.2 that the normal vector $\boldsymbol{\nu}$ coincides with the radius vector. Hence, the components of $\boldsymbol{\nu}$ are

$$\nu_x = \frac{x}{a}, \qquad \nu_y = \frac{y}{a}, \qquad \nu_z = 0. \tag{12.4–9}$$

Consequently, the boundary conditions on the circumference C are

$$\begin{aligned} x\sigma_{xx} + y\sigma_{xy} &= 0, \\ x\sigma_{yx} + y\sigma_{yy} &= 0, \\ x\sigma_{zx} + y\sigma_{zy} &= 0, \end{aligned} \tag{12.4–10}$$

which are again satisfied by Eq. (12.4–4).

It remains to check the condition that the stresses acting on the ends $z = -L$ and $z = L$ are equipollent to a torque. Referring to Fig. 12.3 and using Eq.

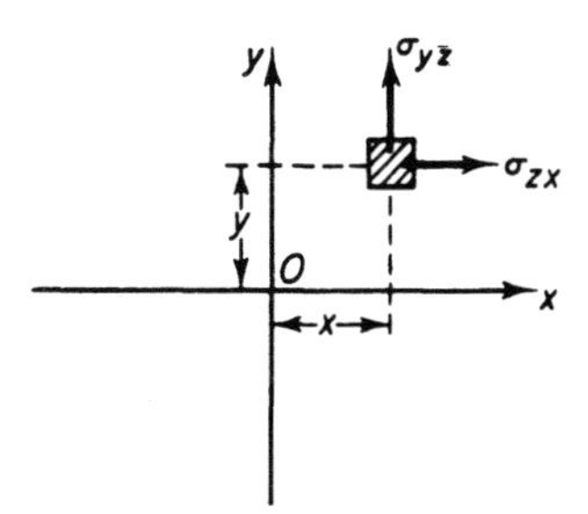

Figure 12.3 Stresses in a twisted shaft.

(12.4–4), we see that the resultant of the stresses acting on the end cross sections are

$$\iint \sigma_{xz}\, dx\, dy = -G\alpha \iint y\, dx\, dy = -G\alpha \int_{-a}^{a} dx \int_{-\sqrt{a^2-x^2}}^{\sqrt{a^2-x^2}} y\, dy = 0,$$

$$\iint \sigma_{yz}\, dx\, dy = G\alpha \iint x\, dx\, dy = 0, \tag{12.4–11}$$

$$\iint \sigma_{zz}\, dx\, dy = 0.$$

Hence, the resultant force vanishes as desired. The resultant moment about the z-axis is, however,

$$\iint (x\sigma_{yz} - y\sigma_{xz}) dx\, dy. \tag{12.4–12}$$

On substituting from Eq. (12.4–4), we have

$$\begin{aligned} \text{moment} &= G\alpha \iint (x^2 + y^2)\, dx\, dy \\ &= G\alpha \int_0^{2\pi} d\theta \int_0^a r^3\, dr \\ &= \frac{2\pi G\alpha a^4}{4}. \end{aligned}$$

Thus, we see that the resultant moment is indeed a torque of magnitude T:

$$T = \frac{\pi a^4 G\alpha}{2} \tag{12.4–13}$$

The checking is now complete. All the equations of equilibrium and the boundary conditions are satisfied. The solution contained in Eqs. (12.4–1) through (12.4–4) is exact.

PROBLEM

12.1 Consider the torsion of a shaft of square cross section. Write down all the boundary conditions. Show that the solution contained in Eqs. (12.4–1) through (12.4–4) no longer satisfy all the boundary conditions.

12.5 BEAMS

When a structural member is used to transmit bending moment and transverse shear, it is called a *beam*. Beams are used constantly in engineering and, therefore, are important objects for study. The floor we stand on is resting on beams. An airplane wing is a beam. Bridges are made of beams, and so on. An engineer should know the stress and deformation in a beam, how to choose the materials for a beam, how to use the material efficiently by a proper geometric design, how to minimize the beam's weight, how to maximize the stiffness and stability of the beam, how to utilize supports to minimize vibrations, how to calculate the loads that act on the beam (static and moving loads, wind loads on a building, aerodynamic load on an airplane, etc.), how to analyze aeroelastic or hydroelastic interactions in case a beam is used in a fluid flow (such as with an airplane wing, or the structure of a ship), and more.

Beams are classified according to the condition of support at their ends. An end is called *simply supported* when it is free to rotate, but is restrained from lateral translation. An end is said to be *free* when it is free to rotate and deflect. An end is said to be *clamped* when translation and rotation are both prevented.

In Sec. 1.11, we considered the pure bending of a prismatic beam of a homogeneous, isotropic Hookean material. We deduced certain results, but we did not check all the field and boundary conditions. We shall now show that all these conditions are satisfied.

Consider the pure bending of a prismatic beam, as shown in Fig. 1.14 on p. 26. Let the beam be subjected to two equal and opposite couples M acting in a plane of symmetry of the cross sections of the beam. Let the x, y, z-axes of reference be chosen as in Sec. 7.7, with the origin located at the centroid of a cross section. In Sec. 7.7, we were led to the conclusion that the stress distribution in the beam is

$$\sigma_{xx} = \frac{Ey}{R}, \qquad \sigma_{yy} = \sigma_{zz} = \sigma_{xy} = \sigma_{yz} = \sigma_{zx} = 0, \tag{12.5–1}$$

$$\frac{M}{EI} = \frac{1}{R}, \qquad \sigma_{xx} = \sigma_o\frac{y}{c}, \qquad \sigma_o = \frac{Mc}{I}, \tag{12.5–2}$$

where c is the distance from the neutral surface to the "outer fiber" of the cross section, M is the bending moment, E is Young's modulus, I is the areal moment of inertia of the cross section, and σ_o is the outer fiber stress. The strains are, therefore,

$$e_{xx} = \frac{y}{R}, \qquad e_{yy} = -\nu\frac{y}{R} = e_{zz}, \qquad e_{xy} = e_{yz} = e_{zx} = 0. \tag{12.5–3}$$

From these, we see that the equations of equilibrium, Eqs. (12.4–5), are satisfied. The equations of compatibility, Eqs. (6.3–4), are also satisfied. The boundary conditions on the lateral surface of the beam are $\overset{\nu}{T}_i = 0$. Since any normal to the lateral surface is perpendicular to the longitudinal axis x, the direction cosine ν_x vanishes; i.e., $\nu_x = 0$ on the lateral surface. Thus, the following boundary conditions are satisfied:

$$\begin{aligned}
\overset{\nu}{T}_x &= 0 = \sigma_{xx}\nu_x + \sigma_{xy}\nu_y + \sigma_{xz}\nu_z, \\
\overset{\nu}{T}_y &= 0 = \sigma_{yx}\nu_x + \sigma_{yy}\nu_y + \sigma_{yz}\nu_z, \\
\overset{\nu}{T}_z &= 0 = \sigma_{zx}\nu_x + \sigma_{yz}\nu_z + \sigma_{zz}\nu_z.
\end{aligned} \tag{12.5–4}$$

The boundary conditions at the ends of the beam are that the stress system must correspond to a pure bending moment, and without a resultant force. The stress system given by Eq. (12.5–1) does that, as discussed in Sec. 1.11. Hence, the solution is exact if the boundary stresses on the ends of the beam are distributed precisely in the manner specified by Eq. (12.5–1), because then all the differential equations and boundary conditions are satisfied.

One of the restrictions imposed in the derivation given in Sec. 1.11, namely, that the cross section of the beam has a plane of symmetry, can be removed. Let us consider, then, a prismatic beam with an arbitrary cross section, such as the one shown in Fig. 12.4. Assume the same stress and strain distribution as in Eqs. (12.5–1), (12.5–2), and (12.5–3). Suppose the boundary conditions, Eqs. (12.5–4), are also satisfied. The resultant axial force is zero again when the origin is taken at the centroid of the cross section. The resultant moment about the z-axis by the traction acting on the end section is given by the surface integral over the cross-section A:

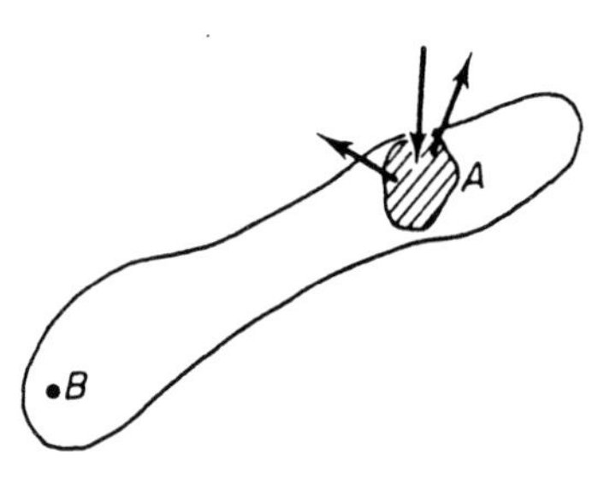

Figure 12.4 An unsymmetric cross section.

$$M_z = \int_A \sigma_{xx} y \, dA = \frac{E}{R}\int_A y^2 \, dA = \frac{EI_z}{R}.$$

It is the same as before, except that we have added the subscript z to show that the bending moment and the areal moment of inertia of the cross section are both taken about the z-axis. The resultant moment about the y-axis, however, is a new element to consider. It is given by an integration of the traction $\sigma_{xx} dA$ acting on an element of area dA situated at a distance z from the y-axis. Hence,

$$M_y = \int_A \sigma_{xx} z \, dA. \tag{12.5–5}$$

On substituting Eq. (12.5–1) into this equation, we obtain

$$M_y = \frac{E}{R}\int_A yz \, dA. \tag{12.5–6}$$

The integral is the negative of the *product of inertia of the cross-sectional area*:

$$P_{yz} = -\int_A yz \, dA \equiv -\iint_A yz \, dy \, dz. \tag{12.5–7}$$

Hence,

$$M_y = \frac{-EP_{yz}}{R}. \qquad (12.5\text{–}8)$$

In case the beam cross section has a plane of symmetry in which the bending moment acts, we choose the xy-plane as that plane of symmetry; then $P_{yz} = 0$. It follows that $M_y = 0$, which shows that our solution in Sec. 1.11 is satisfactory. In the general case, we now choose the coordinate axes in such a way that the product of inertia vanishes. Then

$$P_{yz} = 0, \qquad M_y = 0, \qquad (12.5\text{–}9)$$

and the moment vector is parallel to the z-axis with a magnitude equal to M_z.

The product of inertia vanishes if the y- and z-axes are the principal axes of inertia. Hence, in order that a moment acting in a plane produce bending in the same plane, it is necessary that the plane be a principal plane, i.e., one containing a principal axis of inertia of every cross section. Combining the requirements given by Eqs. (1.11–27) and (12.5–9), we see that *the coordinate axes y, z must be chosen as centroidal axes in the direction of the principal axes of the areal moments of inertia.*

Our verification is now complete. We have found that the stress system given by Eq. (12.5–1) is exact if y is measured from the neutral axis in the direction of a principal axis. The stress system satisfies the equations of equilibrium, the equations of compatibility, and the boundary conditions. Our intuitive assumption that plane sections remain plane is verified in this case.

More refined theories of bending can be found in many books, e.g., Sokolnikoff's *Mathematical Theory of Elasticity*.

12.6 BIOMECHANICS

Continuum mechanics can be applied to biology. Most biological materials can be considered as continua at suitable scales of observation. We have discussed the constitutive equations of a few biological tissues in Chap. 9. Most biological fluids and solids have nonlinear constitutive equations. The bone seems to be an exception, which functions in the small strain range and obeys Hooke's law. But the shape and internal structure of bones are very complex.

In important biological problems, fluid mechanics and solid mechanics are usually coupled together. For example, blood flows in blood vessels which have elastic walls. The heart pumps blood with a muscle. Hence the equations used in Chaps. 11 and 12 are usually coupled together in biomechanics.

Living tissues have a unique feature which is unmatched by nonliving materials. This is the feature of tissue remodeling under stress. By remodeling, the zero-stress state of the material changes, the constitutive equation changes, the mechanics changes. The following chapter is devoted to consider this new aspect of continuum mechanics.

PROBLEMS

12.2 An elementary theory of a circular cylindrical shaft subjected to torsion is given in Sec. 12.4. Let **z** be the axis of the shaft. Let the ends be $z = 0$ and $z = L$. A rectangular Cartesian frame of reference x, y, z is used. The displacement components in the x-, y-, and z-directions are u, v, and w, respectively. The elementary theory gives

$$u = -\alpha zy, \qquad v = \alpha zx$$

where α is the rate of twist angle per unit length of the shaft. The elementary theory does not say anything about the third component of displacement in the axial direction, w, which in general will not vanish if the shaft is non-circular. Let this unknown displacement be

$$w = \alpha\phi(x, y).$$

Using the equation of equilibriun (Navier's equation), find the equation satisfied by the function $\phi(x, y)$. This function is known as the *warping function*.

12.3 The following situation reminds a composite-material designer to pay attention to the question of the stability of a structure in operational condition. Consider a cantilever beam with a rectangular cross section. The beam is made of two strong rods embedded in a matrix. It is loaded by a force P parallel to the line joining the two rods in the cross section, as shown in Fig. P12.3. In practical application under a load, there is a probability that the beam will be twisted to failure. Twist will occur when the load P exceeds a critical value. Formulate a theory that will determine the critical value of the load P that will cause a torsional instability. According to your theory, how should such a composite beam be designed?

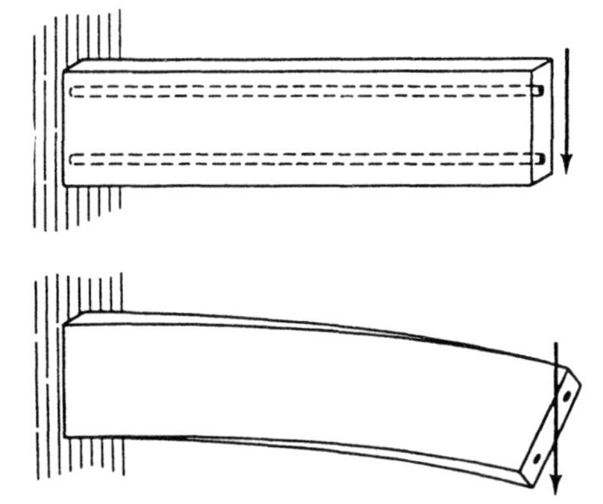

Figure P12.3 A beam of very narrow width may twist under load. The twist could be a critical problem for a beam reinforced with high-strength rods.

12.4 Consider a string of uniform density and material, stretched tightly between two posts (e.g., a violin string or a piano wire). The string is struck at a point. Vibration ensues. Formulate the problem mathematically. Give both the differential equations and the boundary conditions.

12.5 Consider a gong used in an orchestra. Formulate the mathematical problem of gong vibration.

12.6 Formulate a mathematical description of the clouds floating in the sky. How do they move about? Include enough parameters so that the great variety of things you see daily can be described and deduced.

12.7 An airplane flies in the air at a forward speed V relative to the ground. How does the wing maintain this flight? To answer this question, write down the field equations for the air and the wing, and the boundary conditions at the interface between the air and the wing. Present a full set of equations that would be sufficient to furnish a mathematical theory in principle.

12.8 The elastic waves in the rails as a train approaches are typical of waves in many dynamics problems. We can easily hear the impact of the wheels of the train (if we put our ears to the rail) long before the train can be seen. Then, as the train comes by, we can see the deflection of the rails under the wheels. Formulate the problem mathematically so that both of these features can be exhibited.

12.9 Feel the pulse on your wrist. It is a composite elastic wave in your artery. The most important component is undoubtedly the elastic response of the artery to the pressure wave in the blood. To a lesser extent, there must be other waves that are propagated along the arterial wall and caused by disturbances further upstream or downstream. Our arteries are elastic. Formulate a mathematical theory of pulse propagation. Leonhard Euler (1707–1783) formulated the problem and presented an analysis as early as 1775.

12.10 Galileo (1564–1642) proposed the following method for measuring the frequency of vibration of a gong. Attach a small, sharp, pointed knife to a slender rod. Pull the rod over the gong at a constant speed. The vibration of the gong will cause the rod to chatter. Examine the metal surface of the gong; and measure the spacing of the marks, from which the frequency may be calculated.

Explain whether this method will work. How would you compute the frequency? Formulate the problem mathematically from the point of view of the theory of elasticity. Assume a good musical gong to assure that the material is a linear elastic solid that obeys Hooke's law.

12.11 The phenomenon of chatter in machine tools is not unlike that in Galileo's gong experiment. Consider the problem of a high-speed lathe. Formulate the problem of chatter, which ruins a good machine's operation. Propose ways to alleviate the problem.

12.12 A beam vibrates. Write down the differential equation and boundary conditions for a vibrating beam and a method of determining the frequency of vibration of the beam.

12.13 A circular cylindrical shaft spins about its longitudinal axis at an angular speed ω radians/sec. The shaft is simply supported at both ends. Lateral vibrations are always possible when the shaft spins. However, if the rate of spin reaches a critical value, the lateral deformation becomes excessive, and so-called whirling sets in. Describe the phenomenon mathematically. Formulate the equations with which the critical whirling speed can be determined.

12.14 The shaft of an airplane propeller is subjected to both a tension and a torque. How would you propose to measure the stresses in the shaft in flight? How would you measure the power delivered to the propeller in flight?

FURTHER READING

FUNG, Y. C., *Foundations of Solid Mechanics*, Englewood Cliffs, N.J.: Prentice-Hall (1965).

FUNG, Y. C., *Biomechanics: Motion, Flow, Stress, and Growth*, New York: Springer-Verlag (1990).

LOVE, A. E. H., *The Mathematical Theory of Elasticity*, 1st ed. 1892, 4th ed. 1927, New York: Dover Publications (1944).

SOKOLNIKOFF, I. S., *Mathematical Theory of Elasticity*, 2nd ed., New York: McGraw-Hill (1956).

TIMOSHENKO, S., AND N. GOODIER, *Theory of Elasticity*, New York: McGraw-Hill, 1st ed. 1934, 2nd ed. 1951.

13 STRESS, STRAIN, AND ACTIVE REMODELING OF STRUCTURES

We use biological examples to bring out some fundamental issues of continuum mechanics: the zero-stress state, the changes in the zero-stress state and the constitutive equation due to remodeling of a material, the effect of stress and strain on remodeling, and the feedback dynamics of growth and resorption. Nonliving physical systems have these features also.

13.1 INTRODUCTION

In this last chapter, we discuss the mechanics of changes in materials. From the point of view of mechanics, there are three aspects of a solid body in which change plays a fundamental role: the zero-stress state, the constitutive equation, and the overall geometry of the body. Our discussion will focus on these aspects.

The mechanics of flow and deformation is called *rheology*. The literature on rheology is usually concerned only with flow and change in a given material or a given set of materials. The science in which growth and change in materials is a central concern is biology. In continuum mechanics rheology and biology are united. To illustrate the material-change aspects of continuum mechanics, examples can be picked from biology, because they are ubiquitous. In the discussion that follows, we often use the blood vessel as an example.

13.2 How to Discover the Zero-Stress State of Materials in a Solid Body

In the preceding chapters, it is assumed that when there is no external load acting on a body, the stress in the body is zero everywhere. We know, however, that this does not have to be the case; for example, we can sit, but tense up our muscles and create a lot of forces in our muscles and bones. Generally, the stress in a body when there is no external force is called *residual stress*. The effect of residual stress and strain can be dramatic, e.g., the relaxation of residual stress in the earth can cause an earthquake, and an unwanted thermal strain in a nuclear reactor might cause a meltdown.

The simplest way to discover residual stress in a solid body is to cut the body up. Cutting is introducing new surfaces on which the traction is zero. Cutting an unloaded body without residual stress will cause no strain. If strain changes by cutting, then there is residual stress.

Take a blood vessel as an example. If we cut an aorta twice by cross sections perpendicular to the longitudinal axis of the vessel, we obtain a ring. If we cut the ring radially, it will open up into a sector (see Fig. 13.1). By using equations of static equilibrium, we know that the stress resultants and stress moments are zero in the open sector. Whatever stress remains in the vessel wall must be locally in equilibrium. If we cut the open sector further and can show that no additional strain results, then we say that the sector is in the zero-stress state. For the rat artery Fung and Liu (1989) reported that this is the case.

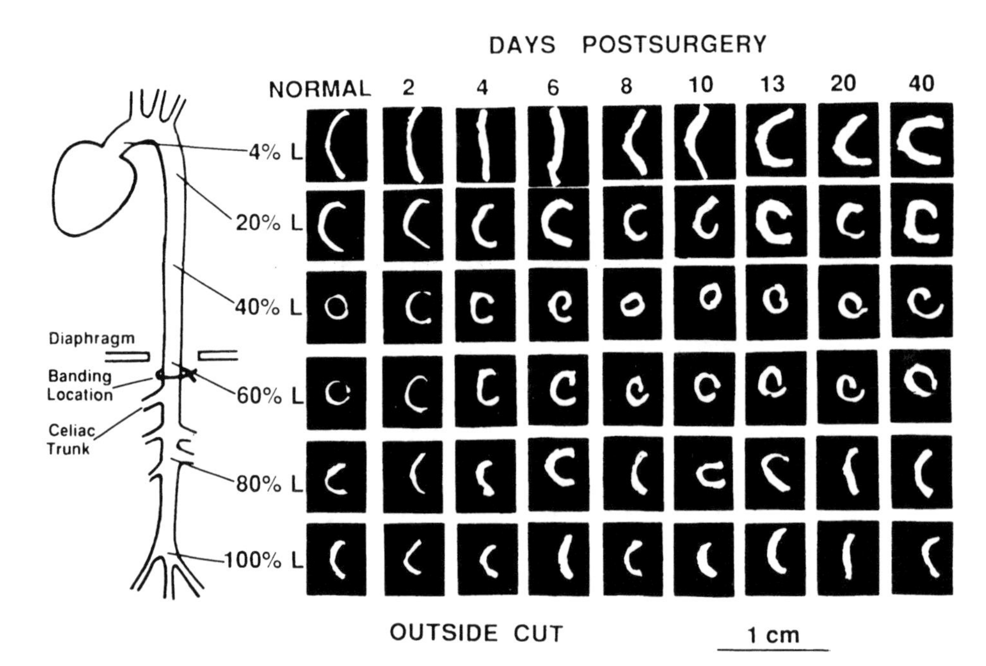

Figure 13.1 Photographs of the cross sections of a rat aorta at zero-stress state. The first column shows the zero-stress state of the aorta of normal rat. The rest shows the change of zero-stress state due to vessel remodeling after a sudden onset of hypertension. The photos are arranged from left to right according to days after surgery, and from top to bottom according to location on the aorta, expressed as distance from the heart in percentage of the total length of the aorta from the aortic valve to iliac bifurcation. The location of the metal clip used to induce hypertension is shown in the sketch at left. The arcs of the blood vessel wall do not appear smooth because of some tissue attached to the wall. In reading these photographs, one should mentally delete these tethered tissues. From Fung and Liu (1989).

Thus assured that the open sectors represent the zero-stress state of a blood vessel, we characterize each sector with an *opening angle*, which is defined as the angle subtended by two radii drawn from the mid-point of the inner wall to the tips of the inner wall of the open sector (see Fig. 13.2). A more complete picture of the zero-stress state of a normal young rat aorta is shown in the first column of Fig. 13.1 (Fung and Liu, 1989). The entire aorta was cut successively into many segments approximately one diameter long. Each segment was then cut radially. It was found that the opening angle varied along the rat aorta: It was about 160° in the ascending aorta, 90° in the arch, 60° in the thoracic region, 5° at the diaphragm level, and 80° toward the iliac bifurcation point.

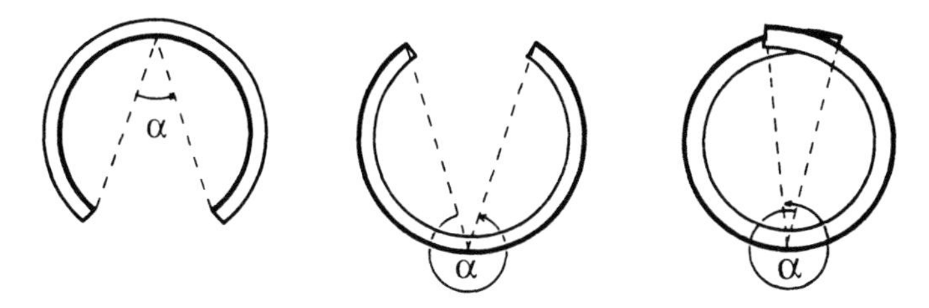

Figure 13.2 Definition of the opening angle. Sector represents a circumferential cross section of a blood vessel at zero-stress state. Angle subtended between two lines originating from the midpoint to the tips of the inner wall is the opening angle.

Following the common iliac artery down a leg of the rat, we found that the opening angle was around 100° in the iliac artery, dropped down in the popliteal artery region to 50°, and then rose again to about 100° in the tibial artery. In the medial plantar artery of the rat, the microarterial vessel of 50 mm diameter had an opening angle on the order of 100° (Liu and Fung, 1989).

There are similar, although not identical, spatial variations of opening angles in the aortas of the pig and dog (Han and Fung, 1991a). Also, there are significant opening angles in the pulmonary arteries (Fung and Liu, 1991), systemic and pulmonary veins (Xie et al., 1991), and trachea (Han and Fung, 1991b) of rats. Thus, we conclude that the zero-stress state of blood vessels and the trachea is shaped as sectors whose opening angles vary with their location on the vessel or trachea and with animal species. In other words, the zero-stress state in a body may vary from place to place. It then follows that the residual stress also varies spatially.

In industrial engineering, residual stresses are usually introduced into a solid body in the manufacturing process. Welding or riveting of metal parts under strain is a common cause of residual stress in airplanes, bridges, and machinery. Plastic deformation or creep in the metal-forming and machining process causes residual stress. Forced fitting is also a common cause. Straining steel rods is the way pre-stressed reinforced concrete beams are made. Heating an outer cylindrical shell to a higher temperature, fitting it to an inner shell, and then cooling the combination down to room temperature is the way gun barrels are made. The purpose is to induce a compressive residual stress in the inner wall and a tensile stress in the outer wall, so that when the gun is fired, the stress concentration at the inner wall of the barrel can be reduced. Shot peening to introduce a compressive residual stress on the outer surface of a metallic body is a way to increase the fatigue life of the body. Techniques using an ion or molecular beam to impregnate matter into the surface of a metallic or ceramic body can similarly introduce a compressive residual stress into a thin layer of the surface of the body to promote a longer service life. Most articles of industrial engineering have residual stress in them.

In living tissues, growth and change are natural. Every cellular or extracellular growth or resorption changes the zero-stress state of the tissue and introduces residual stress. In biological studies, it is easier to measure changes in the zero-stress state than to measure cellular activities in the tissue; hence, observed changes in residual strain are often used as a quantitative tool to study such activities.

Out of the great variety of examples in nature and industry, we shall use a few biological cases to illustrate the long-term effect of the stress in a body on the materials of the body. What in the short term is described by the stress-strain relationship becomes, in the long term, features associated with aging, remodeling, wear, tear, growth, and resorption. Biologists use the term *homeostasis* to describe the condition of normal life. They describe the state of a living organism at normal living conditions by a set of *homeostatic set points*. In a homeostatic condition, there is a certain range of stress in the body of the living organism. When the environment changes, the range of stress changes, the cells in the body respond by modifying themselves, and the tissue is remodeled. In effect, the zero-stress state of the body changes. In time, the mechanical properties of the tissue are also remodeled. We are familiar with these features in our own bodies. We know that homeostasis is not static, but a certain normal mode exists in a dynamic environment. The quantitative aspects will become clearer as we proceed in the sections that follow.

In machines and in nonliving physical objects, analogous features of homeostasis and remodeling may exist. These features are worthy of scientific study.

13.3 REMODELING THE ZERO-STRESS STATE OF A STRUCTURE: A BIOLOGICAL EXAMPLE OF ACTIVE REMODELING DUE TO CHANGE IN STRESS

In one study, hypertension was created in rats by constricting the abdominal aorta with a metal clip placed right above the cœliac trunk. (See Fig. 13.3.) The clip severely constricted the aorta locally and reduced the normal cross-sectional area of the lumen by 97% (Fung and Liu, 1989; Liu and Fung, 1989). This caused a 20% step increase in blood pressure in the upper body and a 55% step decrease in blood pressure in the lower body immediately following surgery. Later, the blood pressure increased gradually, following the course shown in Fig. 13.4. In the upper body, the blood pressure rose rapidly at first and then more gradually, tending to an asymptote at about 75% above normal. In the lower body, the blood pressure rose to normal in about four days and then gradually increased further to an

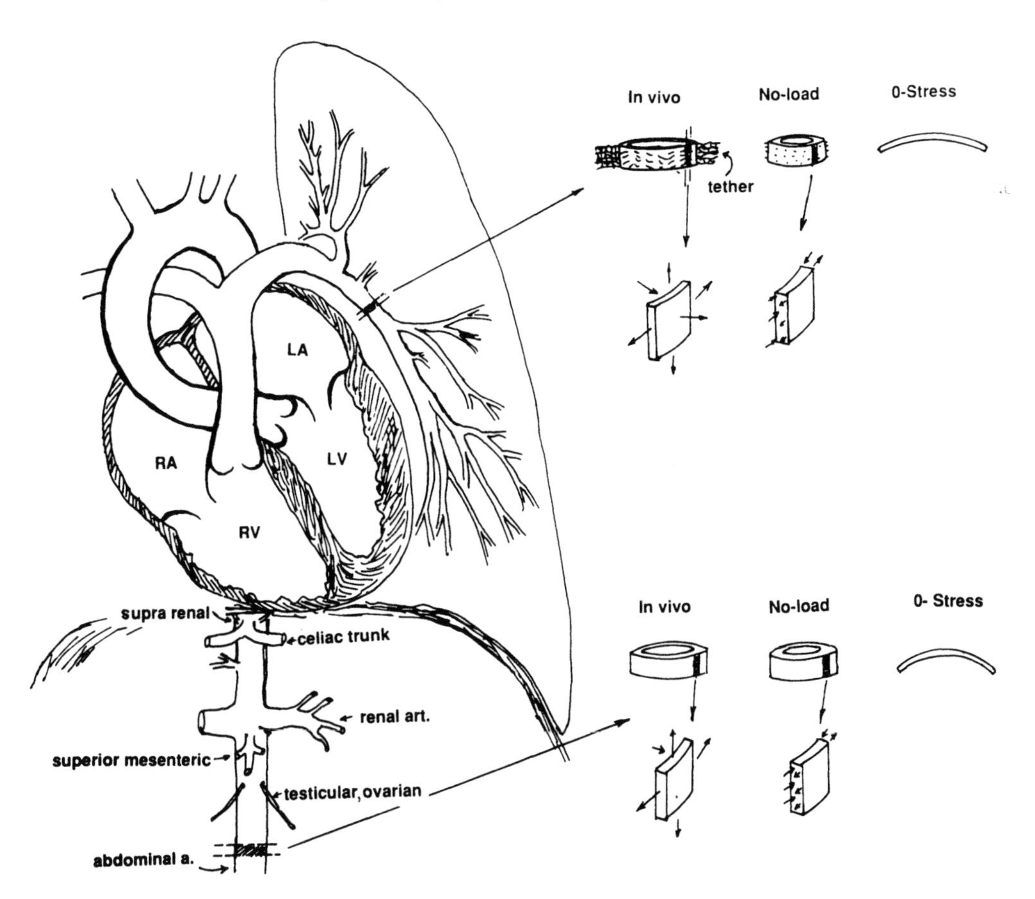

Figure 13.3 A sketch of the heart, aorta, and pulmonary arteries, the stresses in them, the zero-stress state, and the nomenclature of vessels mentioned in the text with regard to control of blood pressure by aortic constriction.

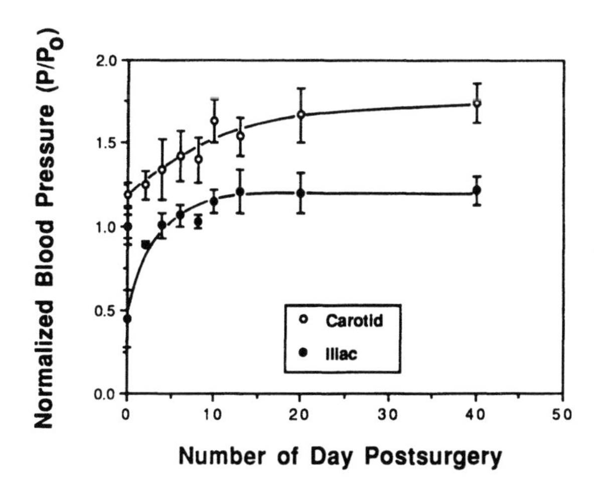

Figure 13.4 The course of change of blood pressure (normalized with respect to that before surgery) when a constriction is suddenly imposed on the aorta at a site below the diaphragm and above the celiac trunk shown in Fig. 13.3. From Fung and Liu (1989).

asymptotic value of 25% above normal. Parallel with these changes in blood pressure, the zero-stress state of the aorta changed as well. The changes are illustrated in Fig. 13.1, in which the location of any section on the aorta is indicated by the percentage distance of that section to the aortic valve measured along the aorta, divided by the total length of the aorta. Successive columns show the zero-stress configurations of the rat aorta at 0, 2, 4, . . . , 40 days after surgery. Successive rows refer to successive locations on the aorta.

The figure shows that following a sudden increase in blood pressure, the opening angles increased gradually, peaked in two to four days, and then decreased gradually to an asymptotic value. Variation with the location of the section on the aorta was great. The maximum change in the opening angle occurred in the ascending aorta, where the total swing of the opening angle was as large as 88°.

Thus, the blood vessel changed its opening angle in a few days following the change in blood pressure. Similar changes were found in the pulmonary arteries of rats after the onset of pulmonary hypertension by exposure to hypoxic gas containing 10% oxygen and 90% nitrogen at atmospheric pressure (Fung and Liu, 1991).

Thus, the zero-stress state of the blood vessel may be remodeled by an active biological process under the influence of changes in homeostatic stress.

13.4 CHANGE OF ZERO-STRESS STATE WITH TEMPERATURE: MATERIALS THAT "REMEMBER" THEIR SHAPES

The mechanical properties of a material may depend on many physical, chemical, and biological factors. We have illustrated a change in a material's zero-stress state due to a biological reaction to stress in the preceding section. Let us now consider a physical factor: temperature. It is well known that at any given state of stress, a change in temperature changes the strain, so that thermal stress may be regarded as caused by changing the zero-stress state through temperature variation.

There is, however, a more dramatic phenomenon in some materials. A hat made of a certain polymer can be folded for carrying and returned to good shape by heating. A medical device made of the same material has been used in Japan to close a patent ductus arteriosus in a young child. Ductus arteriosus is a vessel connecting a fetal heart to fetal lung, allowing blood to flow from the aorta to the pulmonary artery before birth. Normally it is closed immediately after birth. But sometimes it remains open and needs surgery. The device named above is shaped like a tiny umbrella, folded up, threaded to the duct by an endoarterial catheter, then opened up with a little squirt of fairly hot water from the catheter. The opened umbrella closes the ductus arteriosus.

Materials such as these, which appear to "remember" their shape, are materials whose zero-stress state changes with temperature. Alloys of copper-aluminum-nickel, copper-zinc-aluminum, iron-manganese-silicon, nickel-titanium, and polymers like polynorbornene have this property. For example, one nickel-titanium

(Ni-Ti) alloy, composed of equal number of atoms of nickel and titanium, and made into something at a higher temperature, can be deformed into some other shape at a lower temperature. If the deformed body is heated beyond a critical temperature, the Ni-Ti alloy will return to its original shape as manufactured and, if resisted, can generate a stress as high as 700 MPa (10^5 psi). This change is produced in the alloy by a change of crystalline phase known as a *martensitic transformation*. Martensite has a low yield stress, and can be deformed easily and reversibly by a crystalline process called *twining* of the atomic lattice. (See Fig. 13.5.) The martensitic transformation occurs over a range of temperature, above which the material is in the *austenitic phase*. When an austenite is cooled down, random twining occurs in the metal by random internal residual shear stress. Under an external

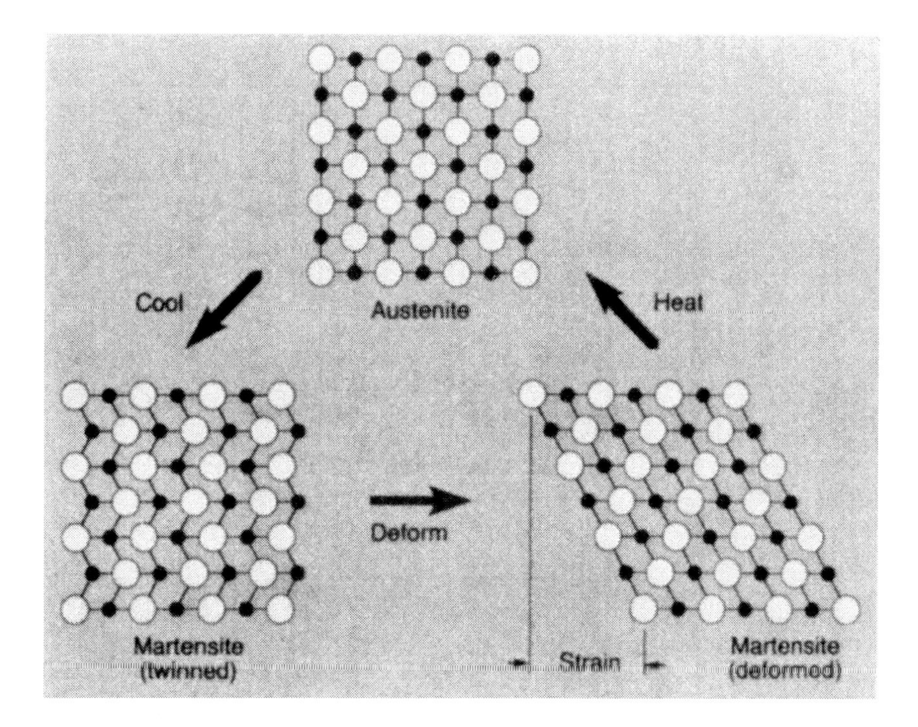

Figure 13.5 The mechanism of shape memory of a nickel-titanium alloy in the austenitic state at a higher temperature. The alloy is deformed at a lower temperature when the crystal structure is martensitic. The deformation in the martensite crystal is by *twining* which occurs under suitable shear stress and is reversible when the shear stress is reversed. If the temperature of the deformed martensite is raised to a level above a critical value, the crystalline structure of the alloy reverts to austenite and to the original shape of the body. From Tom Borden, "Shape-memory alloys: Forming a tight fit," *Mechanical Engineering*, Oct. 1991, p. 68. Reproduced by permission of the author and publisher.

shear load, a martensitic body can deform substantially and reversibly by twining. On heating the deformed martensite to a temperature at which the martensite crystal is transformed into austenite, the crystal reverts to its original shape, because austenite cannot accommodate the twining type of deformation.

The stress-strain curves for martensite and austenite are illustrated in Fig. 13.6. Deformations of martensite at strains greater than about 7% and austenite at strains greater than about 1% are plastic and irreversible. So for practical applications, one has to know the stress-strain curves, the ranges of elasticity and plasticity, the temperature at which austenite is first formed in martensite when heated, and the temperature at which martensite is first formed in austenite when cooled. With this knowledge, people have used Ni-Ti alloy for fastening machine parts, wiring teeth for orthodontic purposes, simulating the erection of an organ, and other phenomena.

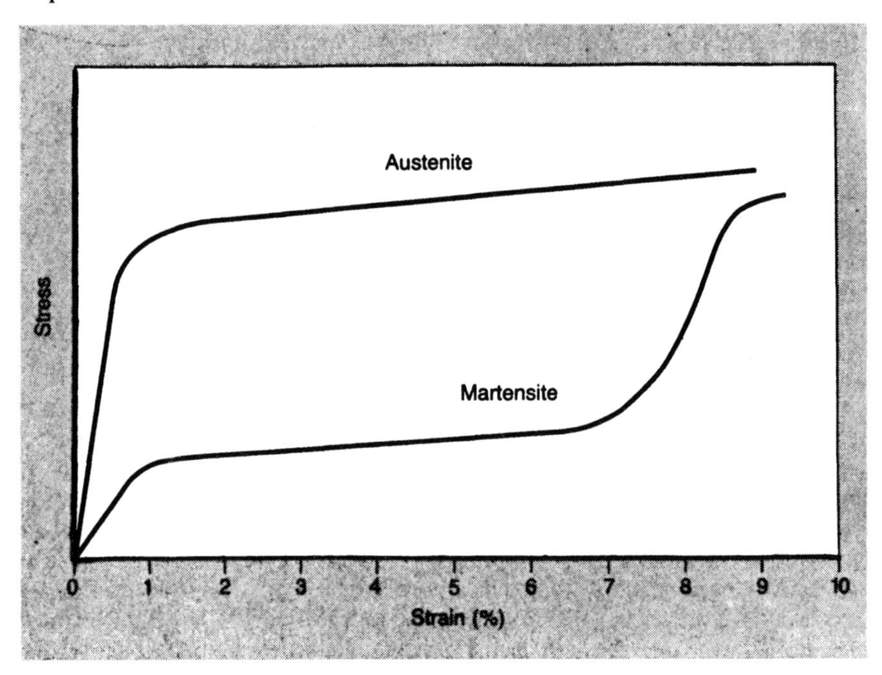

Figure 13.6 The stress-strain relationship of martensite and austenite crystals (tested at different temperature). From Tom Borden. *loc. cit.* By permission.

13.5 MORPHOLOGICAL AND STRUCTURAL REMODELING OF BLOOD VESSELS DUE TO A CHANGE IN BLOOD PRESSURE

The pressure of circulating blood varies from time to time and from place to place. What is normally referred to as the systemic blood pressure is the difference between the pressure of the blood in the aorta at the aortic valve and that in the right atrium. This is the pressure difference that drives the entire "systemic" circulation throughout the body (the peripheral circulation system). The corresponding driving pressure for the pulmonary circulation is the difference between the pressure in the pulmonary artery at the pulmonic valve and the pressure in the left atrium. Both the systemic and the pulmonary circulation are characterized by systolic (in period of contraction of the heart) and diastolic (in period of dilatation of the heart) pressures. When these pressures change, the blood pressure in every vessel of the body changes. When the blood pressure changes, the stress in the blood vessel wall changes.

As is sketched in Fig. 13.3, in the in vivo condition at normal blood pressure, the circumferential stress is usually tensile and is the largest stress component in the vessel wall. The longitudinal stress components exist because the vessel is normally stretched in the axial direction. The radial stress component is compressive at the inner wall, where it is equal to the blood pressure, and gradually decreases to the pressure acting on the outer wall.

The systemic blood pressure can be changed in a number of ways: by drugs, by a high-salt diet, by constricting the flow of blood to the kidneys, etc. If the aorta is constricted severely by a stenosis above the renal arteries (Fig. 13.4), the aorta above the stenosis will become hypertensive. The aorta below the stenosis will become hypotensive at first, but the reduced blood flow to the kidneys will cause the kidneys to secrete more of the enzyme renin into the bloodstream and will raise the blood pressure. If the stenosis is below the kidney arteries and is sufficiently severe, then the lower body will become hypotensive. The pulmonary blood pressure can also be changed in a number of ways. A most convenient way is to change the oxygen concentration of the gas breathed by the animal. If the oxygen concentration is reduced from normal (i.e., if it becomes hypoxic), the smooth muscle cells in the pulmonary blood vessels contract, the vessel diameters are reduced, and the pulmonary blood pressure goes up. This is the reaction human beings who live at sea level encounter when they go to a high altitude.

The hypoxic hypertensive reaction occurs quite fast. If a rat is put into a low-oxygen chamber containing 10% oxygen and 90% nitrogen at atmospheric pressure at sea level, the systolic blood pressure in its lung will shoot up from the normal 2.0 kPa (15 mm Hg) to 2.9 kPa (22 mm Hg) within minutes, become further elevated to 3.6 kPa in a week, and then gradually rise to 4.0 kPa in a month. (The rat's systemic blood pressure remains essentially unchanged in the meantime.) Under such a rise in blood pressure in the lung, the pulmonary blood vessel remodels itself.

Figure 13.7 shows how fast this remodeling proceeds. In the figure, the photographs in each row refer to that segment of the pulmonary artery indicated by the leader line. The first photograph in the top row shows a cross section of the arterial wall of the normal three-month-old rat. The specimen was fixed at the no-load condition. In the figure, the endothelium is facing upward, with the vessel lumen on top. The endothelium is very thin—on the order of a few micrometers. The scale of 100 μm is shown at the bottom of the figure. The dark lines are elastin layers. The upper, darker half of the vessel wall is the media, the lower, lighter half of the vessel wall the adventitia. The second photograph in the first row shows

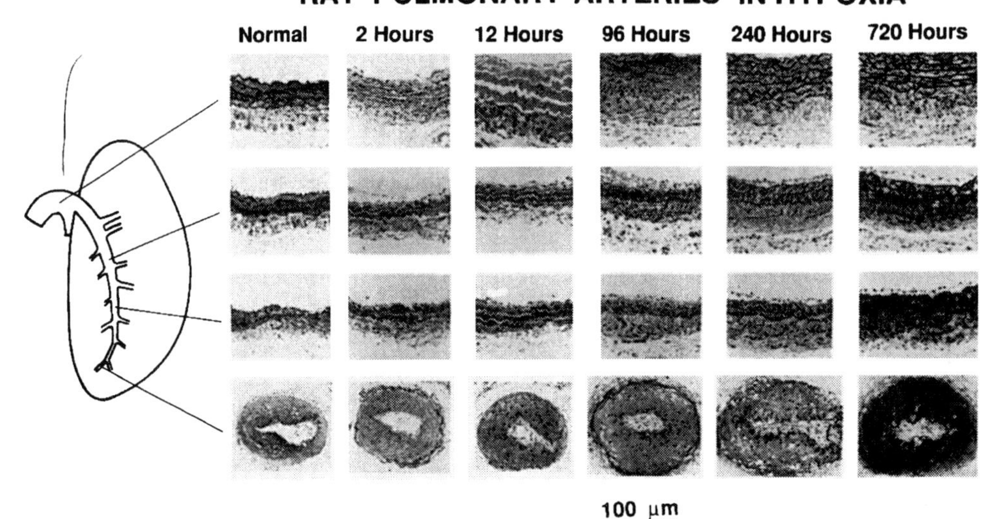

Figure 13.7 Photographs of histological slides from four regions of the main pulmonary artery of a normal rat and several hypertensive rats with different periods of hypoxia. Specimens were fixed at no-load condition. From Fung and Liu (1991).

a cross section of the main pulmonary artery two hours after exposure to lower oxygen pressure. There is evidence of small fluid vesicles and some accumulation of fluid in the endothelium and media. There is also a biochemical change of elastin staining on the vessel wall at this time. The third photograph shows the wall structure 10 hours later. Now the media is greatly thickened, while the adventitia has not changed very much. The fourth photograph shows that at 96 hours of exposure to hypoxia, the adventitia has thickened to about the same thickness as the media. The next two photos show the pulmonary arterial wall structure when the rat's lung is subjected to 10 and 30 days of lowered oxygen concentration. The major change in these later periods is the continued thickening of the adventitia.

The photographs in the second row show the progressive changes in the wall of a smaller pulmonary artery. The third and fourth rows are photographs of arteries of even smaller diameter. The inner diameter of the arteries in the fourth row is on the order of 100 μm, approaching the range of sizes of the arterioles. The remodeling of the vessel wall is evident in pulmonary arteries of all sizes. The maximum rate of change occurs in a day or two.

13.6 REMODELING OF MECHANICAL PROPERTIES

When the material in a blood vessel is changed during remodeling, its mechanical properties change. The mechanical properties of soft biological tissues can be described by the constitutive equations discussed in Secs. 9.5 and 9.7. Hence, we expect that the constitutive equation, or at least its coefficients, will change with tissue remodeling. This is indeed the case, as we shall illustrate with an example.

For the blood vessel, the pseudoelasticity formulation of the constitutive equation, described in Secs. 9.4 and 9.5, applies. We assume that a *pseudoelastic strain energy function* exists, denoted by the symbol $\rho_0 W$ and expressed as a function of the nine components of strain E_{ij} ($i = 1, 2, 3, j = 1, 2, 3$), that is symmetric with respect to E_{ij} and E_{ji}, so that the stress components can be derived by a differentiation, namely,

$$S_{ij} = \frac{\partial \rho_0 W}{\partial E_{ij}} \tag{13.6–1}$$

Here, ρ_0 is the density of the material at the zero-stress state, W is the strain energy per unit mass, $\rho_0 W$ is the strain energy per unit volume, and E_{ij} are strains measured with respect to the material configuration in the zero-stress state.

With regard to the determination of $\rho_0 W$, two approaches may be taken. One is to regard the blood vessel wall as an incompressible material and derive $\rho_0 W$ as a function of E_{ij} in *three dimensions* (Chuong and Fung, 1983). The other is to assume that the blood vessel is a cylindrical body with axisymmetry in mechanical properties and limit oneself to axisymmetric loading and deformation. Then one would be concerned only with two strain components: the circumferential strain E_{11} and the longitudinal strain E_{22}. The radial strain is easily computed from the condition of incompressibility. This technique may be called a *two-dimensional approach*.

For the analytical representation of $\rho_0 W$ for arteries in the two-dimensional approach, a polynomial form has been used by Patel and Vaishnav (1972), a logarithmic form by Hayashi et al. (1971), and an exponential form by Fung et al. (1973, 1979, 1981), see references at the end of Chap. 9. According to Fung et al. (1979),

$$\rho_0 W = C \exp(a_1 E_{11}^2 + a_2 E_{22}^2 + 2a_4 E_{11} E_{22}) \tag{13.6–2}$$

where C, a_1, a_2, and a_4 are material constants, E_{11} is the circumferential strain, and E_{22} is the longitudinal strain, the last two referred to the zero-stress state.

Experiments have been done on rat arteries during the course of development of diabetes after a single injection of streptozocin. The results with the vessel wall treated as one homogeneous material are presented in Table 13.1, from Liu and

TABLE 13.1 COEFFICIENTS C, a_1, a_2, AND a_4 OF THE STRESS-STRAIN RELATIONSHIP OF THE THORACIC AORTA OF 20-DAY DIABETIC AND NORMAL RATS. a_4 WAS FIXED AS THE MEAN VALUE FROM THE NORMAL RATS.*

Group	C (n/cm²)	a_1	a_2	a_4
Normal Rats				
Mean ± SD	12.21 ± 3.32	1.04 ± 0.35	2.69 ± 0.95	0.0036
20-day Diabetic Rats				
Mean ± SD	15.32 ± 9.22	1.53 ± 0.92	3.44 ± 1.07	0.0036

*From Liu, S. Q., and Fung, Y. C. (1992).

Fung (1992). Clearly, the material constants change with the development of diabetes.

13.7 STRESS ANALYSIS WITH THE ZERO-STRESS STATE TAKEN INTO ACCOUNT

If the zero-stress state of a solid body is known, if the strain is infinitesimal, and if the constitutive equation is linear, then the principle of superposition applies, and the mathematical problem of the stress analysis of a body with residual stress is simply a sum of two linear problems: finding the residual stress without an external load and finding the stress under an external load but without residual strain. In this category fall the important classical theories of dislocation and thermal stress.

Nonlinearity introduced by a finite strain or constitutive equation makes the analysis of bodies with residual stress a distinctive subject. The nonlinear analysis is often very difficult. But if we know the zero-stress state and how it is related to the present state, then the analysis of stress in the body could be quite simple.

For example, consider an ileal artery whose cross section in vivo at a blood pressure of 16 kPa (120 mm Hg) is shown in Fig. 13.8. The cross sections under the no-load and zero-stress conditions are also shown in Fig. 13.8. From these figures, we can measure the length of the circumference of the inner wall of the vessel. Let the lengths at the zero-stress state, the no-load state, and the homeostatic (normal, in vivo) state be $L^{i\theta}_{\text{0-stress}}$, $L^{i\theta}_{\text{no-load}}$, and $L^{i\theta}_{\text{hom}}$, respectively, with the superscripts i indicating "inner" and θ indicating "circumferential," and the subscripts "0-stress," "no-load," and "hom" indicating the states of zero stress, no load, and homeostasis, respectively. Similarly, we can measure the circumferential length at

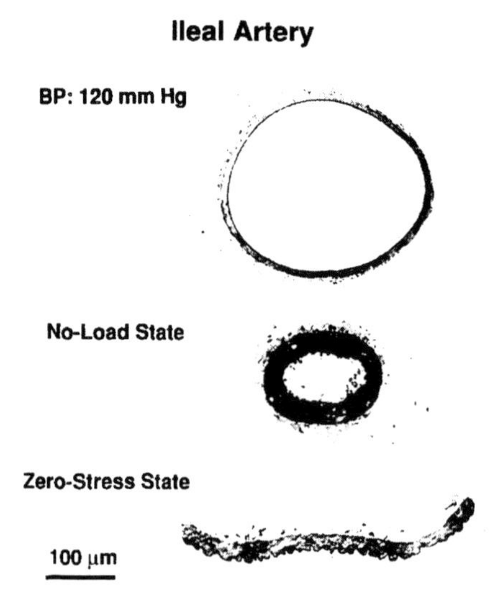

Figure 13.8 The shape of the cross section of an ileal artery of the rat at normal blood pressure (top), no load (middle), and zero stress (bottom).

the outer wall and obtain $L^{o\theta}_{0\text{-stress}}$, $L^{o\theta}_{\text{no-load}}$, and $L^{o\theta}_{\text{hom}}$, with the superscript o indicating "outer." From these, we obtain the stretch ratios

$$\lambda^{(i\theta)}_{\text{no-load}} = \frac{L^{(i\theta)}_{\text{no-load}}}{L^{(i\theta)}_{0\text{-stress}}}, \qquad \lambda^{(i\theta)}_{\text{hom}} = \frac{L^{(i\theta)}_{\text{hom}}}{L^{(i\theta)}_{0\text{-stress}}} \tag{13.7–1}$$

on the inner wall and

$$\lambda^{(o\theta)}_{\text{no-load}} = \frac{L^{(o\theta)}_{\text{no-load}}}{L^{(o\theta)}_{0\text{-stress}}}, \qquad \lambda^{(o\theta)}_{\text{hom}} = \frac{L^{(o\theta)}_{\text{hom}}}{L^{(o\theta)}_{0\text{-stress}}} \tag{13.7–2}$$

on the outer wall.

Typical raw data of the L's of an ileal artery, a medial plantar artery, and a pulmonary artery (branch 1) are given in Table 13.2. The computed stretch ratios are also listed in the table. These results may be compared with those obtained by a theoretical calculation of a hypothetical case in which the *no-load* and *homeostatic* configurations are identical with the real ones, but the residual strains are zero, so that the opening angle is zero and the zero-stress configuration is the same as the no-load configuration. In that case, the stretch ratios of the *no-load* case are unity, but those of the *homeostatic* vessel are

$$\lambda^{(i\theta)}_{\text{hom}} = \frac{L^{(i\theta)}_{\text{hom}}}{L^{(i\theta)}_{\text{no-load}}}, \qquad \lambda^{(o\theta)}_{\text{hom}} = \frac{L^{(o\theta)}_{\text{hom}}}{L^{(o\theta)}_{\text{no-load}}} \tag{13.7–3}$$

These are listed in the last two columns of Table 13.2.

The distribution of the circumferential residual stretch ratio in the vessel wall under the no-load condition is illustrated in the case of an ileal vessel (branch 1) in Fig. 13.9(a). It is seen that the residual stretch ratio is compressive in the inner wall and tensile in the outer wall. Under the conventional assumption that plane sections remain plane in bending, the stretch ratio distribution in the vessel wall is a straight line. In Fig. 13.9(b), the thick, nearly horizontal line shows the actual circumferential stretch ratio distribution in the blood vessel wall when the blood

TABLE 13.2 MEASURED CIRCUMFERENTIAL LENGTHS OF THE INNER AND OUTER WALLS OF RAT ILEAL ARTERY IN THE ZERO-STRESS STATE, IN THE NO-LOAD STATE, AND AT 80 AND 120 μm Hg; AND COMPARISON OF CIRCUMFERENTIAL STRETCH RATIOS COMPUTED ON TWO BASES: (A) RELATIVE TO THE ZERO-STRESS STATE, AND (B) RELATIVE TO THE NO-LOAD STATE.*

	Length, μm		Circumferential Stretch Ratio			
			Re zero-stress state		Re no-load state	
States	Inner Wall	Outer Wall	Inner Wall	Outer Wall	Inner Wall	Outer Wall
Zero Stress	743	963				
No load	590	1,091	0.79	1.13	1.0	1.0
80 mm Hg	1,017	1,281	1.37	1.33	1.72	1.17
120 mm Hg	1,023	1,286	1.38	1.34	1.73	1.18

*Data from Fung and Liu (1992).

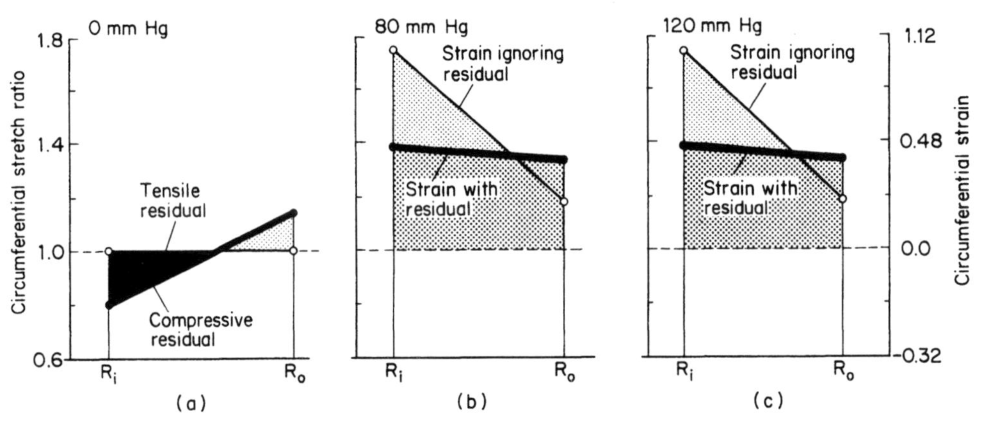

Figure 13.9 Circumferential stretch ratio distribution in an ileal artery (branch 1) whose dimensions are listed in Table 13.2. (a) Measured residual stretch ratio at no-load state. Residual strain can be read from nonlinear scale shown on right. R_i and R_o, inner and outer radius of vessel wall, respectively. Strain is compressive in inner wall region and tensile in outer wall region. (b) Thick, nearly horizontal line joining the solid dots shows measured circumferential stretch ratio (relative to the zero-stress state) at a blood pressure of 80 mm Hg; thinner inclined line joining open circles shows computed hypothetical circumferential stretch ratio when opening angle was ignored. (c) Corresponding strains at blood pressure 120 mm Hg. These curves show that huge errors result if residual strain is ignored. From Fung and Liu (1992). By permission.

pressure is 80 mm Hg, whereas the thinner, inclined line shows the hypothetical circumferential stretch ratio distribution at 80 mm Hg under the assumption that the opening angle is zero. The corresponding strains are all positive (tensile), but the great error caused by ignoring the residual strain (opening angle) is seen. The corresponding stretch ratio distributions in the vessel wall at a blood pressure of 120 mm Hg are illustrated in Fig. 13.9(c). It is clear from Fig. 13.9 that the errors caused by ignoring the residual strains are enormous. It is important to know the zero-stress state of a blood vessel.

The longitudinal stretch ratios from the no-load to the homeostatic condition measured on the specimens of Fig. 13.9 is about 1.35. No change in the ratio was detected experimentally upon cutting open a vessel segment under the no-load condition to the zero-stress state. Hence, the longitudinal stretch from the zero-stress state to the homeostatic state of the ileal vessel is also about 1.35. Finally, the radial stretch ratio can be computed from the condition of incompressibility of the vessel wall:

$$\lambda_r \lambda_\theta \lambda_z = 1. \tag{13.7–4}$$

Thus, the strain state of the vessel is completely determined experimentally.

For arteries, the stresses increase as exponential functions of strains. Hence, if stresses were plotted in graphs corresponding to the strain distributions plotted in Fig. 13.9, a much greater error in stress would be seen as a consequence of ignoring the opening angle.

13.8 STRESS-GROWTH RELATIONSHIP

Biological tissue growth can be affected by many things: nutrition, growth factors (enzymes), the physical and chemical environment, and diseases, as well as stress and strain. If other things were equal, then a stress-growth law will emerge.

A stress-growth relationship has clinical applications in the understanding of diseases, healing, and rehabilitation. If a stress-growth law is known for certain organs, then surgeons can use it to plan surgery on those organs, engineers can use it for tissue engineering, manufacturers of prostheses will have guidance, and physical therapists, athletes, and educators will know the relation between exercise and body development.

Tissue engineering is a field dedicated to making artificial substitutes for living tissues. It is a technology based on molecular biology, cell biology, and organ physiology. To master tissue engineering, one must know how the health of tissues is maintained, improved, or failed in relation to stress and strain.

Machines, in general, do not have the ability to remodel themselves, but such an ability is clearly desirable in some circumstances. It is not beyond the engineer's imagination to conceive of machines with the ability to remodel themselves, but the direction is a totally new one for engineers to think about.

Readers interested in this subject may find the references listed at the end of this chapter helpful. A fairly comprehensive introduction to the mechanics of tissue remodeling is given in Fung (1990), which contains an extensive list of references. In the medical field, bone remodeling has been studied for a long time. Meyer's paper was dated 1867. Wolff's law was proposed in 1869. Papers by Carter and Wong (1988), Cowin (1986), and Fukada (1977) may serve as entry to the current literature. In the preceding sections we used blood vessels to illustrate the features of tissue remodeling: changes in the zero-stress state, structure and arterial composition, constitutive equations, and stress and strain distributions. We could have used bone for this purpose; but changes in soft tissues are more visible and take place faster than those in bone. The getting together of the time constants for tissue remodeling, stress relaxation, strain creep, fluid movement, and mass transport serves to bring biology and mechanics closer together. The papers by Chuong and Fung (1986), Fung (1991), Hayashi and Takamizawa (1989), Takamizawa and Hayashi (1987), Vaishnav and Vossoughi (1987), and Omens and Fung (1990) are relevant to soft tissue mechanics. The book edited by Skalak and Fox (1988) is a collection of papers presented at a tissue engineering conference. There is a large amount of literature on the biology and medicine of tissue remodeling. The papers by Cowan and Crystal (1975), and Meyrick and Reid (1980) are excellent examples.

PROBLEM

13.1 Membranes within Living Cells. Within a cell, membranes are ubiquitous, but their mechanical properties are virtually unknown. As a theoretical concept, intracellular membranes may be assumed to have surface tension, stretching elasticity, shear elasticity, and bending rigidity. The tension and shear are associated with membrane area and deformation, the bending rigidity is associated with the change in curvature of the membrane.

A surface in three-dimensional Euclidean space has two principal curvatures at every point. The sum of the two principal curvatures is called the *mean curvature* of the surface; the product of the two principal curvatures is called the *Gaussian curvature*. One may assume that the energy state of the membrane depends on the mean and Gaussian curvatures. Now, propose a strain energy function for an intracellular membrane. Then solve a mathematical problem: find a *minimal* surface of finite area but zero mean curvature everywhere.

An answer given by Reinhard Lipowsky, in *Nature*, vol. 349, p. 478, Feb. 1991 is shown in Fig. P13.1. Do you think Lipowsky's surface is minimal? What kind of energy state would the surface have? If one wants to claim that a minimal surface has a minimum energy level, how should energy of the membrane be related to the surface area, and the mean and Gaussian curvatures?

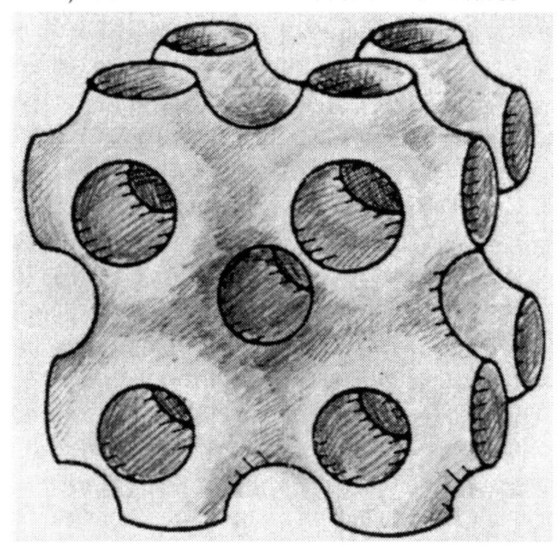

Figure P13.1 Lipowsky's surface.

What kind of surface has zero Gaussian curvature everywhere? Is a developable surface one of zero Gaussian curvature? Are all surfaces with zero Gaussian curvature developable?

REFERENCES

BORDEN, T. (1991). "Shape-Memory Alloys: Forming a Tight Fit." *Mechanical Engineering*, Oct. 1991: 67–72.

CARTER, D. R., AND WONG, M. (1988). "Mechanical Stresses and Endochondral Ossification in the Chondroepiphysis." *J. Orthop. Res.* 6: 148–154.

CHUONG, C. J., AND FUNG, Y. C. (1983). "Three-dimensional Stress Distribution in Arteries." *J. Biomech. Eng.* 105: 268–274.

CHUONG, C. J., AND FUNG, Y. C. (1986). "On Residual Stresses in Arteries." *J. Biomech. Eng.* 108: 189–199.

COWAN, M. J., AND CRYSTAL, R. G. (1975). "Lung Growth after Unilateral Pneumonectomy: Quantitation of Collagen Synthesis and Content." *Am. Rev. Respir. Disease* 111: 267–276.

Cowin, S. C. (1986). "Wolff's Law of Trabecular Architecture at Remodeling Equilibrium." *J. Biomech. Eng.* 108: 83–88.

Fukada, E. (1974). "Piezoelectric Properties of Biological Macromolecules." *Adv. Biophys.* 6: 121.

Fung, Y. C., and Liu, S. Q. (1989). "Change of Residual Strains in Arteries Due to Hypertrophy Caused by Aortic Constriction." *Circ. Res.* 65: 1340–1349.

Fung, Y. C. (1990). *Biomechanics: Motion, Flow, Stress, and Growth.* New York: Springer-Verlag.

Fung, Y. C. (1991). "What Are the Residual Stresses Doing in Our Blood Vessels?" *Annals of Biomedical Engineering*, 19: 237–249.

Fung, Y. C., and Liu, S. Q. (1991). "Changes of Zero-Stress State of Rat Pulmonary Arteries in Hypoxic Hypertension." *J. Appl. Physiol.* 70(6): 2455–2470.

Fung, Y. C., and Liu, S. Q. (1992). " Strain Distribution in Small Blood Vessels with Zero-Stress State Taken into Consideration." *American J. Physiol: Heart and Circulatory Physiology*, 262(2): H544–H552.

Han, H. C., and Fung, Y. C. (1991a). Species dependence on the zero-stress state of aorta: pig vs. rat. *J. Biomech. Eng.* 113: 446–451.

Han, H. C., and Fung, Y. C. (1991). "Residual Strains in Porcine and Canine Trachea. *J. Biomechanics*, 24: 307–315.

Hayashi, K., and Takamizawa, K. (1989). "Stress and Strain Distributions and Residual Stresses in Arterial Walls." In Fung, Y. C., Hayashi, K., and Seguchi, Y., eds., *Progress and New Directions of Biomechanics*. Tokyo: MITA Press, pp. 185–192.

Liu, S. Q., and Fung, Y. C. (1989). "Relationship between Hypertension, Hypertrophy, and Opening Angle of Zero-Stress State of Arteries following Aortic Constriction." *J. Biomech. Eng.* 111: 325–335.

Liu, S. Q., and Fung, Y. C. (1992). Influence of STZ-induced Diabetes on Zero-Stress State of Rat Pulmonary and Systemic Arteries." *Diabetes*. 41: 136–146.

Matsuda, T., Echigo, S., and Kamiya, T. (1991). "Shape-Memory Polymer: Its Application to Cardiovascular Device." (Abstract). *Medical and Biological Engineering and Computing*, Vol. 29, 1991 Supplement, p. 46

Meyer, G. H. (1867). "Die Architektur der spongiosa." *Archiv. für Anatomie, Physiologie, und wissenschaftliche Medizin* (Reichert und Du Bois-Reymonds Archiv) 34: 615–628.

Meyrick, B., and Reid, L. (1980). "Hypoxia-Induced Structural Changes in the Media and Adventitia of the Rat Hillar Pulmonary Artery and Their Regression. *Am. J. Pathol.* 100: 151–178.

Omens, J. H., and Fung, Y. C. (1990). "Residual Strain in the Rat Left Ventricle." *Circ. Res.* 66: 37–45.

Skalak, R., and Fox, D. F. (eds.). (1988). *Tissue Engineering*. New York: Alan Liss.

Takamizawa, K., and Hayashi, K. (1987). "Strain Energy Density Function and Uniform Strain Hypothesis for Arterial Mechanics." *J. Biomech.* 20: 7–17.

Vaishnav, R. N., and Vossoughi, J. (1987). "Residual Stress and Strain in Aortic Segments." *J. Biomechanics* 20: 235–239.

Wolff, J. (1869). "Über die innere Architektur der spongiösen Substanz." *Zentralblatt für die medizinische Wissenschaft* 6: 223–234.

Xie, J. P., Yang, R. F., Liu, S. Q., and Fung, Y. C. (1991). "The Zero-Stress State of Rat Vena Cava." *J. Biomech. Eng.* 113: 36–41.

INDEX

Author Index

R

S

T

U

V

W

X

Y

Subject Index

D

E

F

G

H

I

J

K

L

M

N

T